CAMBRIDGE ENGINE TECHNOLOGY SERIES

General Editors: J. E. Ffowcs Williams, E. Greitzer

VORTEX ELEMENT METHODS FOR FLUID DYNAMIC ANALYSIS OF ENGINEERING SYSTEMS

T0243078

CAMBRIDGE ENGINE TECHNOLOGY SERIES

1 Vortex element methods for fluid dynamic analysis of engineering systems

VORTEX ELEMENT METHODS FOR FLUID DYNAMIC ANALYSIS OF ENGINEERING SYSTEMS

R. I. LEWIS

Professor of Fluid Mechanics and Thermodynamics
University of Newcastle upon Tyne

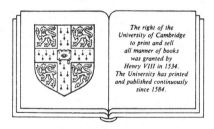

The right of the
University of Cambridge
to print and sell
all manner of books
was granted by
Henry VIII in 1534.
The University has printed
and published continuously
since 1584.

CAMBRIDGE UNIVERSITY PRESS
Cambridge
New York   Port Chester
Melbourne   Sydney

CAMBRIDGE UNIVERSITY PRESS
Cambridge, New York, Melbourne, Madrid, Cape Town, Singapore, São Paulo

Cambridge University Press
The Edinburgh Building, Cambridge CB2 2RU, UK

Published in the United States of America by Cambridge University Press, New York

www.cambridge.org
Information on this title: www.cambridge.org/9780521360104

First published 1991
This digitally printed first paperback version 2005

A catalogue record for this publication is available from the British Library

ISBN-13 978-0-521-36010-4 hardback
ISBN-10 0-521-36010-2 hardback

ISBN-13 978-0-521-01754-1 paperback
ISBN-10 0-521-01754-8 paperback

To
Daphne

Contents

Contents

Contents

Contents

Part 2   Free shear layers, vortex dynamics and vortex cloud analysis

Contents

xiii

Contents

Contents

Contents

Preface

Only thirty years have elapsed since E. Martensen published his well known paper proposing the surface vorticity boundary integral method for potential flow analysis. Generally regarded as the foundation stone, this paper has led to the establishment of a considerable volume of numerical methodology, applicable to a wide range of engineering problems, especially in the fields of aerodynamics and turbomachines. During this period we have also witnessed a technological transformation in the engineering world of immense proportions and of great historical significance. This has been based upon parallel advances in both theoretical and practical engineering skills which have been breath-taking at times. Theoretical methods, to which this book is dedicated, have undergone a renaissance spurred on by the rapid growth of computing power in response to the ever increasing demands of engineering hardware. The main characteristic of this new-birth has been a shift from the pyramid of classical methods to a whole host of numerical techniques more suited to direct modelling of real engineering problems. The explosion of this activity has been damped down only by the difficulties of transferring and absorbing into normal practice a technology which can, as in the case of many numerical methods, become highly personalised. After three decades there is a need for books which sift and catalogue and which attempt to lay out the new fundamental methodologies to suit the needs of engineers, teachers and research workers.

This is indeed the main purpose behind the present book which attempts to lay out a systematic treatment of the surface vorticity method in relation primarily to the author's special field of interest of turbomachinery fluid dynamics. Martensen's paper was published in the early days of first generation computers when there were already several successful computational schemes available for cascade analysis, notably those of Scholz (1951) and Schlichting (1955a). Taking advantage of many of the brilliant insights and ideas bred by classical modelling, this generation of research

xvii

engineers used mathematical methods of reduction to bring computations within the range of desk calculators. The first use of digital computers by mechanical engineers was largely concerned with the programming of these established methods spread across such disciplines as fluid dynamics, heat transfer, stress analysis and vibration analysis. Much of this pioneering work is indeed still viable today. In the field of turbomachines however there was soon a diversion of effort towards the development of new numerical methods matched both to engineering need and also to emerging computer capability.

The field of aerodynamics comprises two categories of flow regime designated 'internal' and 'external'. External aerodynamics, typified for example by the flow past an aircraft, led to the development in the late 1950s and early 1960s of the surface source boundary integral method for three-dimensional flow analysis, sometimes known alternatively as the Douglas–Neumann or panel method. On the other hand turbomachines involve mainly internal flows, often of even greater complexity. In this field of study, because of the need to solve the equations of motion throughout the fluid, there has tended to be a greater emphasis upon mesh methods such as the streamline curvature, finite difference, matrix through-flow and time marching techniques. Nevertheless, in parallel with this there has also been a steady development of boundary integral techniques employing the much more natural model of surface vorticity analysis. Indeed, without the introduction of prescribed bound vorticity, source panel methods cannot reproduce lift forces. The attraction of the surface vorticity method is its physical basis which attempts to model directly the vortical nature of fluid flows. All real flows involving interaction of bodies with uniform streams, are controlled primarily by the vorticity created at the surface and contained within the boundary layer. Its convection and diffusion both within the boundary layer and further downstream in the wake of a body are of course crucial to the motion, bringing in also the influence of Reynolds number, the balance between dynamic and viscous action within the fluid. However it is the surface vorticity which exercises the primary control over fluid flows.

For many years surface vorticity theory has tended to be presented and regarded as just yet another numerical tool for solving potential flow problems. Since 1979 however advantage has at last been taken of the far greater scope of surface vorticity methods for direct physical modelling of complex real flows. As a technique for simulation of vortex streets and bluff body wakes, 'vortex dynamics' began to emerge in the 1970s as a new analytical tool, usually assuming prescribed separation points and often relying on conformal transformations to simplify body shapes. However, in parallel with this surface vorticity modelling was extended both to the same class of problems and to more complex situations with the same aim of attempting to solve the two-dimensional Navier–Stokes equations. In these works fluid rotation is discretised into a cloud of vortex elements. Computation proceeds either in a Lagrangian manner, maintaining space–time records of each vortex element, or in an Eulerian manner employing grid structures to capture and re-order the vorticity in order to gain economies in computation. Recent developments tend towards a mixture of these approaches and this will probably be the way forward in application of vortex cloud analysis to turbomachine cascade and meridional flow problems. Very recently successful attempts have been made to extend vortex cloud modelling to acoustic excitation due to bluff body wakes, Hourigan *et al.* (1986), and to convective heat transfer, Smith & Stansby (1989), in both cases by coupling in the additional physical equations. Undoubtedly there is scope for much more extensive application of these powerful new techniques in the years ahead.

The strategy of this book is to present an unfolding methodology designed to lead easily into computing schemes, beginning with potential flow modelling in Part I and progressing right through to full vortex cloud modelling in Part II. The field of coverage is strongly linked to rotodynamic machines and bluff bodies but the reader should be able reasonably easily to extend this work to a wider range of applications to suit particular interests. A number of useful computer programs used in the text have been included as an appendix. These and a number of other engineering design and analysis programs are available on magnetic disc from the author.

Acknowledgements

I would like to acknowledge the outstanding contributions to the subject matter of this book made by the authors quoted and very many others. In particular I would like to pay tribute to my own post graduate students who, over many years, have made an impressive and cumulative impact upon vorticity modelling, in most cases feeding directly through into the design process. Diagrams re-traced from publications have been acknowledged in the relevant figure titles. The writing of a book in the midst of a busy academic life would be impossible without a generous attitude and the moral support of colleagues and the continual encouragement of my wife and family, to all of whom I am extremely grateful. Finally I must record a deep debt of gratitude to Roberta Stocks who has completed to perfection the daunting task of converting my scribbles into the finished manuscript.

R. I. Lewis

PART I

The surface vorticity method
for inviscid ideal fluid flow

CHAPTER 1

The basis of surface singularity modelling

1.1 Introduction

The principal aims of this book are to outline the fundamental basis of the surface vorticity boundary integral method for fluid flow analysis and to present a progressive treatment which will lead the reader directly to practical computations. Over the past two and a half decades the surface vorticity method has been developed and applied as a predictive tool to a wide range of engineering problems, many of which will be covered by the book. Sample solutions will be given throughout, sometimes related to Pascal computer programs which have been collated for a selection of problems in the Appendix. The main aims of this introductory chapter are to lay down the fundamental basis of both source and vorticity surface panel methods, to explain the fluid dynamic significance of the surface vorticity model and to introduce a few initial applications to potential flow problems.

As numerical techniques, surface singularity methods were not without progenitors but grew quite naturally from the very fertile field of earlier linearised aerofoil theories. Such methods, originally contrived for hand calculations, traditionally used internal source distributions to model profile thickness and vortex distributions to model aerodynamic loading, a quite natural approach consistent with the well known properties of source and vortex singularities. On the other hand it can be shown that the potential flow past a body placed in a uniform stream can be modelled equally well by replacing the body surface with either a source or a vortex sheet of appropriate strength, Fig. 1.1. Integral equations can then be written expressing the Neumann boundary condition of zero normal surface velocity for the source model or the Dirichlet condition of zero parallel surface velocity for the vorticity model. Whichever type of singularity is chosen the final outcome is the same, namely a prediction of the potential flow velocity close to the body profile. The numerical strategy is also fairly similar as we shall see from

3

The basis of surface singularity modelling

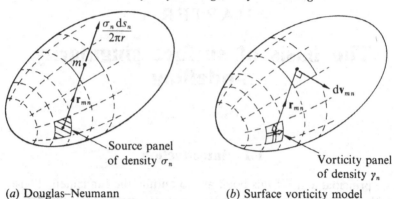

(a) Douglas–Neumann   (b) Surface vorticity model
source panel model

Fig. 1.1. Surface source and vorticity panel models for three-dimensional potential flow.

Sections 1.6 and 1.7. Lifting bodies form an important exception to this remark, since lift forces normal to a uniform stream cannot be simulated by sources alone but require also the introduction of vorticity distributions, a matter which will be taken up in Chapter 2. On the other hand the surface vorticity model is capable of handling potential flows for any situation including lifting bodies. We shall begin in Sections 1.2 and 1.3 with a presentation of these basic surface singularity models and their associated integral equations.

Surface vorticity modelling offers the additional advantage over source panels that it actually represents a direct simulation of an ideal fluid flow. In all real flows a boundary viscous shear layer exists adjacent to the body surface. Inviscid potential flow is akin to the case of the flow of a real fluid at infinite Reynolds number for which the boundary layer is of infinitesimal thickness. In this situation the boundary layer vorticity is squashed into an infinitely thin vorticity sheet across which the velocity parallel to the surface changes discontinuously from zero in contact with the wall to the potential flow value just outside the vorticity sheet. Thus surface vorticity modelling is the most natural of all boundary integral techniques. Further discussion of its physical significance will be given in Section 1.4. Part I of this book is concerned with such ideal flows for a range of applications especially in the fields of aerodynamics and rotodynamic machines including also some situations involving rotational main stream flow. In the early sections of Part II further consideration will be given to the physical sig-

4

nificance of the surface vorticity model including its extension to the simulation of real boundary layer flows and the establishment of wake eddies behind bluff bodies.

Surface source modelling by contrast is capable of no direct physical interpretation but is purely a vehicle, albeit a powerful one, for analysing three-dimensional potential flows. Historically it predated the surface vorticity method by half a decade or so as an emerging practical tool for numerical analysis at a time of intense pressure for the creation of flexible computational procedures for design use in the aeronautical field. To some extent surface vorticity methods were thus upstaged and have consequently received but a small fraction of the attention paid to the surface source panel technique. This book is aimed at redressing the balance. Although most of the text will in consequence be concerned with surface vorticity theory and applications, we will devote some of chapter 1 to consideration in parallel of both source and vorticity panel methods. Following the introductory Sections 1.2 to 1.5 on fundamentals we will move on quickly to numerical models for plane two-dimensional flows in Sections 1.6 and 1.7, leading to comparisons between the source and vorticity schemes for flow past a circle and an ellipse. One or two other plane two-dimensional problems will then be dealt with in Section 1.9 involving bodies with sharp corners and simplifications for symmetrical bodies. The chapter is concluded with a summary of the surface vorticity equations expressed in curvilinear coordinates.

1.2 The source panel or Douglas–Neumann method

Many of today's established numerical methods for engineering design and analysis find their origins in classical mathematics predating the age of the digital computer. Surface singularity methods are no exception. For example in 1929 Kellogg wrote a comprehensive book dealing with potential theory by the use of integral equations, including the treatment of volume and surface singularities. Such works tended to concentrate upon solutions to incompressible inviscid flows, expressed for example by Laplace's equation for the velocity potential ϕ

$$\nabla^2 \phi = \frac{\partial^2 \phi}{\partial x^2} + \frac{\partial^2 \phi}{\partial y^2} + \frac{\partial^2 \phi}{\partial z^2} = 0 \qquad (1.1)$$

The basis of surface singularity modelling

where (x, y, z) are Cartesian coordinates. A well known elementary solution to this equation is given by $\phi = 1/r$ where r is the radial distance between (x_n, y_n, z_n) and some other point of fluid action (x_m, y_m, z_m). The physical interpretation of this solution is that of flow from a point source in three-dimensional space. Thus, the velocity potential at m due to a point source of unit strength at n (where point source strength is defined here as the volume of fluid emitted in unit time) is given by

$$\phi = -\frac{1}{4\pi r_{mn}} \tag{1.2}$$

where

$$r_{mn} = \{(x_m - x_n)^2 + (y_m - y_n)^2 + (z_m - z_n)^2\}^{\frac{1}{2}} \tag{1.3}$$

As shown by Kellog (1929) and elaborated by A. M. O. Smith (1962), the flow past a body immersed in a uniform stream W_∞ may then be expressed by the following integral equation,

$$\tfrac{1}{2}\sigma_m - \frac{1}{4\pi} \iint_S \frac{\partial}{\partial n}\left(\frac{1}{r_{mn}}\right)\sigma_n \, dS_n + \mathbf{i}_m \cdot \mathbf{W}_\infty = 0 \tag{1.4}$$

where $\mathbf{i}_m$ is a unit vector normal to the body surface S, and σ_n is the source density per unit area. This equation represents the earliest form of surface singularity model in which the body surface is replaced by a surface source distribution σ_n, Fig. 1.1(a). Equation (1.4) then states the Neumann boundary condition that for all points m on the body the velocity normal to the surface is zero. If this equation is satisfied then the body surface becomes a stream surface of the flow. For computation we may complete the normal derivative inside the Kernel resulting in

$$\tfrac{1}{2}\sigma_m + \frac{1}{4\pi} \iint_S \frac{\sigma_n}{r_{mn}^3}\mathbf{r}_{mn} \cdot \mathbf{i}_m \, dS_n + \mathbf{i}_m \cdot \mathbf{W}_\infty = 0 \tag{1.5}$$

This equation states that the sum of three velocity components normal to the surface at point m, when combined, comes to zero. The last term is the component of the uniform stream resolved along the surface normal $\mathbf{i}_m$. The second term accounts for the influence of all surface source elements $\sigma_n \, dS_n$. Here we note that the actual velocity at m due to one such element is given by

$$dv_{mn} = \frac{\sigma_n \, dS_n}{4\pi r_{mn}^2} \tag{1.6}$$

6

and has the vector direction of $\mathbf{r}_{mn}$, which can be represented by the unit vector $\mathbf{r}_{mn}/r_{mn}$. Since the integal is taken actually on the surface S, we must introduce also the first term of (1.5), $\frac{1}{2}\sigma_m$, which represents the velocity discontinuity stepping onto the outside of the source sheet.

The numerical strategy of the panel method involves the representation of the body surface by a finite distribution of source panels defined geometrically by a suitable grid, Fig. 1.1(*a*). One control point m is chosen for each source panel for application of the Neumann boundary condition through (1.5). The surface integral then becomes a summation for all panels, resulting in a set of M linear equations for M unknown values of σ_m. Solution is straightforward usually and yields the necessary surface source strength to ensure that the flow remains parallel to the body surface. Following on from this the local potential flow velocity parallel to the surface can be evaluated directly by means of a second integral equation of the form,

$$\mathbf{v}_m = \mathbf{i}_m X \left\{ \frac{1}{4\pi} \int\!\!\int_S \frac{\sigma_n}{r_{mn}^3} \mathbf{r}_{mn} X \mathbf{i}_m \, \mathrm{d}S_n + \mathbf{W}_\infty X \mathbf{i}_m \right\} \qquad (1.7)$$

This has been expressed in vector form, reminding us that for three-dimensional bodies the surface potential flow is of course two-dimensional. Reduction to Cartesian or other coordinate systems is necessary for numerical computations but is soon accomplished. Later, in Section 1.7, we will illustrate this by a simple numerical example, but for the present our aim is to draw out some of the fundamental equations and models of surface singularity methods. In the case of the source panel method, which is actually not to be the main substance of this book, it remains only to point out that two integral equations must be solved, one indirect and the other direct, using the source 'singularity' distribution σ_m as an intermediate parameter for reaching the solution. Unlike the use of surface vorticity, source panels provide no ready physical interpretation or special advantage as a physical model except in very special cases such as surface transpiration or change in fluid volume due to evaporation or condensation at a surface. Nevertheless, as a computational method for potential flows the source panel method has been widely used with great success since about 1953, notably in the field of aeronautics. The literature is extensive and mention will be made here only of representative early work by A. M. O. Smith (1962), A. M. O. Smith & Hess (1966), A. M. O. Smith & Pierce

(1958) and Hess (1962) covering basic theory with a range of applications. A more recent survey of models and formulations was also given by Hunt (1978). Discussions of the relationships between volume surface and line distributions of vorticity, sources and doublets have been given by Semple (1977), Hunt (1978) and R. Rohatynski (1986).

1.3 The surface vorticity or Martensen method

Although no doubt early theorems related to surface vorticity distributions, such as those of Kellogg previously referred to, could be located in older texts, the seed corn publication in this field was undoubtedly that of Martensen (1959). Martensen not only laid out the basis of a powerful new computational technique, but he also extended his new boundary integral theory to deal with turbo-machine cascades, a subject which we will deal with in some detail in Chapters 2 and 3. However, Jacob & Riegels (1963) would seem to have been the first contributors of a practical working scheme designed for digital computers, taking 15 minutes to execute on a IBM 650 computer for analysis of an aerofoil with 36 surface vorticity elements; no mean achievement at that time. Numerical modelling often offers great scope for ingenuity and inventiveness and several good ideas put forward by Jacob and Riegels have stood the test of time. However there were many problems to be identified and solved before the method could progress to accept-ability as a reliable engineering predictive tool. D. H. Wilkinson pioneered many of these problems of modelling and practical methodology, publishing a most significant paper in 1967 which formed an important foundation stone for computer applications. He also extended his work to mixed-flow cascades, Wilkinson (1969), another very important and far reaching contribution. In parallel with this Nyiri (1964), (1970) independently produced an extension of Martensen's method to mixed-flow pump cascades, later updated as a practical numerical scheme, Nyiri & Baranyi (1983). There are of course many other important publications covering a range of applications. These will not be reviewed here but referred to in relevant parts of the text.

The surface vorticity model is illustrated in Fig. 1.1(*b*) for a three-dimensional body. In this scheme the body surface is covered with a finite number of surface vorticity panels initially of unknown strength. Following a similar procedure to the source panel method, one control point m is chosen for each panel for application of the surface flow boundary condition, taking account of the influence of all other surface vorticity panels and of the mainstream flow. In this case on the other hand, it is appropriate to adopt the boundary condition of zero velocity on, and parallel to the body surface†, (we shall consider why in more detail later in Section 1.4). The actual velocity induced at m by a small line vortex element at n of strength Γ_n per unit length* and of length dl_n is given by the Biot–Savart law, namely, with reference to Fig. 1.2,

$$d\mathbf{v}_{mn} = \frac{\Gamma_n \, dl_n X \mathbf{r}_{mn}}{4\pi r_{mn}^3} \tag{1.8}$$

By taking the cross product of $d\mathbf{v}_{mn}$ with the unit vector $\mathbf{i}_m$ normal to the surface at m twice, we obtain the velocity parallel to the surface at m induced by the line vortex element. Thus

$$d\mathbf{v}_{smn} = \mathbf{i}_m X (d\mathbf{v}_{mn} X \mathbf{i}_m)$$

$$= \frac{\mathbf{i}_m X ((\Gamma_n X \mathbf{r}_{mn}) X \mathbf{i}_m) \, dl_n}{4\pi r_{mn}^3} \tag{1.9}$$

In reality the surface is to be covered not with concentrated line vortices but with an area density of distributed sheet vorticity which

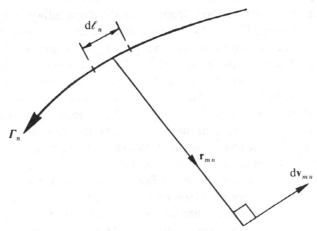

Fig. 1.2. Velocity induced by a line vortex element.

† Since we are addressing the Dirichlet problem for $\mathbf{q}$ this will be termed the Dirichlet boundary condition throughout this book.

* Throughout this book vortex strength is defined as positive according to the right hand corkscrew rule.

we will denote here by the symbol γ_m. Making use of (1.9) the Dirichlet boundary condition of zero velocity on (and parallel to) the body surface at m may then be expressed

$$-\tfrac{1}{2}\gamma_m + \frac{1}{4\pi} \iint_S \frac{\mathbf{i}_m X((\gamma_n X \mathbf{r}_{mn}) X \mathbf{i}_m)\, \mathrm{d}S}{r_{mn}^{\,3}} + \mathbf{i}_m X(\mathbf{W}_\infty X \mathbf{i}_m) = 0 \quad (1.10)$$

The last term is the component of the mainstream velocity $\mathbf{W}_\infty$ resolved parallel to the body surface. The first term is the velocity discontinuity experienced if we move from the centre of the vorticity sheet onto the body surface beneath.

As it stands this integral equation, like that for the source panel method (1.5), is of little practical use and is recorded here only in view of its importance as a general statement of the problem. We will later on in Section 1.10 express it in curvlinear coordinates which are of much more value for setting up computational schemes in various coordinate systems. At this point however it will be much more helpful to move on to a simple physical interpretation of the surface vorticity method followed by practical application to a numerical scheme for solving a simple problem.

1.4 Physical significance of the surface vorticity model

In all real flows a boundary shear layer develops adjacent to the surface of a body, Fig. 1.3(a). Sufficient vorticity is present in this layer to reduce the fluid velocity from v_s just outside the shear layer to a value of zero on the body surface. The action of viscosity is to cause the vorticity in this shear layer to diffuse normal to the surface, resulting in the familiar viscous boundary layer. The vorticity itself however is the product of the dynamic behaviour of the outer flow and we will show later that the rate of vorticity production adjacent to the surface is directly related to the pressure gradient. Traditionally a real flow is usually regarded as comprising a largely irrotational inviscid outer flow in the bulk of the domain, separated from the body surface by a thin but highly active viscous shear layer. These regimes are of course frequently treated separately for analytical expediency, with suitable matching conditions at the outer edge a–b of the boundary layer. In reality, as we have just pointed out, vorticity creation is largely attributable to the outer flow, a fact which is underlined if we consider in particular the special case of infinite Reynolds number, or inviscid potential flow.

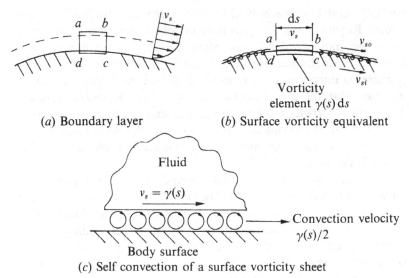

(a) Boundary layer (b) Surface vorticity equivalent

(c) Self convection of a surface vorticity sheet

Fig. 1.3. Boundary layer and surface vorticity equivalent in potential flow.

Suppose that we were able gradually to reduce the fluid viscosity to zero in a real fluid flow. In the limit, due to progressive reduction of viscous diffusion, the boundary layer would approach infinitesimal thickness. As the viscosity approached zero and the Reynolds number approached infinity, the body surface would be covered with an infinitely thin vorticity sheet $\gamma(s)$, Fig. 1.3(b), across which the fluid velocity would change discontinuously from zero beneath the sheet on the body surface to v_s parallel to the surface just above the sheet. In the case of a real flow with extremely high Reynolds number, we are aware that the boundary layer may separate spontaneously with rising static pressure in the direction of the mainstream flow. Furthermore the boundary layer will normally become turbulent at very high Reynolds numbers. Both phenomena are connected with the interrelationship between the viscous diffusion and convection processes in the boundary layer which the Reynolds number symbolises. Leaving aside these additional features of a real flow connected with instabilities of the shear layer itself, we see that inviscid potential flows can be thought of as a special type of infinite Reynolds number flow. An irrotational potential flow thus comprises a surface vorticity sheet covering the body surface, separating the irrotational flow of the outer domain from a motionless flow in the inner domain. In this sense the surface

11

vorticity model is precisely true to the physical reality of a real infinite Reynolds number (but fully attached) flow and is therefore the most natural of all numerical methods for potential flow analysis. Furthermore, as we shall see in Chapters 9–11, it is also possible to introduce models to simulate viscous diffusion, so that we may relax the present constraint of infinite Reynolds number. The surface vorticity method, unlike the source panel method, thus offers special attractions as a route towards the simulation of real fluid flows because the model truly reflects the physical reality, Lewis & Porthouse (1983a).

To decide upon an appropriate boundary condition (which we have already asserted to be the Dirichlet condition) let us consider the flow induced by such a surface vorticity sheet in closer detail, Fig. 1.3. First let us define the contour *abcd* surrounding a small vorticity element $\gamma(s)\,\mathrm{d}s$ where *ab* and *dc* are parallel to the streamlines, while *da* and *cb* are normal to them. Now $\gamma(s)$ is defined as the vorticity strength per unit length at point s. The circulation around *abcd*, defined clockwise–positive, may be equated to the total amount of vorticity enclosed by the contour, that is

$$(v_{so} - v_{si})\,\mathrm{d}s = \gamma(s)\,\mathrm{d}s$$

where v_{so} and v_{si} are the fluid velocities just outside and inside the sheet, which must be parallel to the surface. Our boundary condition of zero velocity on the body surface is thus satisfied if we specify

$$v_{si} = 0 \qquad\qquad\qquad\qquad (1.11)$$

whereupon

$$v_{so} = v_s = \gamma(s) \qquad\qquad\qquad\qquad (1.12)$$

The neatness of Martensen's method lies in these two equations. Equation (1.11) is the basis of Martensen's boundary integral equation as summarised previously by (1.10). The solution of this equation yields the surface vorticity distribution of the potential flow. The second equation (1.12) then tells us that the potential flow velocity close to the body surface v_s is now immediately known, being exactly equal to the surface vorticity $\gamma(s)$. The surface vorticity method, in addition to its direct simulation of physical reality, thus offers the additional attraction compared with the source panel method that no second integral equation is required to derive v_s from the surface singularity distribution.

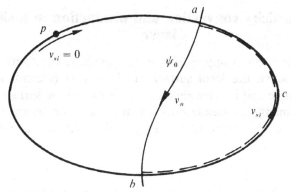

Fig. 1.4. Check for leakage flow with the Dirichlet boundary condition in Martensen's method.

The reader may feel that the Dirichlet boundary condition stated might be insufficient to ensure flow parallel to the body surface and that the Neumann boundary condition should be imposed either in addition or instead. To counter this view let us first assume that Dirichlet is inadequate and that consequently there is a leakage velocity v_n normal to the body at a, Fig. 1.4. However if there are no sources present inside the body contour the only possibility is that the streamline ψ_0 will cross the body a second time at point b. If we now apply the circulation theorem around the contour abc just inside the surface vorticity sheet, then

$$\oint_{abc} \mathbf{v} \cdot \mathbf{ds} = \int_a^b v_n \, ds + \int_b^a v_{si} \, ds = 0$$

$$\underset{A}{} \quad \underset{B}{}$$

assuming also that there is no vorticity contained within the body profile. Since zero v_{si} has been enforced by the Dirichlet boundary condition, term B and therefore term A are independently zero. Since v_n is undirectional along the supposed streamline ψ_0, the only possibility is that it must also be zero throughout. The Dirichlet condition is thus totally adequate provided there are no vortex or source distributions within the body profile. The reader is referred to Martensen (1959) for a rigorous proof.

13

1.5 Vorticity convection and production in a shear layer

Most surface vorticity applications in the past have dealt with steady flows, for which the local surface vorticity $\gamma(s)$ is constant with respect to time and is often regarded as bound to the surface. Thus in a two-dimensional steady flow situation the total bound circulation would be directly calculable through the contour integral

$$\Gamma = \oint \gamma(s) \, \mathrm{d}s \tag{1.13}$$

In reality however we know that the actual boundary layer vorticity is continuously being convected downstream, Fig. 1.3(c). Let us now consider the case of a vortex sheet of strength $\gamma(s)$ coincident with the x axis and stretching between $x = \pm\infty$. Applying the circulation theorem again to an element $\gamma(s) \, \mathrm{d}s$ and taking into consideration symmetry about the x axis, Fig. 1.5(a), we see that the velocity above and below the sheet are given respectively by $\pm\frac{1}{2}\gamma(s)$. If we now superimpose a uniform stream of strength $\frac{1}{2}\gamma(s)$ over the whole (x, y) plane we have a correct surface vorticity model for flow past a plane wall, Fig. 1.5(b). From this simple study we observe that for such a flow the vortex sheet, like the boundary layer which it represents, is also convected downstream, in this case with a velocity exactly equal to half of its strength, $\frac{1}{2}\gamma(s)$.

The foregoing argument was based upon the special case of flow past a plane wall, for which the surface vorticity is identical for all points on the wall. In fact the same principle applies to potential flow past bodies of arbitrary shape. Locally at point s the velocity changes from zero on the surface just beneath the vortex sheet, to the sheet convection velocity $v_d = \frac{1}{2}\gamma(s)$ at the centre of the vorticity sheet, to $v_s = \gamma(s)$ just outside the sheet. In this case of course $\gamma(s)$ varies in magnitude along the wall, a fact which at first sight seems to be at odds with Kelvin's theorem of the constancy of circulation. Thus if the vorticity at s_2 has been convected from s_1 somewhere upstream, how is it possible that $\gamma(s_2)$ does not equal $\gamma(s_1)$ bearing in mind Kelvin's theorem? The simple answer to this seeming dilemma is that vorticity is continually being created or destroyed at a body surface in an inviscid flow whether the motion is steady or unsteady. Thus if we define $\mathrm{d}\gamma(s)$ as the net vorticity per unit length generated at point s in time $\mathrm{d}t$, Fig. 1.3(b), then the net vorticity flux leaving the control volume $abcd$ can be related to

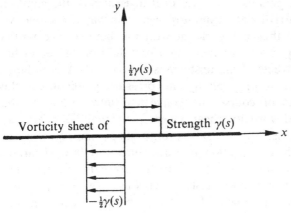

(*a*) Vorticity sheet alone

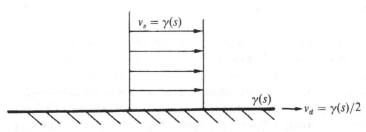

(*b*) Vorticity sheet $\gamma(s)$ with uniform stream $\frac{1}{2}\gamma(s)$

Fig. 1.5. Surface vorticity model of a uniform stream past a plane wall.

$d\gamma(s)$ through

$$\frac{\text{Vorticity created}}{\text{in time } \Delta t} = \frac{\text{Net vorticity flux,}}{\text{crossing } abcd}$$

that is

$$d\gamma(s) \cdot ds = \{\tfrac{1}{2}(v_s + dv_s)(\gamma(s) + d\gamma(s)) - \tfrac{1}{2}v_s\gamma(s)\} \, dt$$

Neglecting second-order products of infinitesimal quantities and introducing (1.12) we have finally

$$\frac{d\gamma(s)}{dt} = \frac{d}{ds}\left(\frac{v_s^2}{2}\right) = -\frac{1}{\rho}\frac{dp}{ds} \tag{1.14}$$

This equation reveals the influence of the surface pressure gradient upon vorticity production in a potential flow. Surface vorticity is spontaneously generated if the pressure falls and is

15

destroyed if the pressure rises. In ideal inviscid flow the convected vorticity sheet arriving at s from upstream, is always in contact with the surface and thus is coincident with the new surface vorticity created by the pressure gradient. In a real fluid on the other hand the vorticity convected from upsteam is not coincident with the new vorticity created at the wall by the pressure gradient but flows outside it. It is of course the convective interactions of these successively laid vorticity shear layers in a real boundary layer which characterises the velocity profile shape and gives rise to instabilities such as transition to turbulence and flow separation. Equation (1.14) will be derived directly from the Navier–Stokes equations in Chapter 10 for unsteady viscous flow for which case it provides a useful technique for calculating the surface pressure distribution on a body through

$$p = p_1 - \int_{s_1}^{s} \rho \frac{d\gamma(s)}{dt} \, ds \qquad (1.15)$$

In the vortex cloud method for simulating the Navier–Stokes equations, to be dealt with in Part II, the local vorticity production rate $d\gamma(s)/dt$ is a by-product of the Martensen analysis over successive time steps. Equation (1.15) is then extremely useful for calculating the surface pressure distribution. In the present context of steady potential flows the second two factors of (1.14) may be combined instead to give local pressure from Bernoulli's equation,

$$p = p_0 - \tfrac{1}{2}\rho v_s^2 \qquad (1.16)$$

where p_0 is stagnation pressure, which is then constant throughout the fluid.

To summarise a few of the significant points which we have just made, potential flows do in fact comprise surface vorticity sheets covering all body surfaces, providing the necessary discontinuity in velocity from zero on the body to the potential flow velocity at the edge of the flow domain. These vorticity sheets convect downstream, but vorticity is also created by the surface pressure gradient imposed by the outer flow. The surface vorticity numerical model is thus a direct simulation of the physical 'reality' for an ideal inviscid or infinite Reynolds number flow, offering the most direct and natural boundary integral representation. Finally it is of interest to note that there is no such thing as bound vorticity in fluid flow since all surface vorticity, once created under the influence of the surface pressure gradient, is free to convect and diffuse. Equation (1.13) and the notion of a total bound vortex strength are nevertheless very useful for calculating lift forces and especially so of course in the case of steady flow past aerofoils or cascades. In this special case the local surface vorticity is continually replenished at the same level, giving the impression therefore of being bound.

1.6 Surface vorticity model for plane two-dimensional flow

Consider the flow past a two-dimensional body in the (x, y) plane, immersed in a uniform stream W_∞ inclined at an angle α_∞ to the x axis, Fig. 1.6. Applying the foregoing arguments, we may represent this flow by a distributed vorticity sheet $\gamma(s)$ clothing the whole body but initially of unknown strength. The distance s is measured

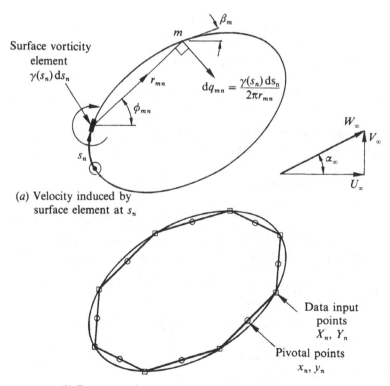

(a) Velocity induced by surface element at s_n

(b) Representation of body surface by straight line elements

Fig. 1.6. Discrete surface vorticity model for a two-dimensional body.

17

clockwise around the body perimeter from some zero datum 0 such as the leading edge in the case of an aerofoil. The velocity dq_{mn} induced at s_m due to a small vorticity element $\gamma(s_n)\,ds_n$ located at s_n elsewhere on the body, follows from the Biot–Savart law (1.8) which in this case reduces to the velocity induced by a rectilinear vortex, namely

$$dq_{mn} = \frac{\gamma(s_n)\,ds_n}{2\pi r_{mn}} \tag{1.17}$$

We need to resolve dq_{mn} parallel to the body surface at m where the profile slope is defined as β_m. For computational convenience the (x, y) components of dq_{mn} may first be expressed in terms of coordinate locations through

$$\left. \begin{aligned} dU_{mn} &= \frac{\gamma(s_n)\,ds_n}{2\pi r_{mn}} \sin \phi_{mn} = \left(\frac{y_m - y_n}{2\pi r_{mn}^{\,2}}\right)\gamma(s_n)\,ds_n \\ dV_{mn} &= -\frac{\gamma(s_n)\,ds_n}{2\pi r_{mn}} \cos \phi_{mn} = -\left(\frac{x_m - x_n}{2\pi r_{mn}^{\,2}}\right)\gamma(s_n)\,ds_n \end{aligned} \right\} \tag{1.18}$$

Resolving dq_{mn} parallel to s_m we then obtain

$$dv_{smn} = \frac{1}{2\pi}\left\{\frac{(y_m - y_n)\cos\beta_m - (x_m - x_n)\sin\beta_m}{(x_m - x_n)^2 + (y_m - y_n)^2}\right\}\gamma(s_n)\,ds_n$$

Stating the Dirichlet boundary condition at s_m, (1.10) becomes, for plane two-dimensional flow,

$$-\tfrac{1}{2}\gamma(s_m) + \oint k(s_m, s_n)\gamma(s_n)\,ds_n$$

$$+ W_\infty(\cos \alpha_\infty \cos \beta_m + \sin \alpha_\infty \sin \beta_m) = 0 \tag{1.19}$$

where the last term is the component of W_∞ resolved parallel to the body surface at m and the coupling coefficient $k(s_m, s_n)$ is given by

$$k(s_m, s_n) = \frac{1}{2\pi}\left\{\frac{(y_m - y_n)\cos\beta_m - (x_m - x_n)\sin\beta_m}{(x_m - x_n)^2 + (y_m - y_n)^2}\right\} \tag{1.20}$$

Equation (1.19) is Martensen's boundary integral equation for plane two-dimensional flow and is a Fredholm integral equation of the second kind. Compared with some other singularity methods it offers the special advantage that its Kernel is non-singular. Derivation of numerical solutions is thus extremely straightforward and this we shall consider next.

Equation (1.19) is to be satisfied at all points on the body surface. A practical approach to approximate this would be to select a finite number M of so-called 'pivotal points' representative of the surface, Fig. 1.6(b). This can be achieved most simply if the surface is broken down into a finite number M of straight line elements of length Δs_n, whereupon (1.2) may be expressed as a linear equation of the form

$$\sum_{n=1}^{M} K(s_m, s_n)\gamma(s_n) = -U_\infty \cos \beta_m - V_\infty \sin \beta_m \qquad (1.21)$$

U_∞ and V_∞ are the components of W_∞ parallel to the x and y axes and $K(s_m, s_n)$ are coupling coefficients linking elements m and n given by

$$K(s_m, s_n) = \frac{\Delta s_n}{2\pi} \left\{ \frac{(y_m - y_n) \cos \beta_m - (x_m - x_n) \sin \beta_m}{(x_m - x_n)^2 + (y_m - y_n)^2} \right\}$$

$$= k(s_m, s_n) \, \Delta s_n \qquad (1.22)$$

Several comments are needed at this point. Firstly the summation in (1.21) represents evaluation of the contour integral in the previous equation by the trapezium rule. Secondly one such equation must be written for the centre or 'pivotal' point of each element resulting in a set of M equations for the M unknown values of surface vorticity, $\gamma(1), \gamma(2) \ldots \gamma(M)$. Thirdly $k(s_m, s_n)$ is exactly equal to the velocity parallel to the body surface at s_m induced by a two-dimensional point vortex of unit strength located at s_n.

It will be observed also that $K(s_m, s_n)$ is finite but indeterminate as written for the special case of $n = m$ since both numerator and denominator are then zero. Also implicit in (1.21) is the absorption of the term $-\frac{1}{2}\gamma(s_m)$ into $K(s_m, s_m)$ which may then be written

$$K(s_m, s_m) = -\tfrac{1}{2} + K_{mm}' \qquad (1.23)$$

where $K_{mm}' \cdot \gamma(s_m)$ is the self-induced velocity of element m and the self-inducing coupling coefficient is given by

$$K_{mm}' = \frac{\Delta s_m}{2\pi} \mathop{\mathscr{L}_t}_{s_m \to s_n} \left\{ \frac{(y_m - y_n) \cos \beta_m - (x_m - x_n) \sin \beta_m}{(x_m - x_n)^2 + (y_m - y_n)^2} \right\} \qquad (1.24)$$

For the moment we will assume that the self-induced velocity of each vorticity element is zero, which would be true if the actual body surface were polygonal as depicted in Fig. 1.6(b). If we had chosen to use curved elements then K_{mm}' would be non-zero and we will return to this matter again in the next section to provide a suitable correction for the effect of local surface curvature. For the moment we will write approximately

$$K(s_m, s_m) \approx -\tfrac{1}{2} \tag{1.23a}$$

Equations (1.21) then have the matrix form

$$
\begin{vmatrix}
-0.5 & K_{12} & K_{13} & \cdots & K_{1M} \\
K_{21} & -0.5 & K_{23} & \cdots & K_{2M} \\
K_{31} & K_{32} & -0.5 & \cdots & K_{3M} \\
\cdot & \cdot & \cdot & \cdots & \cdot \\
\cdot & \cdot & \cdot & \cdots & \cdot \\
K_{M1} & K_{M2} & K_{M3} & \cdots & -0.5
\end{vmatrix}
\begin{vmatrix}
\gamma(s_1) \\
\gamma(s_2) \\
\gamma(s_3) \\
\cdot \\
\cdot \\
\gamma(s_M)
\end{vmatrix}
=
\begin{vmatrix}
rhs_1 \\
rhs_2 \\
rhs_3 \\
\cdot \\
\cdot \\
rhs_M
\end{vmatrix}
\tag{1.25}
$$

with simplified notation $K_{mn} \equiv K(s_m, s_n)$ and right hand sides

$$rhs_m \equiv -U_\infty \cos \beta_m - V_\infty \sin \beta_m$$

Example – Flow past a circular cylinder (zero SIV term)
  If the coordinates of a cylinder of radius a are expressed by

$$
\left.
\begin{aligned}
x &= a(1 - \cos \phi) \\
y &= a \sin \phi
\end{aligned}
\right\}
\tag{1.26}
$$

then the exact solution for the surface velocity due to a uniform stream U_∞ parallel to the x axis, Batchelor (1970), is

$$v_s = 2U_\infty \sin \phi \tag{1.27}$$

  Solution of the above linear equations by matrix inversion and multiplication for the 18 element case depicted in Fig. 1.7, produced the results shown in column 2 of Table 1.1.

  Compared with the exact solution v_s is of the correct form but underestimated by a scaling factor of 0.94686 for all elements due to neglecting the element self-inducing velocity term $K_{mm}{}'$. In the next section a correction for this will be derived, resulting in the excellent results recorded in column 3.

Table 1.1. *Plane potential flow past a circle by the surface vorticity method*

Pivotal point	v_s exact solution	$v_s \approx$ surface vorticity method	
		No curvature correction	Including curvature correction
1	0..347 296	0.328 842	0.347 091
2	1.000 000	0.946 858	0.999 420
3	1.532 089	1.450 669	1.531 214
4	1.879 385	1.779 509	1.878 311
5	2.000 000	1.893 713	1.998 856
6	1.879 385	1.779 508	1.878 303
7	1.532 089	1.450 670	1.531 212
8	1.000 000	0.946 857	0.999 420
9	0.347 296	0.328 839	0.347 088
10	0.347 296	0.328 839	0.347 104
11	1.000 000	0.946 856	0.999 432
12	1.532 089	1.450 667	1.531 221
13	1.879 385	1.779 507	1.878 318
14	2.000 000	1.893 711	1.998 865
15	1.879 385	1.779 507	1.878 313
16	1.532 089	1.450 668	1.531 216
17	1.000 000	0.946 853	0.999 435
18	0.347 296	0.328 837	0.347 100

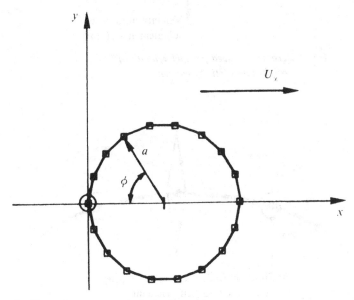

Fig. 1.7. Plane flow past a circle in the z plane.

1.6.1 Self-induced velocity of a surface vorticity element due to curvature

The self-inducing coupling coefficient equation (1.24) may be rewritten

$$K_{mm}' = \frac{\Delta s_m}{2\pi} \mathscr{L}_{\substack{t \\ s_m \to s_n}} \left\{ \frac{(y_m - y_n)\dfrac{\mathrm{d}x_m}{\mathrm{d}s_m} - (x_m - x_n)\dfrac{\mathrm{d}y_m}{\mathrm{d}s_m}}{(x_m - x_n)^2 - (y_m - y_n)^2} \right\} \qquad (1.28)$$

where

$$\left. \begin{aligned} \cos \beta_m &= \frac{\mathrm{d}x_m}{\mathrm{d}s_m} \\[2mm] \sin \beta_m &= \frac{\mathrm{d}y_m}{\mathrm{d}s_m} \end{aligned} \right\} \qquad (1.29)$$

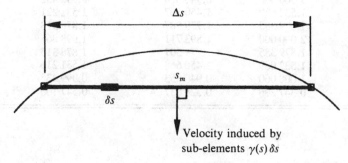

(a)   *Zero self-induced parallel velocity of straight line vorticity element*

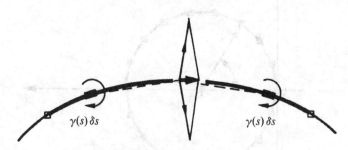

(b) Finite self-induced parallel velocity of a curved vorticity element

Fig. 1.8. Self-induced velocity of a surface vorticity element.

22

Surface vorticity model for plane two-dimensional flow

As already shown, K_{mm}' actually represents the velocity parallel to the body surface at m induced by element m itself with unit strength vorticity. If the element were straight, Fig. 1.8(a), contributions from sub-elements δs would always be normal to the element at s_m resulting in zero self-induced velocity. For a curved element on the other hand, Fig. 1.8(b), there will always be a net induced velocity parallel to the surface. Any finite value possessed by K_{mm}' then is due entirely to element curvature.

Since the coordinates (x_m, y_m) represent local geometry of element m, we can now apply L'Hospital's rule to (1.28) treating them as variables while holding (x_n, y_n) constant. Thus differentiating both numerator and denominator with respect to s_m we obtain

$$K_{mm}' = \frac{\Delta s_m}{4\pi} \mathop{\mathscr{L}}_{s_m \to s_n} \left\{ \frac{(y_m - y_n)\dfrac{d^2 x_m}{ds_m^{\,2}} - (x_m - x_n)\dfrac{d^2 y_m}{ds_m^{\,2}}}{(x_m - x_n)\dfrac{dx_m}{ds_m} + (y_m - y_n)\dfrac{dy_m}{ds_m}} \right\}$$

Since this expression is still indeterminate we must apply L'Hospital's rule a second time resulting in

$$K_{mm}' = \frac{\Delta s_m}{4\pi} \mathop{\mathscr{L}}_{s_n \to s_m}$$

$$\times \left\{ \frac{(y_m - y_n)\dfrac{d^3 x_m}{ds_m^{\,3}} + \dfrac{dy_m}{ds_m}\dfrac{d^2 x_m}{ds_m^{\,2}} - (x_m - x_n)\dfrac{d^3 y_m}{ds_m^{\,3}} - \dfrac{dv_m}{ds_m}\dfrac{d^2 y_m}{ds_m^{\,2}}}{(x_m - x_n)\dfrac{d^2 x_m}{ds_m^{\,2}} + \left(\dfrac{dx_m}{ds_m}\right)^2 + (y_m - y_n)\dfrac{d^2 y_m}{ds_m^{\,2}} + \left(\dfrac{dy_m}{ds_m}\right)^2} \right\}$$

$$= \frac{\Delta s_m}{4\pi} \left\{ \frac{\dfrac{dy_m}{ds_m}\dfrac{d^2 x_m}{ds_m^{\,2}} - \dfrac{dx_m}{ds_m}\dfrac{d^2 y_m}{ds_m^{\,2}}}{\left(\dfrac{dx_m}{ds_m}\right)^2 + \left(\dfrac{dy_m}{ds_m}\right)^2} \right\} \qquad (1.30)$$

K_{mm}' may now be evaluated from the prescribed surface geometry, the second order derivatives indicating the significance of element curvature. Most early workers in this field introduced curve fitting and smoothing procedures for processing tabulated input profile data in order to estimate the first and second order derivatives in (1.30). For example Jacob & Riegels (1963) proposed a Fourier series method to evaluate x_m', x_m'' etc. As pointed out by Wilkinson (1967a) however, Fourier series methods suffer from the fact that while the ordinates of the series converge absolutely to the ordinates of the real curve as the number of terms in the series tends to infinity, the derivatives of the series do not necessarily converge to the derivatives of the real curve and may even diverge from them. Thus curve fitting procedures themselves often have intrinsic difficulties. For example fitting a polynomial exactly through say five successive body surface locations may yield good interpolation of (x, y) values in between but erroneous higher order derivatives.

Added to this is the problem of inaccurate initial profile data. For example, given a tabulated set of data $y_i = f(x_1, x_2, \ldots, x_N)$ where y_i are subject to error, evaluation of y_i' and y_i'' by finite differencing leads to increasing error, suggesting the need for initial data smoothing. The most usual approach is to deal with smoothing and curve fitting simultaneously by adopting a least-squares polynomial procedure. Wilkinson (1967a) investigated the comparative errors in various alternative procedures advocating finally least-squares fitting of a parabola through five points to smooth the initial input data and for subsequent evaluation of the higher order derivatives. D. G. Graham (1972) applied a similar least-squares approach to smoothing with both second- and third-order polynomials followed by the use of a third-order polynomial curve fit of the smoothed data to obtain profile derivatives by a finite differencing technique. The application of this method to aerofoil cascade flows has been reported by D. G. Graham and Lewis (1970).

Curve fitting in practice tends to demand subjective interpretation and can in fact be virtually eliminated if we take the previous analysis one stage further. Bearing in mind the simplicity of the final results it is extraordinary that the following treatment was apparently not explored prior to the report by Lewis (1980). First we introduce the transformations

$$\frac{dy_m}{ds_m} = \frac{dy_m}{dx_m} \frac{dx_m}{ds_m}$$

$$\frac{d^2 y_m}{ds_m{}^2} = \frac{d^2 y_m}{dx_m{}^2} \left(\frac{dx_m}{ds_m}\right)^2 + \frac{dy_m}{dx_m} \frac{d^2 x_m}{ds_m{}^2}$$

K_{mm}' then reduces to

$$K_{mm}' = -\frac{\Delta s_m}{2\pi} \left\{ \frac{\dfrac{dx_m}{ds_m} \dfrac{d^2 y_m}{dx_m{}^2}}{1 + \left(\dfrac{dy_m}{dx_m}\right)^2} \right\}$$

24

dx_m/ds_m may be eliminated from this since

$$ds_m{}^2 = dx_m{}^2 + dy_m{}^2$$

and hence

$$\frac{dx_m}{ds_m} = \frac{1}{\sqrt{\left[1 + \left(\dfrac{dy_m}{ds_m}\right)^2\right]}}$$

Finally we have the simple result that

$$K_{mm}{}' = \frac{\Delta s_m}{4\pi}\left\{\frac{-\dfrac{d^2y_m}{dx_m{}^2}}{\left\{1 + \left(\dfrac{dy_m}{dx_m}\right)^2\right\}^{\frac{3}{2}}}\right\}$$

$$= \frac{\Delta s_m}{4\pi r_m} \approx -\frac{\Delta \beta_m}{4\pi} \tag{1.31}$$

where r_m is the internal radius of curvature of element m and $\Delta\beta_m$ the change of profile slope from one end of the element to the other. As advocated by Lewis (1984a), extremely good potential flow predictions can be obtained with straight line elements provided (1.31) is adopted to correct for element curvature. This is demonstrated by the modified results for circular cylinder flow given in column 3 of Table 1.1. Surface vorticity analysis with curvature correction agreed with the exact solution to well within 0.1% even with so few pivotal points. This method of treatment is similar to circular arc spline fitting as illustrated by Fig. 1.9 which itself is quite a reasonable method for profile generation.

1.6.2 Computational scheme for surface vorticity analysis

All of the equations are now available for preparing a computational scheme and a sample Pascal program circle.pas is included in the Appendix (Program 1.1). The computational stages are as follows.

(i) Input data
Specify $M + 1$ input data coordinates (X_n, Y_n) as illustrated in Fig. 1.6 moving clockwise around the profile from the leading edge.

25

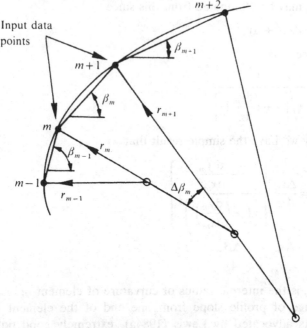

Fig. 1.9. Circular arc spline fit related to use of straight line elements.

Point $M + 1$ coincides with point 1, ensuring profile closure. In Program 1.1 the profile data points are calculated from equations (1.26) within the program.

(ii) Data preparation
Straight line elements are obtained by joining successive data points. Element lengths are then given by

$$\Delta s_n = \sqrt{[(X_{n+1} - X_n)^2 + (Y_{n+1} - Y_n)^2]} \tag{1.32}$$

and the profile slopes follow from equations (1.29) in finite difference form

$$\left. \begin{aligned} \cos \beta_n &= (X_{n+1} - X_n)/\Delta s_n \\ \sin \beta_n &= (Y_{n+1} - Y_n)/\Delta s_n \end{aligned} \right\} \tag{1.33}$$

These values may be used to evaluate the profile slope β_n taking care to select the correct quadrant.

Pivotal points (x_n, y_n) are then located at the centre of each element through

$$\left. \begin{aligned} x_n &= \tfrac{1}{2}(X_{n+1} + X_n) \\ y_n &= \tfrac{1}{2}(Y_{n+1} + Y_n) \end{aligned} \right\} \tag{1.34}$$

(iii) Coupling coefficients

$K(s_m, s_n)$ values are now calculable from (1.22) for $m \neq n$. $K(s_m, s_m)$ values follow from (1.31) where $\Delta\beta_m$ can be estimated as half of the change in slope between the two adjacent elements. Thus

$$K(s_m, s_m) = -\frac{1}{2} - \frac{1}{8\pi}(\beta_{m+1} - \beta_{m-1}) \qquad (1.35)$$

(iv) Matrix inversion

The coupling coefficient matrix has finite values throughout with a dominant leading diagonal, offering no difficulties for solution by matrix inversion. A Pascal procedure economic in memory requirements is included in Program 1.1. This, together with the profile data handling procedure just outlined, is used throughout this text.

(v) Right hand sides

Right hand side values follow directly from (1.21).

(vi) Solution

The solution is obtained by multiplying the inverted matrix by the right hand side column vector rhs.

1.7 Comparison of surface vorticity analysis with Douglas–Neumann Scheme

At this point it would be of interest to apply the Douglas–Neumann method to plane two-dimensional flows for comparison. Replacing the vorticity sheet by a source sheet $\sigma(s)$, introducing the Neumann boundary condition of zero velocity *normal* to the body and using the same line element model, the resulting boundary integral equation is quite similar to (1.9), namely

$$\tfrac{1}{2}\sigma(s_m) + \frac{1}{2\pi}\oint\left\{\frac{(y_m - y_n)\cos\beta_m - (x_m - x_n)\sin\beta_m}{(x_m - x_n)^2 + (y_m - y_n)^2}\right\}\sigma(s_n)\,\mathrm{d}s_n$$

$$+ W_\infty(\sin\alpha_\infty\cos\beta_m - \cos\alpha_\infty\sin\beta_m) = 0 \quad (1.36)$$

This is the plane two-dimensional flow equivalent form of (1.5). We observe that the Kernel is in fact identical to that for the surface vorticity model so that we could express this in numerical form as

27

The basis of surface singularity modelling

before through

$$\sum_{n=1}^{M} \bar{K}(s_m, s_n)\sigma(s_n) = U_\infty \sin \beta_m - V_\infty \cos \beta_m \qquad (1.37)$$

where $\bar{K}(s_m, s_n)$ has the same form as $K(s_m, s_n)$, (1.22).

There are two differences in recommended practice for the source panel method. Firstly, following Chuen-Yen Chow (1979), no account will be taken of element curvature so that

$$\bar{K}(s_m, s_m) = -\tfrac{1}{2} \qquad (1.38)$$

Secondly it is usual to obtain the average value of the coupling coefficient over the straight line source elements, defined by

$$\bar{K}(s_m, s_n) = \frac{1}{\Delta s_m} \int_0^{\Delta s_m} k(s_m, s_n) \, ds_n \qquad (1.39)$$

As shown by Chuen-Yen Chow, this integral can actually be reduced to closed form for plane two-dimensional flow. Alternatively numerical integration may be used based upon the notion of introducing N equal length sub-elements, Fig. 1.10, whereupon

$$\bar{K}(s_m, s_n) = \frac{1}{2\pi N} \sum_{i=1}^{N} \left\{ \frac{(y_m - y_i) \cos \beta_m - (x_m - x_i) \sin \beta_m}{(x_m - x_i)^2 + (y_m - y_i)^2} \right\} \qquad (1.40)$$

where the centre-location of the ith sub-element is

$$\left. \begin{aligned} x_i &= x_n + (i - \tfrac{1}{2}(1 + N)) \frac{\Delta s_n \cos \beta_n}{N} \\ y_i &= y_n + (i - \tfrac{1}{2}(1 + N)) \frac{\Delta s_n \sin \beta_n}{N} \end{aligned} \right\} \qquad (1.41)$$

The solution of equation (1.37) provides the source strength distribution required to satisfy the Neumann boundary condition. Surface velocity may then be calculated by resolving both the uniform stream and the source-induced velocities parallel to the body surface as described by the second integral equation (1.7), which in the present case of plane two-dimensional flow becomes

$$v_{sm} = \frac{1}{2\pi} \oint \left\{ \frac{(x_m - x_n) \cos \beta_m + (y_m - y_n) \sin \beta_m}{(x_m - x_n)^2 + (y_m - y_n)^2} \right\} \sigma(s_n) \, ds_n$$
$$+ U_\infty \cos \beta_m + V_\infty \sin \beta_m \qquad (1.42)$$

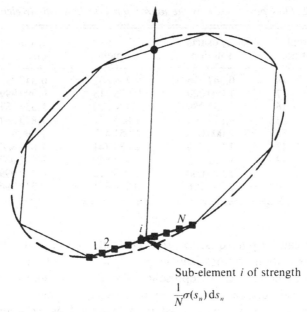

Sub-element i of strength

$$\frac{1}{N}\sigma(s_n)\,\mathrm{d}s_n$$

Fig. 1.10. Use of sub-elements in source panel method.

Following the same numerical strategy, this may be expressed

$$v_{sm} = \sum_{\substack{n=1 \\ n\neq m}}^{N} \bar{L}(s_m, s_n)\sigma(s_n) + U_\infty \cos \beta_m + V_\infty \sin \beta_m \qquad (1.43)$$

where the average source velocity coupling coefficient may be obtained as before by the use of sub-elements,

$$\bar{L}(s_m, s_n) = \frac{1}{\Delta s_n} \int_0^{\Delta s_n} L(s_m, s_i)\,\mathrm{d}s_i$$

$$= \frac{\Delta s_n}{2\pi N} \sum_{i=1}^{N} \left\{ \frac{(x_m - x_i)\cos \beta_m + (y_m - y_i)\sin \beta_m}{(x_m - x_i)^2 + (y_m - y_i)^2} \right\} \qquad (1.44)$$

There are three points to raise here. Firstly the source panel method requires the calculation of two sets of coupling coefficients as compared with only one for the surface vorticity method. Some savings can be made if $\bar{K}(s_m, s_n)$ and $\bar{L}(s_m, s_n)$ are evaluated simultaneously but at the expense of extra memory to hold the $\bar{L}(s_m, s_n)$ matrix. Secondly (1.42) is singular and one might expect possible inaccuracies in its numerical equivalent, (1.43). Since the self-induced velocity of a line source element is zero the term

29

Table 1.2. *Flow past a circle by the source panel method with 18 elements*

Pivotal point no.		v_s exact solution	1 sub-element	200 sub-elements
1	18	0.347 296	0.338 862	0.347 296
2	17	1.000 000	0.975 713	0.999 999
3	16	1.532 089	1.494 879	1.532 087
4	15	1.879 385	1.833 741	1.879 384
5	14	2.000 000	1.951 427	1.999 998
6	13	1.879 385	1.833 741	1.879 384
7	12	1.532 089	1.494 879	1.532 087
8	11	1.000 000	0.975 713	0.999 999
9	10	0.347 296	0.338 861	0.347 296

$\bar{L}(s_m, s_m)$ can be put equal to zero. The outcome of (1.43) is then an approximation to the Cauchy principal value of the singular integral equation which it represents. Thirdly it is usual to ignore element local surface curvature in both $\bar{K}(s_m, s_m)$ and $\bar{L}(s_m, s_m)$. Pascal program source.pas, listed in Appendix I as Program 1.2, undertakes this computation for flow past a circular cylinder for a chosen number of sub-elements. The output for 18 elements is compared with the exact solution in Table 1.2 above. Because of symmetry v_s values for opposite elements, e.g. 1 and 18, 2 and 17 etc., are identical.

Two cases are shown in Table 1.2 for 1 and 200 sub-elements respectively. While reasonable results were obtained with only single element representation, it is of considerable interest to note that the use of 200 sub-elements to evaluate $\bar{K}(s_m, s_n)$ and $\bar{L}(s_m, s_n)$ resulted in an exact representation of the flow past a circular cylinder. If the program is re-run with less surface elements, even with the minimum possible of $M = 4$, the exact solution is still reproduced by the source panel method as presented here. This is extraordinary bearing in mind that the following approximations have been made:

(i) Straight line elements to model the circular body shape.
(ii) Self-induced velocity due to body curvature has been ignored in the governing equation (1.37) for Neumann boundary conditions.
(iii) Sub-elements are assumed also to lie along the straight line elements.

(iv) The source strength is assumed constant along each element when progressing from the integral equation to its numerical equivalent. Only the coupling coefficients are averaged.

Although analysis might reveal why this approximate model yields the exact solution, presumably due to cancellation of error terms, it is not immediately obvious why this is so. Unfortunately introduction of sub-elements into the surface vorticity model and removal of the correction for element curvature self-induced velocity has only a ruinous effect and is not recommended. On the other hand this acceptable but fortuitous result for source panels is true only for circular cylinder flows as illustrated by the following example for inclined flow of a uniform stream $W_\infty = 1.0$ past an ellipse of ratio minor axis/major axis = 0.2, with an angle of attack $\alpha_\infty = 30°$, Fig. 1.11 and Table 1.3.

From this comparison for a body with a widely varying surface radius of curvature, the source panel method no longer gives exact results but is subject to errors which are small and of similar magnitude to those of the surface vorticity model. With 40 elements an accuracy well within 1% was obtained by both methods for the whole surface and increased accuracy can be obtained with yet more elements.

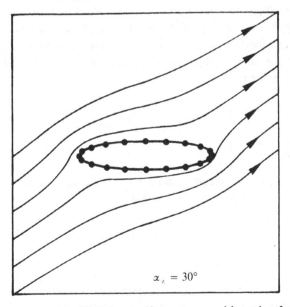

$\alpha_r = 30°$

Fig. 1.11. Flow past an ellipse in a uniform stream with angle of attack 30°.

31

Table 1.3. *Flow past ellipse by surface vorticity and source panel methods.* (*Fig. 1.11*)

Pivotal point no.		Surface vorticity method	Source panels 200 sub-elements	Exact v_s
1	11	3.871 989	3.829 432	3.852 799
2	12	2.402 106	2.367 435	2.340 044
3	13	1.682 961	1.675 428	1.664 101
4	14	1.281 192	1.279 385	1.274 276
5	15	0.982 305	0.984 278	0.982 429
6	16	0.707 595	0.713 117	0.713 777
7	17	0.401 048	0.410 387	0.414 037
8	18	−0.023 193	0.009 192	0.000 000
9	19	−0.810 255	−0.786 361	−0.760 326
10	20	−2.834 843	−2.824 572	−2.799 222

1.8 Calculation of streamlines and velocities within the flow field

Once the vortex element strengths $\gamma(s_n)$ have been determined, it is a simple matter to calculate streamline patterns and velocities throughout the flow domain. Two methods are available for locating streamlines. The first method involves calculation of the stream function ψ throughout the field of interest, followed by interpolation for prescribed streamlines ψ_1, $\psi_2 \dots$ etc. The stream functions for a uniform stream $W_\infty = U_\infty + iV_\infty$ and a point vortex Γ at the origin, Batchelor (1970), are

$$\left. \begin{array}{l} \psi_{W_\infty} = U_\infty y - V_\infty x \\ \\ \psi_\Gamma = \dfrac{\Gamma}{2\pi} \ln(r) \end{array} \right\} \tag{1.45}$$

If we give ψ_{W_∞} the datum value zero at the origin, then the stream function at any other point (x_m, y_m) will be

$$\psi_m = U_\infty y_m - V_\infty x_m + \frac{1}{2\pi} \sum_{n=1}^{M} \gamma(s_n)\, \Delta s_n \ln(r_{mn}) \tag{1.46}$$

A second and in some ways more convenient method involves tracing out the coordinates of each streamline by moving along the local velocity vector q_m in discrete steps. Thus the velocity

32

components at (x_m, y_m), making use of equations (1.18), are

$$
\left.
\begin{aligned}
u_m &= U_\infty + \frac{1}{2\pi} \sum_{n=1}^{M} \gamma(s_n)\, \Delta s_n \left(\frac{y_m - y_n}{r_{mn}^{2}} \right) \\
v_m &= V_\infty - \frac{1}{2\pi} \sum_{n=1}^{M} \gamma(s_n)\, \Delta s_n \left(\frac{x_m - x_n}{r_{mn}^{2}} \right)
\end{aligned}
\right\}
\tag{1.47}
$$

For a discrete displacement $\Delta \ell$, a first order or forward differencing estimate of the next location on the streamline would be

$$
\left.
\begin{aligned}
x_{m+1} &= x_m + (u_m/q_m)\, \Delta \ell \\
y_{m+1} &= y_m + (v_m/q_m)\, \Delta \ell
\end{aligned}
\right\}
\tag{1.48}
$$

A useful check upon the accuracy of this numerical process can be obtained by applying it to a single vortex Γ at the origin with no uniform stream, whereupon, at radius r

$$
\left.
\begin{aligned}
x_{m+1} &= x_m - \frac{y_m}{r_m}\, \Delta \ell \\
y_{m+1} &= y_m + \frac{x_m}{r_m}\, \Delta \ell
\end{aligned}
\right\}
\tag{1.49}
$$

In reality streamlines are circular. However, even with as many as 50 time steps, Fig. 1.12, we see that the predicted streamline spirals outward significantly due to the errors of the forward differencing scheme, which implies always proceeding tangential to the stream-line, Fig. 1.12(*a*). A considerable improvement is possible by implementing the central difference type of procedure illustrated in Fig. 1.12(*b*). The forward difference equations are used for the first step leading to initial estimates, point *b*

$$
\left.
\begin{aligned}
x_{m+1}' &= x_m + (u_m/q_m)\, \Delta \ell \\
y_{m+1}' &= y_m + (v_m/g_m)\, \Delta \ell
\end{aligned}
\right\}
\tag{1.48a}
$$

For these positions equations (1.47) are again evaluated resulting in velocities u_m' and v_m' applicable at point *b*. Taking the average velocity components, as shown vectorially in Fig. 1.12(*b*), results in

$$
\left.
\begin{aligned}
x_{m+1} &= x_m + \tfrac{1}{2}(u_m/q_m + u_m'/q_m')\, \Delta \ell \\
y_{m+1} &= y_m + \tfrac{1}{2}(v_m/q_m + v_m'/q_m')\, \Delta \ell
\end{aligned}
\right\}
\tag{1.50}
$$

This process can be repeated several times for further improvement. Almost perfect closure was obtained with three iterations for twenty steps and with ten iterations for only six steps. Inviscid

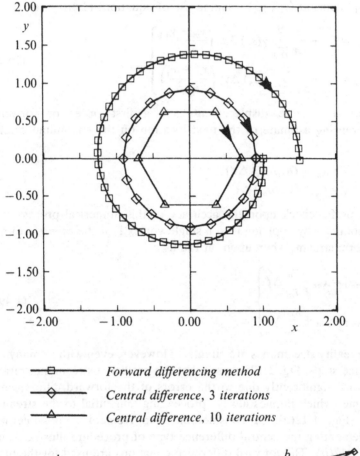

---□--- *Forward differencing method*

---◇--- *Central difference, 3 iterations*

---△--- *Central difference, 10 iterations*

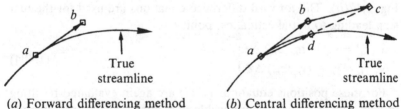

(a) Forward differencing method   (b) Central differencing method

Fig. 1.12. Forward and central differencing estimates of streamline path.

potential flows are thermodynamically reversible, which means in this context that the flow is actually physically reversible if run backwards in time. For the flow under consideration this will be true if the predicted streamlines form a closed circle. It is recommended that, say, five iterations are performed and that occasionally a check on reversibility is made by retracting a streamline with steps length changed to $-\Delta \ell$.

A special attraction of this technique is its suitablity to computer plotting. In addition the velocity components u_m, v_m are also known along the streamlines without further effort. The predicted flow pattern for an ellipse in a uniform stream with 30° angle of attack are shown in Fig. 1.11 using the central difference method with five iterations. These results were obtained from Program No. 1.3 of the Appendix. The only other refinements not included in this program are the introduction of sub-elements to avoid slight undulation in streamlines close to the body surface and special consideration to detect arrival at the body surface should the selected starting location (x_1, y_1) lie on the stagnation streamline.

1.9   Flows with symmetry about the x axis

Referring back to the tabulated output for flow past a circular cylinder in a uniform stream parallel to the x axis, Table 1.1, we observe a four-fold redundancy in the statement of the problem in Section (1.6) in view of symmetry of the cylinder and flow about both x and y axes. At the outset we could have observed that elements reflected in these axes would have the same vortex strength (although those reflected in the x axis would have the opposite sign since $\gamma(s)$ is anticlockwise). For example we can state that

$$\gamma(s_1) = \gamma(s_9) = -\gamma(s_{10}) = -\gamma(s_{18})$$
$$\gamma(s_2) = \gamma(s_8) = -\gamma(s_{11}) = -\gamma(s_{17}) \tag{1.51}$$
etc.

Taking advantage of this we can reduce the number of columns in the coupling coefficient matrix, (1.25), to five involving only $\gamma(s_1)$ to $\gamma(s_5)$ by combining columns with identical element strengths. For example column 1 would become

$$(K_{1,1} + K_{1,9} - K_{1,10} - K_{1,18})$$
$$(K_{2,1} + K_{2,8} - K_{2,11} - K_{2,17})$$
etc.

To retain a square matrix we need keep only the first five equations since these are the ones which state the boundary conditions at our remaining elements $\gamma(s_1)$ to $\gamma(s_5)$.

35

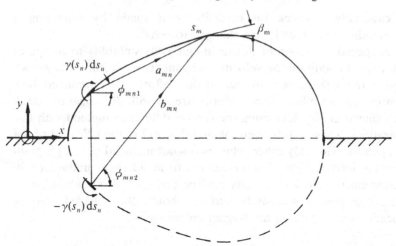

Fig. 1.13. Mirror image system for modelling flow past a body with symmetry about the x axis. Reproduced by courtesy of the Institution of Mechanical Engineers.

Alternatively we could have eliminated half of the redundancy for symmetry about the x axis only by retaining $\gamma(s_1)$ to $\gamma(s_9)$, still saving computational effort while permitting asymmetry about the y axis. This approach opens the way to analysis of flow past two-dimensional bodies attached to an infinite plane wall, Fig. 1.13.

For such problems a better initial approach is to consider first the mirror image reflection system required to produce symmetry. Each element $\gamma(s_n)\,ds_n$ can then be combined with its reflection in the x axis, $-\gamma(s_n)\,ds_n$, to form a vortex pair, for which the combined coupling coefficient can be written as follows

$$K(s_m, s_n) = \frac{\Delta s_n}{2\pi}\left\{\frac{1}{a_{mn}}\sin(\phi_{mn1} - \beta_m) - \frac{1}{b_{mn}}\sin(\phi_{mn2} - \beta_m)\right\}$$

$$= \frac{\Delta s_n}{2\pi}\left[\left\{\frac{y_m - y_n}{a_{mn}{}^2} - \frac{y_m + y_n}{b_{mn}{}^2}\right\}\cos\beta_m\right.$$

$$\left. - \left\{\frac{1}{a_{mn}{}^2} - \frac{1}{b_{mn}{}^2}\right\}(x_m - x_n)\sin\beta_m\right] \qquad (1.52)$$

where

$$\left.\begin{aligned}a_{mn} &= \sqrt{\{(x_m - x_n)^2 + (y_m - y_n)^2\}}\\b_{mn} &= \sqrt{\{(x_m - x_n)^2 + (y_m + y_n)^2\}}\end{aligned}\right\} \qquad (1.53)$$

In this special situation we should observe that the self-induced coupling coefficient equation (1.35) must be modified to include the mirror image vortex, resulting in

$$K(s_m, s_m) = -\frac{1}{2} - \frac{\Delta \beta_m}{4\pi} - \frac{\cos \beta_m}{4\pi y_m} \tag{1.54}$$

As a preparation for extending this method to flow separation from sharp edged bluff bodies, to be considered in Chapter 7, calculations were undertaken by Lewis (1981) for flow over a ridge, Fig. 1.14, subjecting the surface vorticity method to the extreme

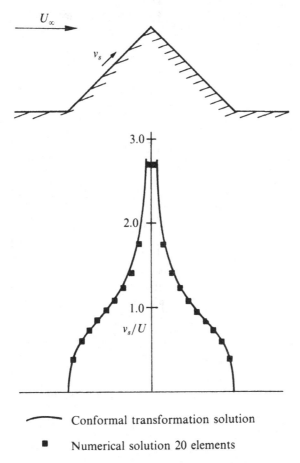

——— Conformal transformation solution

■   Numerical solution 20 elements

Fig. 1.14. Potential flow over a ridge. Reproduced by courtesy of the Institution of Mechanical Engineers.

conditions of modelling a potential flow singularity. A Schwarz–Christoffel conformal transformation is available to deal with this case. Flow over the ridge involves acceleration from a stagnation point at the base to infinite velocity over the crest. With just twenty evenly spaced pivotal points a quite creditable solution was ob-

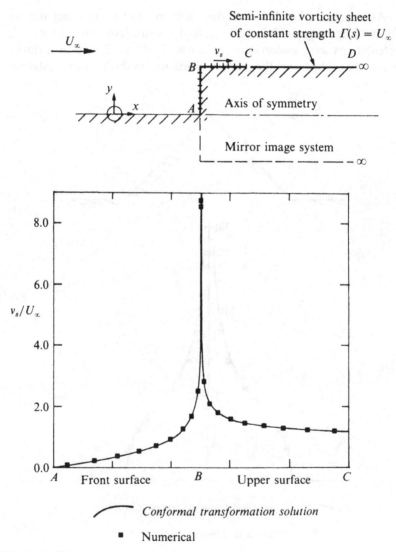

Fig. 1.15. Flow over a step in a plane wall. Reproduced by courtesy of the Institution of Mechanical Engineers.

tained, the worst errors occurring naturally enough in the neighbourhood of the flow singularity at the crest.

An equally rigorous test of vorticity modelling is presented by the problem of the potential flow past a step discontinuity of height $y = 1.0$ in a plane wall. As illustrated in Fig. 1.15, this can be treated as the flow past a body which is symmetrical about the x axis but extends all the way to $+\infty$ downstream. In this problem the velocity increases from stagnation at A to infinity at the sharp corner B moving along the front face. Downstream of this the flow decelerates back again to U_∞ as $x \rightarrow \infty$ at D. As shown by the conformal transformation solution, most but not all of this velocity recovery is accomplished by the time the fluid reaches point C where $BC = AB$. Since it is hardly practicable to cover the whole of BD with discrete elements, a semi-infinite vortex sheet of prescribed constant strength $\Gamma(s) = U_\infty$ was imposed between C and D together with its mirror image reflection. Ten elements were chosen on both AB and BC, with pivotal points concentrated towards point B using a square law distribution to emphasise the more rapid velocity variation around the sharp corner. Comparison with the exact solution was excellent even though the velocity recovery is still incomplete at point C.

1.10 Generalised equations for surface vorticity modelling in curvilinear coordinates

When modelling a particular flow problem it is normally feasible to proceed quickly to a simple geometrical discretisation of the boundary into elements as we have just illustrated by several examples. For three-dimensional bodies this may be less straightforward. To conclude this chapter we shall express some of the basic equations of the surface vorticity method in generalised curvilinear coordinates and through this treatment bring out some additional features of three-dimensional modelling.

Adopting curvilinear coordinates (u_1, u_2), Fig. (1.16), the surface velocity in potential flow past a three-dimensional body will in general have two components v_{s1} and v_{s2}. If the surface vorticity likewise is resolved into two components $\mathbf{i}_1 \gamma_1(u_1, u_2)$ and $\mathbf{i}_2 \gamma_2(u_1, u_2)$, then from consideration of circulations about elements $ds_1 = h_1 \, du_1$ and $ds_2 = h_2 \, du_2$ we have, (by analogy with (1.12) for two-dimensional flow).

39

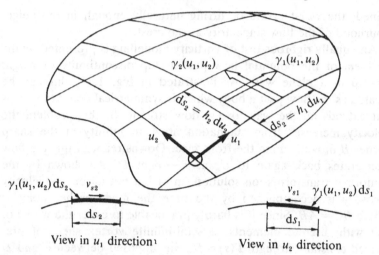

Fig. 1.16. Surface vorticity components of a vortex sheet. Reproduced by courtesy of the Institution of Mechanical Engineers.

$$v_{s1} = \gamma_2(u_1, u_2)$$
$$v_{s2} = -\gamma_1(u_1, u_2) \qquad\qquad (1.55)$$

$\mathbf{i}_1$ and $\mathbf{i}_2$ are unit vectors along the coordinate axes (u_1, u_2) and $\mathbf{i}_3$ is the unit vector normal to the surface. The condition of irrotationality on the body surface can be stated

$$\operatorname{curl} \mathbf{v}_s = 0$$

which, in curvilinear coordinates becomes

$$\frac{1}{h_1 h_2}\left\{ \frac{\partial}{\partial u_1}(h_2 v_{s2}) - \frac{\partial}{\partial u_2}(h_1 v_{s1}) \right\} = 0 \qquad (1.56)$$

From this a continuity equation for the vorticity sheet may be derived by substitution from (1.55), namely

$$\operatorname{div} \boldsymbol{\gamma} = 0$$

or in curvilinear coordinates

$$\frac{1}{h_1 h_2}\left\{ \frac{\partial}{\partial u_1}[h_2 \gamma_1(u_1, u_2)] + \frac{\partial}{\partial u_2}[h_1 \gamma_2(u_1, u_2)] \right\} = 0 \qquad (1.57)$$

This equation, which constitutes an application of Helmholtz's vortex theorem to the vortex sheet, expresses the interdependency of the vorticity components $\gamma_1(u_1, u_2)$ and $\gamma_2(u_1, u_2)$. In a three-dimensional problem only one of these is truly independent. For purposes of illustration $\gamma_1(u_1, u_2)$ could be thought of as 'bound' vorticity giving rise to 'shed' vorticity of strength

$$d[h_1\gamma_2(u_1, u_2)] = -\frac{\partial}{\partial u_1}[h_2\gamma_1(u_1, u_2)]\, du_2 \qquad (1.58)$$

$\gamma_2(u_1, u_2)$ can therefore always be expressed in terms of $\gamma_1(u_1, u_2)$ for a three-dimensional body surface grid, thereby reducing by one half the number of unknowns in a numerical scheme to equal that of the source panel method.

Applying the Biot–Savart law, (1.8), the induced velocity at m due to a surface vorticity element at n at a radial distance of r_{mn}, is then

$$d\mathbf{v}_{mn} = \frac{[\mathbf{i}_{1n}\gamma_1(u_{1n}, u_{2n}) + \mathbf{i}_{2n}\gamma_2(u_{1n}, u_{2n})]X\mathbf{r}_{mn}}{4\pi r_{mn}^{\,3}} h_{1n}h_{2n}\, du_{1n}\, du_{2n}$$

$$(1.59)$$

The resolved component of this parallel to the surface at point m then follows from

$$d\mathbf{v}_{smn} = \mathbf{i}_{3m}X(d\mathbf{y}_{mn}X\mathbf{i}_{3m}) \qquad (1.60)$$

where the unit vectors $(\mathbf{i}_{1m}, \mathbf{i}_{2m}, \mathbf{i}_{3m})$ are parallel to the curvilinear coordinate directions at m and thus occupy a different spatial orientation from the unit vectors at n on the element, $(\mathbf{i}_{1n}, \mathbf{i}_{2n}, \mathbf{i}_{3n})$. Combining (1.59) and (1.60) we have finally the vector form of Martensen's equation for three-dimensional flows expressed in general curvilinear coordinates as given originally by Lewis & Ryan (1972).

$$-\frac{1}{4\pi} \oint_s \frac{1}{r_{mm}^{\,3}}\mathbf{i}_{3m}X\{[\{\mathbf{i}_{1n}\gamma_1(u_{1n}, u_{2n})$$

$$+\,\mathbf{i}_{2n}\gamma_2(u_{1n}, u_{2n})\}X\mathbf{r}_{mn}]X\mathbf{i}_{3m}\}\, ds_{1n}\, ds_{2n}$$

$$-\tfrac{1}{2}\{\mathbf{i}_{1m}\gamma_2(u_{1m}, u_{2m}) - \mathbf{i}_{2m}\gamma_1(u_{1m}, u_{2m})\}$$

$$+\,\mathbf{i}_{3m}X(\mathbf{W}_\infty X\mathbf{i}_{3m}) = 0 \qquad (1.61)$$

This generalised equation may be used to produce working equations in coordinate systems appropriate to any particular problem. Since the unit vectors $(\mathbf{i}_1, \mathbf{i}_2, \mathbf{i}_3)$ change orientation over the curvilinear grid, they must first be expressed in terms of local panel geometrical variables. For example in Chapter 4 this equation

The basis of surface singularity modelling

will be applied to axisymmetric flow past an annular aerofoil, for which it is convenient to adopt a polar coordinate system (x, r, θ). Since it is then convenient to use unit vectors parallel to these directions, these must first be related to the curvilinear unit vectors through transformations such as (4.3a). There is little merit in proceeding further here with such analysis which is specific to each particular application. It is sufficient to point out that completion of the vector products in (1.61) for three-dimensional problems will result in general in two sets of equations for the Dirichlet boundary condition applicable to coordinate directions u_1 and u_2 respectively of the form

$$
\left.
\begin{aligned}
\tfrac{1}{2}\gamma_{1m} + \frac{1}{4\pi} \oiint_s (\gamma_{1n}K_{nm}' + \gamma_{2n}L_{nm}')\, ds_{1n}\, ds_{2n} + v_{t2m} = 0 \\[2mm]
-\tfrac{1}{2}\gamma_{2m} + \frac{1}{4\pi} \oiint_s (\gamma_{1n}K_{nm}'' + \gamma_{2n}L_{nm}'')\, ds_{1n}\, ds_{2n} + v_{t1m} = 0
\end{aligned}
\right\}
\tag{1.62}
$$

with the simplified notation $\gamma_{1n} \equiv \gamma_1(u_{1n}, u_{2n}) \ldots$ etc. v_{t1n} and v_{t2m} are the components of the uniform stream W_∞ parallel to the surface at m in the directions of the curvilinear coordinates (u_{1n}, u_{2n}).

It will be observed that the kernels of these integrals involve Biot–Savart law contributions from both bound and shed vorticities $\gamma_1(u_{1n}, u_{2n})$ and $\gamma_2(u_{1n}, u_{2n})$ expressed through coupling coefficients K_{nm}', L_{mn}' for the first equation and K_{nm}'', L_{mn}'' for the second equation. If $\gamma_1(u_{1n}, u_{2n})$ and $\gamma_2(u_{1n}, u_{2n})$ were independent there would thus be $2M$ unknown surface vorticity values for a surface representation by M vortex panels for which (1.62) would provide $M + M = 2M$ different linear equations. The greatest merit of such an approach is its simplicity since equations (1.62) provide an explicit statement which remains uncomplicated by the continuity equation (1.58), which, as we have already mentioned, expresses the shed vorticity $\gamma_2(u_1, u_2)$ as a direct function of the bound vorticity $\gamma_1(u_1, u_2)$.

Such a statement of the problem, simple though it is for conversion into a numerical procedure, involves 50% redundancy. The alternative approach normally followed is to solve the first equation only for the Dirichlet boundary condition parallel to the principal curvilinear coordinate direction u_1, thus involving only M equations for the unknown $\gamma_1(u_{1n}, u_{2n})$ values. Use must first be

42

made of the vorticity sheet continuity equation (1.58) in order to eliminate $\gamma_2(u_{1n}, u_{2n})$ from the kernel of the integral. In practice this can lead to extremely complex analysis, examples of which by V. P. Hill (1975, 1978) and Turbal (1973) are referred to in Chapter 6 for flow past an annular aerofoil with angle of attack.

CHAPTER 2

Lifting bodies, two-dimensional aerofoils and cascades

2.1 Introduction

An outline computational scheme was developed in Chapter 1 for application of the surface vorticity method to two-dimensional flow past non-lifting bodies of arbitrary shape. In the fields of aeronautics and engine aerodynamics on the other hand there is a special interest in lifting bodies and control surfaces such as aerofoils, struts and turbine, compressor or fan blades. The objective of this chapter is to extend the analysis to deal with these important applications which exhibit three features not yet considered, namely:

  (i) Such devices are required to generate lift, associated with net bound circulation on the body.
 (ii) In the applications cited the lifting surfaces are normally thin foils for which special computational problems arise due to the close proximity of vorticity elements on opposite sides of the profile.
(iii) A device may involve an assembly of several lifting bodies, taking deliberate advantage of their mutual aerodynamic interference.

We will deal with these matters in turn beginning with an extension of flow past a circular cylinder, Section 1.6, to the case of the Flettner rotor or lifting rotating cylinder, Section 2.2. Progressing to the closely related problem of flow past an ellipse, Sections 2.3 and 2.4, problems of type (ii) will be dealt with for the treatment of thin non-lifting and lifting bodies. This leads naturally into the case of generalised thin aerofoils, Section 2.5, for which comparisons will be provided from Joukowski's exact solutions. These early sections of this chapter lay down very important basic methodology for the surface vorticity method applicable to lifting aerofoils. Comparison with exact solutions is vital in the preliminary stages of program development. For this reason the reader is provided with all the necessary equations to achieve this.

In the remainder of the chapter consideration will be given to the problem of the flow past systems of multiple aerofoils taking advantage of mutual aerodynamic interference. These fall into two categories dealt with in Sections 2.6 and 2.7 respectively. The first category, of great importance in the field of turbomachines, involves the flow through aerofoil cascades, the problem originally addressed by Martensen (1959). The second involves the use of slots and flaps in close proximity to an aerofoil in order to modify its aerodynamic characteristics.

2.2 Circular cylinder with bound circulation – Flettner rotor

The principle of generating lift by spinning a cylinder in a uniform stream has been known since the early days of aeronautics and indeed formed an important part of the foundations of elementary aerofoil theory, Glauert (1926). In potential flow this device may be modelled quite simply by introducing a bound vortex Γ at the centre of the cylinder of radius r_0 in a uniform stream U_∞. Various flow patterns are then generated, depending upon the dimensionless parameter $\Gamma/U_\infty r_0$, Fig. 2.1. Since an exact solution to this flow problem is available, which may be extended by conformal transformation to the flow past an ellipse or a Joukowski aerofoil, it will help the reader in program development and testing if we outline this theory in parallel with surface vorticity analysis.

Following Glauert, the rotating cylinder may be modelled by locating a doublet of strength $\lambda = 2\pi U_\infty r_0^2$ and a point vortex Γ at its centre, assumed to coincide with the origin of the z plane. The complex potential is then

$$\omega = U_\infty\left(z + \frac{r_0^2}{z}\right) + i\frac{\Gamma}{2\pi}\ln z \qquad (2.1)$$

from which we may obtain the velocity v_s on the cylinder surface $(z = r_0 e^{i\theta})$.

$$v_s = \left|\frac{d\omega}{dz}\right|_{z=r_0} = 2U_\infty \sin\theta + \frac{\Gamma}{2\pi r_0} \qquad (2.2)$$

For the general uniform stream $W_\infty = U_\infty + iV_\infty$ with angle of attack $\alpha_\infty = \arctan(V_\infty/U_\infty)$, we can show by rotating coordinates

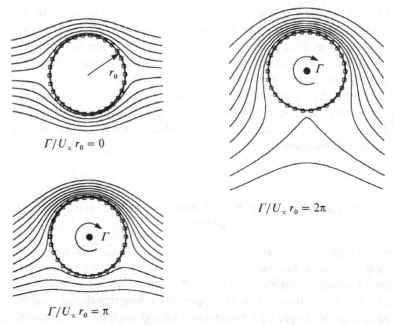

$\Gamma/U_x r_0 = 0$

$\Gamma/U_x r_0 = 2\pi$

$\Gamma/U_x r_0 = \pi$

Fig. 2.1. Flow induced by cylinder with circulation in a uniform derived by the surface vorticity method.

through α_∞ that this expression transforms to

$$v_s = 2W_\infty \sin(\theta - \alpha_\infty) + \frac{\Gamma}{2\pi r_0} \qquad (2.2a)$$

As shown by Glauert by integrating the surface pressure on the cylinder, a lift force L is generated in the direction normal to W_∞ given by the Magnus law.

$$L = \rho W_\infty \Gamma \qquad (2.3)$$

Introducing the usual definition of lift coefficient

$$C_L = \frac{L}{\frac{1}{2}\rho W_\infty^2 \ell} \qquad (2.4)$$

where ℓ is a typical dimension of the body, in this case its diameter $2r_0$, we have

$$C_L = \Gamma/W_\infty r_0 \qquad (2.5)$$

The dimensionless parameter $\Gamma/U_\infty r_0$ previously referred to is in fact the lift coefficient with uniform stream U_∞. To estimate the appropriate rotor speed of rotation Ω for creation of the streamline patterns and related lift coefficients shown in Fig. 2.1, (which have in fact been calculated by the surface vorticity model we are about to derive), we may observe that the fluid immediately in contact with the cylinder surface beneath the surface vorticity sheet, must have velocity $v_{si} = r_0\Omega$. Thus we obtain the relationships

$$\Gamma = v_{si}2\pi r_0 = 2\pi\Omega r_0{}^2 \qquad (2.6)$$

and

$$C_L = \frac{2\pi\Omega r_0}{U_\infty} \qquad (2.7)$$

Although actual streamline patterns quite similar to the theoretical solutions of Fig. 2.1 have been illustrated by Prandtl and Tietjens (1931), Batchelor (1970) and other authors, the corresponding rotor speeds must in practice be very much higher.

As a first step towards a general treatment for lifting aerofoils, let us now extend the surface vorticity analysis of Section 1.6 to deal with the Flettner rotor. Since all of the bound vorticity is now to be found on the cylinder surface, we may relate this to the bound vortex Γ of the classical model through

$$\Gamma = \oint \gamma(s)\,\mathrm{d}s = \sum_{n=1}^{M} \gamma(s_n)\,\Delta s_n \qquad (2.8)$$

Since Γ is to be prescribed as an additional constraint upon the flow regime, we now have an extra equation linking the element strengths $\gamma(s_n)$, in addition to those already stated for the non-lifting surface vorticity scheme, namely

$$\sum_{n=1}^{M} K(s_m, s_n)\gamma(s_n) = -U_\infty \cos\beta_m - V_\infty \sin\beta_m \qquad (1.21)$$

To overcome the problem posed by having $M + 1$ independent linear equations for only M unknowns, three possibilities may be considered. Firstly we might replace one Martensen equation by (2.8), not an attractive option since we would lose control of the boundary condition at one element. Secondly we could add (2.8) to say the first equation. This option, explored by Lewis (1984a) leads to good results except for the element thus compromised. Thirdly and best of all when put to the test, we could add (2.8) to all

Two-dimensional aerofoils and cascades

Table 2.1. *Surface velocities for a circular cylinder with circulation in a uniform stream U_∞*

		Case 1		Case 2	
		$U_\infty = 1.0$	$\Gamma = \pi$	$U_\infty = 0.0$	$\Gamma = \pi$
x	y	numerical	exact	numerical	exact
0.005 463	0.051 978	1.210 296	1.209 056	1.001 265	0.999 999
0.027 077	0.153 662	1.619 229	1.618 033	1.001 265	0.999 999
0.069 359	0.248 630	2.001 131	1.999 999	1.001 258	0.999 999
0.130 463	0.332 733	2.339 376	2.338 260	1.001 271	0.999 999
0.207 717	0.402 293	2.619 093	2.618 033	1.001 260	0.999 999
0.297 746	0.454 270	2.828 127	2.827 090	1.001 262	0.999 999
0.396 614	0.486 395	2.957 322	2.956 294	1.001 264	0.999 999
0.500 000	0.497 261	3.001 016	2.999 999	1.001 263	0.999 999
0.603 386	0.486 395	2.957 316	2.956 294	1.001 263	0.999 999
0.702 254	0.454 270	2.828 130	2.827 090	1.001 262	0.999 999
0.792 283	0.402 293	2.619 100	2.618 033	1.001 264	0.999 999
0.869 537	0.332 732	2.339 360	2.338 260	1.001 262	0.999 999
0.930 641	0.248 630	2.001 139	1.999 999	1.001 262	0.999 999
0.972 923	0.153 662	1.619 217	1.618 033	1.001 261	0.999 999
0.994 537	0.051 978	1.210 293	1.209 056	1.001 262	0.999 999
0.994 537	0.051 978	0.792 233	0.790 942	1.001 263	0.999 999
0.972 923	0.153 662	0.383 301	0.381 965	1.001 261	0.999 999
0.930 641	0.248 631	0.001 379	0.000 001	1.001 260	0.999 999
0.869 537	0.332 733	−0.336 835	0.338 262	1.001 262	0.999 999
0.792 283	0.402 293	−0.616 574	0.618 035	1.001 263	0.999 999
0.702 254	0.454 270	−0.825 604	0.827 092	1.001 260	0.999 999
0.603 386	0.486 395	−0.954 792	0.956 296	1.001 264	0.999 999
0.500 000	0.497 261	−0.998 492	1.000 001	1.001 264	0.999 999
0.396 614	0.486 395	−0.954 793	0.956 296	1.001 260	0.999 999
0.297 746	0.454 270	−0.825 597	0.827 092	1.001 265	0.999 999
0.207 717	0.402 293	−0.616 570	0.618 035	1.001 264	0.999 999
0.130 463	0.332 732	−0.336 834	0.338 262	1.001 261	0.999 999
0.069 359	0.248 630	0.001 387	0.000 001	1.001 261	0.999 999
0.027 077	0.153 662	0.383 305	0.381 965	1.001 263	0.999 999
0.005 463	0.051 978	0.792 238	0.790 943	1.001 266	0.999 999

Martensen equations resulting in the following modified form of (1.21)

$$\sum_{n=1}^{M} (K(s_m, s_n) + \Delta s_n)\gamma(s_m) = -U_\infty \cos \beta_m - V_\infty \sin \beta_m + \Gamma \qquad (2.9)$$

Relevant changes to Program 1.1 to achieve this are introduced into Program 2.1, leading to results for two sample cases shown in Table 2.1 with V_∞ equal to zero. Case 2 ($\Gamma = \pi$, $U_\infty = 0.0$) is of

48

special interest since it corresponds to the case of cylinder rotation in the absence of a uniform stream. For this case the surface velocity $r_0 \Omega$ for $r_0 = 0.5$ is constant and equal to 1.0. Both cases were predicted to within 0.13% using 30 elements. Streamline patterns for Γ values of 0, π (as for case 1) and 2π, calculated by the technique outlined in Section 1.10, are shown also in Fig. 2.1.

2.3   Flow past a thin ellipse

As the next step towards a surface vorticity computational scheme for aerofoils let us reconsider the case of flow past an ellipse introduced in Section 1.9. The non-lifting ellipse analysed there was of 20% thickness/chord ratio and was handled with fair precision by Program 1.3 with only 20 pivotal points. For very thin bodies on the other hand serious errors may arise. Before we investigate this problem in depth let us first extend the previous exact solution for flow past a circle in the z plane to flow past an ellipse in the ζ plane to provide the means of comparison with the numerical method. This can be achieved by introducing Joukowski's transformation

$$\zeta = z + \frac{a^2}{z} \tag{2.10}$$

where a is a constant.

By substitution of the equation for the circle $z = r_0 e^{i\theta}$, the coordinates of the ellipse in the ζ plane become

$$\left.\begin{aligned}
\xi &= \left(r_0 + \frac{a^2}{r_0}\right)\cos\theta \\
\eta &= \left(r_0 - \frac{a^2}{r_0}\right)\sin\theta
\end{aligned}\right\} \tag{2.11}$$

The ellipse thickness ratio λ is then given by

$$\lambda = \frac{\text{minor axis}}{\text{major axis}} = \frac{2\eta_{\max}}{2\xi_{\max}} = \frac{r_0^2 + a^2}{r_0^2 - a^2} \tag{2.12}$$

Now the fluid velocity on the circle in the z plane is already known from (2.2a) and the velocity relationship between the two planes may be obtained by differentiating the complex potential,

$$u_z - iv_z = \frac{d\omega}{dz} = (u_\zeta - iv_\zeta)\frac{d\zeta}{dz}$$

49

The velocity on the surface of the ellipse then follows by taking the modulus of both sides

$$v_{s_\zeta} = v_{s_z} \bigg/ \left| \frac{d\zeta}{dz} \right|$$

Differentiating the transformation equation (2.10) and taking its modulus we have finally, after some analytical reductions, the exact solution for surface velocity on the ellipse, namely

$$v_s = \frac{2W_\infty \sin(\theta - \alpha_\infty) + \Gamma/2\pi r_0}{\sqrt{\left[1 + \left(\dfrac{a}{r_0} \right)^4 - 2 \left(\dfrac{a}{r_0} \right)^2 \cos 2\theta \right]}} \tag{2.13}$$

where the ratio (a/r_0) is a function of the ellipse thickness ratio λ. Rearranging (2.10) we obtain

$$\frac{a}{r_0} = \sqrt{\left(\frac{1-\lambda}{1+\lambda} \right)} \tag{2.10a}$$

where, from (2.12)

$$r_0 = (\text{major axis})(1+\lambda)/4 \tag{2.12b}$$

2.3.1   Reconsideration of non-lifting ellipse

These formulations were in fact implemented in Program 1.3 for the case of non-lifting flow past an ellipse ($\Gamma = 0$). As we have observed, Table 1.3, accurate results were obtained for $\lambda = 0.2$ with just 20 elements. On the other hand if the calculation is repeated for a 5% thick elliptic profile ($\lambda = 0.05$), significant errors occur even with a 40 element representation. This situation is illustrated in Fig. 2.2 where the exact solution equation (2.13) with $\Gamma = 0$ is compared with surface vorticity solutions by Program 1.3 for 60, 40 and 30 element representations. Although excellent precision was obtained with 60 elements, unacceptable errors were creeping in with 40 elements. A further reduction in surface resolution to 30 elements led to disastrous results.

This problem is caused primarily by the inadequacy of the coupling coefficients as given by (1.22) to represent correctly the mutual interference of opposite elements for thin body shapes. The nature of the problem is illustrated by Fig. 2.3 in which the coupling coefficients $k(s_m, s_n)$ have been plotted versus s for element 8 due to

Flow past a thin ellipse

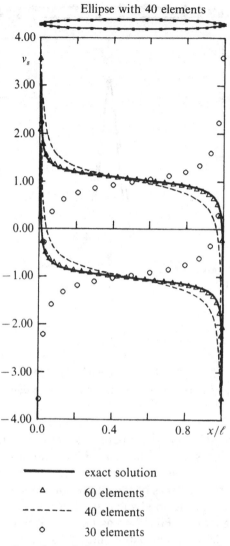

Fig. 2.2. Effect of number of elements upon accuracy of surface vorticity solution for flow past thin ellipse, $\alpha_\infty = 10°$, $W_\infty = 1.0$.

the 15 elements on the opposite surface of an ellipse ($\lambda = 0.05$) with 30 element representation. The curve labelled 'exact' represents the ideal numerical model of an infinite number of surface elements. Bearing in mind the term $\oint k(s_m, s_n)\gamma(s_n)\,ds_n$ in (1.19), the accuracy of the numerical statement of the Dirichlet boundary condition at

51

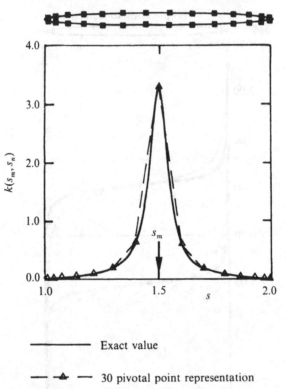

———————— Exact value

— ▲ — 30 pivotal point representation

Fig. 2.3. Coupling coefficients at point s_m on an ellipse due to elements on opposite surface.

s_m is closely related to the area under this curve shown as a dotted curve in accordance with the trapezoidal integration which we have adopted in our numerical representation, (2.9). Due to the close proximity of the two surfaces of the ellipse, the coupling coefficients for the elements opposite to element m, namely element $(M + 1 - m)$ and its neighbours, Fig. 2.4, assume enormous proportions. $k(s_m, s_{M+1-m})$ alone contributes most towards the integral $\oint k(s_m, s_n)\gamma(s_n)\,\mathrm{d}s_n$. Furthermore errors in the representation of this integral for the opposite element, though they appear from Fig. 2.3 to be small in proportion locally, may be as large as or greater than the total contribution to $\oint k(s_m, s_n)\gamma(s_n)\,\mathrm{d}s_n$ due to most of the other elements. This error level of the back diagonal coefficients of the matrix, namely $K(s_m, s_{M+1-m})$, may thus dominate other coefficients and lead to catastrophic results as already illustrated by Fig. 2.2.

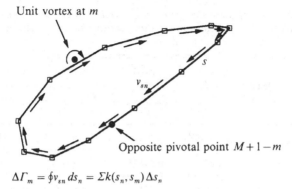

$$\Delta \Gamma_m = \oint v_{sn}\, ds_n = \Sigma k(s_n, s_m)\, \Delta s_n$$

Fig. 2.4. Circulation induced around profile interior due to a unit vortex just outside the profile at element m.

Now, as stated in Chapter 1, for thick bodies such as a cylinder, the matrix has a dominant leading diagonal. As illustrated by equations (1.25), the leading diagonal comprises terms $K(s_m, s_m)$ which are all of order of magnitude 0.5. In these cases all other coefficients are small by comparison and the solution is dominated by the leading diagonal terms. It is for this reason that care must be taken to represent $K(s_m, s_m)$ accurately by including the self-induced velocity due to local profile curvature as dealt with in Section 1.7. For thin profiles, on the other hand, the back diagonal coefficients also begin to grow in size. There is some evidence that numerical troubles may arise when $K(s_m, s_{M+1-m})$ becomes of similar order to $K(s_m, s_m)$ so that the back diagonal of the coupling coefficient matrix begins to dominate. This is borne out by the data presented in Table 2.2 for an ellipse with $\lambda = 0.05$, $W_\infty = 1.0$, and $\alpha_\infty = 10°$.

For various total numbers of elements M, a range of relevant parameters have been tabulated for the mid-element m of the upper surface of the ellipse. Column (1) shows local profile thickness ΔT_m as a fraction of local element length Δs_m. Column (2) lists the opposite point or back diagonal coupling coefficient $K(s_m, s_{M+1-m})$ and column (3) lists the area A under Fig. 2.3 taken for the whole surface perimeter s. We observe that errors in A are negligible for large numbers of elements but with $M = 40$ are beginning to mount up exponentially. The coefficients $K(s_m, s_{M+1-m})$ are also beginning to increase significantly at this stage as M is reduced. Although there is no dramatic event to be observed, it is really the combined

Table 2.2. *Analysis of the influence of total number of elements upon errors and various causal parameters for a thin ellipse with* $\lambda = 0.05$, $\alpha_\infty = 10°$, $W_\infty = 1.0$

Number of Elements M	Element studied m	$\dfrac{\Delta T_m}{\Delta s_m}$ (1)	Matrix coefficient $K(s_m, s_{M+1-m})$ (2)	$A = \oint k(s_m, s_n)\,\mathrm{d}s_n$ (3)	Circulation error $\Delta\Gamma_m$ (4)
10	3	0.153 884	1.034 252	1.625 216	0.618 547
16	4	0.251 354	0.633 158	1.248 836	0.262 450
20	5	0.315 678	0.504 154	1.149 007	0.161 704
26	7	0.411 787	0.386 498	1.070 675	0.082 899
30	8	0.475 718	0.334 558	1.041 023	0.054 295
36	9	0.571 497	0.278 485	1.015 452	0.029 355
40	10	0.635 305	0.250 516	1.005 525	0.019 667
46	12	0.730 974	0.217 739	0.996 062	0.010 912
50	13	0.794 727	0.200 263	0.992 452	0.007 425
60	15	0.954 053	0.166 820	0.988 230	0.002 933
80	20	1.272 582	0.125 064	0.985 729	0.000 591
100	25	1.591 026	0.100 033	0.985 279	0.000 210
∞	∞	∞	0	0.984 830	0.000 000

effects revealed in columns (2) and (3) which indicate likely errors and their predominance in the back diagonal coefficients.

As an alternative method for indicating error we may examine the circulation induced around the profile interior due to a unit vortex located at s_m, Fig. (2.4), namely

$$\Delta\Gamma_m = \oint k(s_n, s_m)\,\mathrm{d}s_n \qquad (2.14)$$

or in numerical form

$$\Delta\Gamma_m = \frac{1}{\Delta s_m} \sum_{n=1}^{M} K(s_n, s_m)\,\Delta s_n \qquad (2.15)$$

As defined, the coupling coefficient $k(s_n, s_m)$ implies the velocity v_{sn} parallel to the surface at n induced by a unit point vortex in contact with the body surface at m but just outside curve s. The circulation around such a contour which does not surround the unit vortex must be zero according to Kelvin's theorem. Consequently the residual value $\Delta\Gamma_m$ represents the circulation error implied by using the coupling coefficients as already calculated for setting up the matrix. As a matter of interest (2.15) involves only the coupling coefficients in a given column m, a very important observation to

54

which we will return later. For the moment we will refer only to the circulation or so-called 'leakage' errors implied by the values of $\Delta\Gamma_m$ given in Table 2.2 with the observation that these do begin to escalate for element numbers less than $M = 40$.

The one remaining observation not yet made is that these troubles arise for $\Delta T_m / \Delta s_m$ ratios below about 0.64. Indeed this parameter is of greater geometrical relevance than simply the total number of elements M. In general when selecting elements it is important that element length Δs_m should always be kept no greater than local profile thickness ΔT_m giving a margin of safety for the reasons just outlined. Having said that, as we shall now see, two alternative methods can be devised to reduce these *opposite point* errors based upon the observations we have just made. These were considered early in this work by Jacob & Riegels (1963) and by Wilkinson (1967a) and the second of the techniques is now generally adopted for lifting body flows.

(i) Use of sub-elements
(ii) Back diagonal correction to enforce zero internal circulation

2.3.2   Use of sub-elements

As stated by (1.21) the standard Martensen numerical model implies the replacement of the line vorticity elements by a concentrated vortex of strength $\gamma(s_n) \Delta s_n$ at each pivotal point*. An improved representation of the true element vorticity influence might be anticipated if $K(s_m, s_n)$ were recalculated as the average value for element n following the strategy adopted for source panels in Section 1.9, equations (1.39) and (1.40). Thus we may write

$$\bar{K}(s_m, s_n) = \frac{1}{\Delta s_n} \int_0^{\Delta s_n} K(s_m, s_n) \, ds_n \tag{2.16}$$

or in numerical form

$$\bar{K}(s_m, s_n) = \frac{1}{N} \sum_{i=1}^N \left\{ \frac{(y_m - y_i) \cos \beta_m - (x_m - x_i) \sin \beta_m}{(x_m - x_i)^2 + (y_m - y_i)^2} \right\} \tag{2.17}$$

---

* The reader is referred to Section 1.6 for a deeper understanding of this point. The coupling coefficients $k(s_m, s_n)$ give the velocity parallel to the profile at s_m due to a unit point vortex at m.

where

$$
\left.
\begin{aligned}
x_i &= x_n + (i - \tfrac{1}{2}(1+N)) \frac{\Delta s_n \cos \beta_m}{N} \\
y_i &= y_n + (i - \tfrac{1}{2}(1+N)) \frac{\Delta s_n \sin \beta_m}{N}
\end{aligned}
\right\}
\tag{2.18}
$$

In effect (2.17) introduces N sub-elements of equal length $\Delta s_n/N$ and mid point coordinates (x_i, y_i).

Curiously enough this technique can produce worse results for thicker aerofoils or bodies such as a circular cylinder. On the other hand application to the thin ellipse presently under consideration leads to considerable improvement for the 30-element case, Fig. 2.5, c.f. Fig. 2.2. For this case with only 2 sub-elements extremely good predictions were obtained and no noticeable improvement resulted from the use of 100 sub-elements. Tests with smaller numbers of elements M reveal that the use of 2 sub-elements generally produces quite reasonable results which can be better than those obtained using more sub-elements. The following rules should therefore be adhered to when selecting pivotal points:

(i) Element lengths Δs_n should be no less than local body thickness ΔT_n

(ii) To economise on matrix size rule (i) may be broken provided
   (a) $\Delta s_n > \Delta T_n > 0.5 \Delta s_n$
   (b) Two sub-elements are used

Rule (ii) results in halving the number of elements and therefore matrix size required to model a thin body.

2.3.3   Back diagonal correction

Jacob and Riegels (1963) first advocated this technique which is based upon the observation made in Section 2.3. From Kelvin's theorem the net circulation $\Delta \Gamma_m$ around the profile interior induced by a surface vorticity element such as $\gamma(s_m)\Delta s_m$ should be zero. If this condition is enforced upon the matrix coefficients, (2.15) becomes

$$
K(s_{\text{opp}}, s_m) = -\frac{1}{\Delta s_{\text{opp}}} \sum_{\substack{n=1 \\ n \neq \text{opp}}}^{M} K(s_n, s_m)\, \Delta s_n \quad (\text{opp} = M + 1 - m)
\tag{2.19}
$$

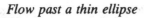

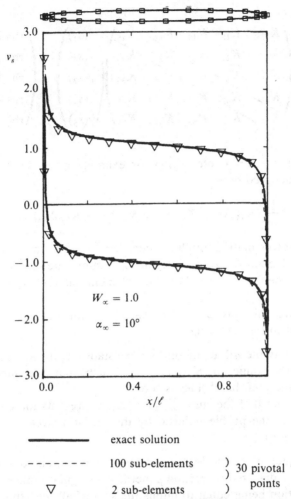

Fig. 2.5. Inclined flow past a thin non-lifting ellipse using sub-elements.

On this basis these authors recommended that the back diagonal matrix coefficients $K(s_{M+1-m}, s_m)$ be replaced by the value given by this equation thereby ensuring that net circulation around the profile interior implied by the numerical model is made to be zero. Upon closer inspection we observe that (2.19) involves the coupling coefficients in column m only. To clarify this let us write out in full the set of linear Martensen equations (1.21) for a simplified case with just $M = 5$ pivotal points, drawing attention to the back diagonal coefficients.

$$
\begin{array}{c}
\begin{array}{cccccc}
m=1 & 2 & 3 & 4 & 5
\end{array}\\
\begin{array}{c}
n=1\\2\\3\\4\\5
\end{array}
\begin{pmatrix}
K_{11} & K_{12} & K_{13} & K_{14} & K_{15}\\
K_{21} & K_{22} & K_{23} & K_{24} & K_{25}\\
K_{31} & K_{32} & K_{33} & K_{34} & K_{35}\\
K_{41} & K_{42} & K_{43} & K_{44} & K_{45}\\
K_{51} & K_{52} & K_{53} & K_{54} & K_{55}
\end{pmatrix}
\begin{pmatrix}
\gamma(s_1)\\
\gamma(s_2)\\
\gamma(s_3)\\
\gamma(s_4)\\
\gamma(s_5)
\end{pmatrix}
=
\begin{pmatrix}
\text{rhs } 1\\
\text{rhs } 2\\
\text{rhs } 3\\
\text{rhs } 4\\
\text{rhs } 5
\end{pmatrix}
\end{array}
$$

$$(2.20)$$

Introducing $m = 4$ into (2.19) for example gives, in the matrix notation adopted here,

$$
K_{24} = -\frac{1}{\Delta s_2}\left(K_{14}\,\Delta s_1 + K_{34}\,\Delta s_3 + K_{44}\,\Delta s_4 + K_{54}\,\Delta s_5\right)
$$

which involves matrix coupling coefficients in column 4 only. Back diagonal element K_{24} is to be replaced by minus the sum of all other column 4 coefficients scaled by their element lengths Δs_n and finally divided by $-\Delta s_2$.

Applying this procedure to each column of the matrix is equivalent to the following;

(i) Multiply the nth equation by a constant Δs_n for $n = 1 \ldots M$

(ii) Fix the value of $K(s_{M+1-m}, s_m)$ such that the sum of all equation left hand sides is zero

(iii) We note that the sum $\sum_{n=1}^{M} \text{rhs}_n\,\Delta s_n$ represents the circulation around the profile induced by the uniform stream W_∞ and is also zero

Consequently for the Martensen method as presented in Chapter 1, the back diagonal correction generates a singular matrix with any one equation being equal to minus the sum of all the others. Jacob and Riegels, who were considering lifting aerofoils, were able to escape this difficulty by prescribing one extra pivotal point at the trailing edge location $n = $ te. Assuming there to be a stagnation point at te for which γ(te) would in any case be zero, they were then able to delete equation and column numbers te from the matrix. In this way they were able both to satisfy the trailing edge Kutta condition and to overcome the problem of an otherwise singular matrix. For non-lifting bodies such as the thin ellipse presently under consideration however, this procedure is inapplicable leaving the use of sub-elements as the only apparent option available at this stage of our argument for increasing the accuracy.

However, as we shall now see, for lifting bodies special treatments other than the use of sub-elements are available to avoid this difficulty.

2.4 Thin ellipse as a lifting aerofoil

We are now in a position to consider the flow past a thin ellipse regarded as a simple form of lifting aerofoil, for which the trailing edge Kutta condition must be imposed. This can be approached in one of two possible ways. Firstly, following Jacob & Riegels (1963) and as already mentioned, the trailing edge could be treated as a stagnation point. This is indeed the true situation for inviscid irrotational flow past an aerofoil with a sharp or rounded trailing edge. Alternatively, in response to the observed behaviour of aerofoils in the flow of a real fluid, we could state that the static pressure approaches the same value moving towards the trailing edge along the upper and lower surfaces. Applying these approaches to the thin ellipse, three methods present themselves for dealing with the trailing edge Kutta condition.

(i) *Method 1. Prescribed bound vorticity* Γ
   With the help of conformal transformation theory, Section 2.3, the required bound circulation Γ may be evaluated sufficient to make the trailing edge of the ellipse a stagnation point
(ii) *Method 2. Unloaded trailing edge*
   Γ may be calculated directly from the surface vorticity analysis to impose the restraint of equal static pressure on the two surface elements adjacent to the trailing edge, te and te + 1
(iii) *Method 3. The Wilkinson trailing edge condition*
   This method, now generally adopted, is a variant on method 2 yielding marginal computational advantages

Back diagonal correction may be introduced into all three of these methods eliminating the need to use sub-elements. We shall deal with them in turn in the following three sub-sections.

2.4.1   Kutta condition, method 1 – prescribed bound circulation Γ

In Section 2.2 the surface vorticity equations were developed for a circular cylinder with prescribed circulation Γ in a uniform stream

W_∞, (2.9). The exact solution of this problem was extended in Section 2.3 to deal with lift-generating ellipses (2.13) although only the non-lifting case was tackled at that point. To deal with these flows by the surface vorticity method, two possible approaches present themselves:

(i) To prescribe several values of Γ and select the one which produces the most suitable looking surface pressure or velocity distribution in the trailing edge region

(ii) To make use of the exact solution to pre-determine Γ

The first of these options is generally applicable to all lifting aerofoils for which there is no exact solution to assist with selection of Γ. It offers the advantage that room is left for human judgment with scope to correlate to typical experimental trailing edge loadings. For this reason the method has proved popular for some industrial applications.

The second option is useful for obtaining a datum check when undertaking program development. From the exact solution for the ellipse (2.13) stagnation conditions at the trailing edge require that $v_s = 0$ when $\theta = 0$, whereupon Γ is determined by

$$\Gamma = 4\pi r_0 W_\infty \sin \alpha_\infty = \pi(1 + \lambda)(\text{major axis}) \sin \alpha \qquad (2.21)$$

Program 2.2, ellipse1.pas, illustrates this for an ellipse with prescribed bound circulation. This program has also been provided with the facility to introduce sub-elements both to demonstrate the technique and to permit some experimentation. Solutions with 30 pivotal points and two sub-elements (in this case without back diagonal correction) are shown in Fig. 2.6 for three prescribed values of Γ, namely 0.4, 0.572 808 and 0.74 for the 5% thick ellipse with $W_\infty = 1.0$, $\alpha_\infty = 10°$. For $\Gamma = 0.572\,808$ obtained from (2.21), the corresponding exact solution is given for comparison and agrees well as expected, with smooth velocity variation progressing towards the trailing edge. The incorrect specification of the Kutta condition resulting from the other two erroneous Γ values is indicated by the consequent velocity peaks due to recirculation around the trailing edge.

2.4.2 Kutta condition method 2 – trailing edge unloading

Wilkinson (1967a) suggested the alternative Kutta condition statement that the static pressure and therefore surface vorticity at

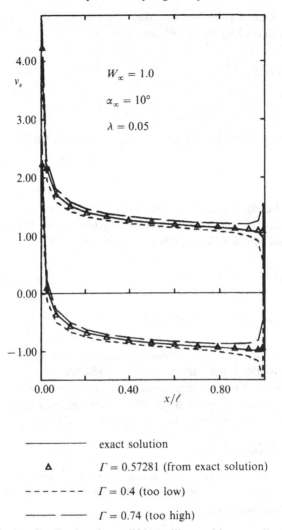

Fig. 2.6. Velocity distribution for a lifting ellipse with prescribed bound circulation.

the two trailing edge elements on the upper and lower surfaces should have the same magnitude. This can be achieved if we impose the constraint

$$\gamma(s_{te}) = -\gamma(s_{te+1}) \qquad (2.22)$$

remembering that for smooth flow leaving the trailing edge $\gamma(s_{te})$ must be clockwise and $\gamma(s_{te+1})$ must be anti-clockwise and therefore negative. One way to achieve this is to replace Γ by a unit bound vortex so that (2.8) and (2.9) become

$$\sum_{n=1}^{M} \gamma(s_n)\, \Delta s_n = 1.0 \tag{2.8a}$$

and

$$\sum_{n=1}^{M} (K(s_m, s_n) + \Delta s_m)\gamma(s_n) = -(U_\infty \cos \beta_m + V_\infty \sin \beta_m) + 1.0$$

$$\qquad\qquad\qquad\text{(rhs I)} \qquad\qquad \text{(rhs II)}$$

$$\tag{2.9a}$$

Since the right hand side has two independent components, we may replace (2.9a) by two separate systems, one for the uniform stream and one for the unit bound vortex strength.

$$\left.\begin{aligned} \sum_{n=1}^{M} (K(s_m, s_n) + \Delta s_m)\gamma_1(s_n) &= -(U_\infty \cos \beta_m + V_\infty \sin \beta_m) \\ \sum_{n=1}^{M} (K(s_m, s_n) + \Delta s_m)\gamma_2(s_n) &= 1 \end{aligned}\right\} \tag{2.23}$$

We note that the same coupling coefficient is applicable to both sets of equations thus demanding no additional major computing requirements. Independent solutions are then obtained for $\gamma_1(s_n)$ and $\gamma_2(s_n)$ which may subsequently be recombined to give the final solution for any particular bound circulation Γ through

$$\gamma(s_n) = \gamma_1(s_n) + \Gamma\gamma_2(s_n) \tag{2.24}$$

Introducing this into the trailing edge equation (2.22), we are able to derive an expression for Γ in terms of the trailing edge element unit solutions.

$$\Gamma = -\frac{\gamma_1(s_{te}) + \gamma_1(s_{te+1})}{\gamma_2(s_{te}) + \gamma_2(s_{te+1})} \tag{2.25}$$

A further advantage can be gained in both methods 1 and 2 by the use of back diagonal correction to the $K(s_m, s_n)$ matrix before introducing the extra Δs_m term in the left hand sides of equations (2.23). The later introduction of Δs_m, representing the addition of (2.8a) to each equation, prevents the matrix from remaining singular. There is therefore no need to make use of sub-elements to gain further accuracy.

2.4.3   Method 3 – Wilkinson's Kutta condition

Wilkinson (1967a), in putting forward (2.22) as a simple means for unloading a lifting aerofoil trailing edge, went one stage further. Since $\gamma(s_{te})$ and $-\gamma(s_{te+1})$ are now equal, we have one less unknown value of surface vorticity to find and may eliminate one equation. In this method there is no need to introduce bound circulation through (2.8) and we shall revert to the normal form of Martensen's equation for element m

$$\sum_{n=1}^{M} K(s_m, s_n)\gamma(s_n) = -(U_\infty \cos \beta_m + V_\infty \sin \beta_m) \qquad (1.21)$$

Introducing the Kutta condition (2.22) we observe that the nth equation may be written

$$K(n, 1)\gamma(s_1) + \ldots + (K(n, \text{te}) - K(n, \text{te} + 1))\gamma(\text{te})$$
$$+ \ldots + K(n, M)\gamma(M) = \text{rhs}_n \quad (2.26)$$

where columns te and te + 1 have been combined. Since there are now only $M - 1$ columns for the $M - 1$ unknown vorticity values, it is mandatory to reduce the number of equations also to $M - 1$. Three possibilities present themselves.

(a) To simply strike out one equation such as that for pivotal point te + 1
(b) To add the two equations representing the trailing edge elements
(c) To subtract these two equations instead

Experimentation with these options shows that option (c) delivers the best results. The physical explanation for this can be seen from Fig. 2.7, which illustrates the nature of the trailing edge streamline flow we are seeking to model. We recall however that each Martensen equation is a statement that the velocity parallel to the inner surface v_{si}, shown as a double headed arrow, is zero, with the sign convention that v_{si} points in the clockwise direction around the profile. At the trailing edge the outcome of this sign convention is that v_{si} points towards the trailing edge at element te on the upper surface and away from the trailing edge at element te + 1 on the lower surface. These two element boundary conditions may be combined to represent the average *downstream* flow in the trailing edge region if we first multiply equation number te + 1 by -1 to reverse the direction of v_{si} which the equation represents. Then we

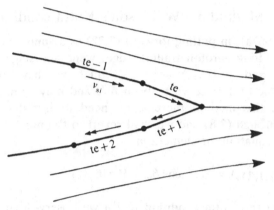

Fig. 2.7. Aerofoil trailing edge flow.

must add it to equation number te as a replacement for the two trailing edge equations.

A special advantage of this technique pointed out by Wilkinson is that back diagonal correction, as explained in Section 2.3.3, may first be applied to the original coupling coefficient matrix. When reduced to size $M - 1$ by the elimination of one column and one row, the matrix ceases to be ill-conditioned but retains the zero internal circulation or leakage flux characteristics endowed by back diagonal correction.

The bound circulation may be calculated by summing all the surface vorticity through (2.8).

These three methods are compared in Table 2.3 with the exact solution for our test case ellipse using only 30 elements and no sub-elements but with back diagonal correction in all three cases. The predicted surface velocity distributions for methods 2 and 3 are almost indistinguishable and agree closely with method 1. Methods 2 and 3 offer the advantage that Γ and C_L may be predicted rather than prescribed and predicted with remarkable accuracy for so few elements. Finally we note from Table 2.2 that body thickness/ element length $\Delta T_m / \Delta s_m$ for this case was only 0.475 718, demonstrating the economy in elements permitted by the extra accuracy afforded by back diagonal correction, evidently a superior technique to that of introducing sub-elements. The reader will find that Program 2.2 delivers quite good results for this 5% thick ellipse with as few as ten elements!

Table 2.3. *Flow past an ellipse acting as a lifting aerofoil with* $\lambda = 0.05$, *$W_\infty = 1.0$ and $\alpha_\infty = 10$*

Element number	Exact solution v_s	Method 1 prescribed Γ v_s	Method 2 unloaded trailing edge v_s	Method 3 Wilkinson method v_s
1	4.075 465	4.192 013	4.183 404	4.183 464
2	2.159 815	2.174 612	2.171 423	2.171 381
3	1.708 121	1.709 796	1.707 811	1.707 807
4	1.506 714	1.507 595	1.506 107	1.506 107
5	1.390 974	1.391 562	1.390 330	1.390 337
6	1.314 487	1.314 915	1.313 825	1.313 823
7	1.259 136	1.259 455	1.258 436	1.258 438
8	1.216 379	1.216 614	1.215 617	1.215 612
9	1.181 630	1.181 803	1.180 784	1.180 783
10	1.152 169	1.152 297	1.151 207	1.151 204
11	1.126 207	1.126 309	1.125 078	1.125 079
12	1.102 340	1.102 475	1.100 988	1.100 989
13	1.078 865	1.079 379	1.077 394	1.077 389
14	1.050 560	1.063 306	1.060 121	1.060 117
15	0.942 400	1.053 155	1.044 540	1.044 542
16	−0.925 145	−1.035 930	−1.044 540	−1.044 542
17	−0.993 476	−1.006 140	−1.009 328	−1.009 344
18	−0.981 519	−0.981 865	−0.983 850	−0.983 846
19	−0.962 575	−0.962 459	−0.963 946	−0.963 951
20	−0.940 526	−0.940 290	−0.941 521	−0.941 532
21	−0.915 415	−0.915 110	−0.916 200	−0.916 202
22	−0.886 350	−0.885 987	−0.887 006	−0.887 008
23	−0.851 718	−0.851 290	−0.852 286	−0.852 285
24	−0.808 843	−0.808 334	−0.809 353	−0.809 357
25	−0.753 097	−0.752 491	−0.753 581	−0.753 586
26	−0.675 759	−0.675 035	−0.676 267	−0.676 272
27	−0.558 201	−0.557 339	−0.558 827	−0.558 832
28	−0.352 263	−0.351 447	−0.353 432	−0.353 451
29	0.115 779	0.105 166	0.101 979	0.101 996
30	2.207 920	2.102 928	2.094 318	2.094 282
Bound vortex Γ	0.572 808	0.572 808	0.569 683	0.569 677
Lift coeff. C_L	1.145 616	1.145 616	1.139 367	1.139 354

2.5 Aerofoils

The preceding analysis may be applied with little further modification to predict the potential flow past aerofoils of arbitrary shape. Program 2.3 completes such a computation portrayed in the flow diagram.

The two new features in this sequence relate to profile specification in box (1) and an economic technique for repeat runs varying W_∞ and α_∞, boxes (7) and (8). These will now be dealt with in reverse order.

Flow diagram for program 2.3

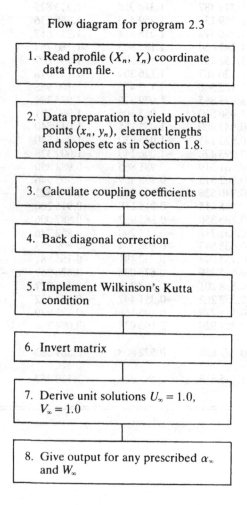

1. Read profile (X_n, Y_n) coordinate data from file.

2. Data preparation to yield pivotal points (x_n, y_n), element lengths and slopes etc as in Section 1.8.

3. Calculate coupling coefficients

4. Back diagonal correction

5. Implement Wilkinson's Kutta condition

6. Invert matrix

7. Derive unit solutions $U_\infty = 1.0$, $V_\infty = 1.0$

8. Give output for any prescribed α_∞ and W_∞

Aerofoils

2.5.1 Unit Solutions

Applying the technique already introduced for other reasons in Section 2.4.2, the linear equation (2.27) may be separated into two equations for the individual components of the right hand side if we introduce the unit vorticities $\gamma_u(s_n)$ and $\gamma_v(s_n)$ defined by

$$\gamma(s_n) = U_\infty \gamma_u(s_n) + V_\infty \gamma_v(s_n)$$

$$= W_\infty(\cos \alpha_\infty \gamma_u(s_n) + \sin \alpha_\infty \gamma_v(s_n)) \tag{2.27}$$

Equations (2.27) then become

$$\left.\begin{array}{l} \sum_{n=1}^{M} K(s_m, s_n)\gamma_u(s_n) = -\cos \beta_m \\[2mm] \sum_{n=1}^{M} K(s_m, s_n)\gamma_v(s_n) = -\sin \beta_m \end{array}\right\} \tag{2.28}$$

where of course β_m is the profile slope at element m. Unit solutions $\gamma_u(s_n)$ and $\gamma_v(s_n)$ may thus be derived which are independent of mainstream velocity W_∞ and angle of attack α_∞. In fact $\gamma_u(s_n)$ and $\gamma_v(s_n)$ represent the solutions for unit uniform streams $U_\infty = 1.0$ and $V_\infty = 1.0$ respectively. Furthermore the same coupling coefficient is involved in both unit equations. Following the solution of equations (2.28), $\gamma(s_n)$ for any choice of W_∞ and α_∞ is obtained by combining the unit solutions through equations (2.27). This process is implemented in Program 2.3 which is re-entrant in W_∞ and α_∞ in the last procedure 'input_flow_data'.

Surface velocity follows directly since $v_s = \gamma(s)$, (1.12). For aerofoil applications a more usual practice is to focus on surface pressure distribution by defining the surface pressure coefficient

$$C_p = \frac{p - p_\infty}{\frac{1}{2}\rho W_\infty^2}$$

$$= 1 - \left\{\frac{\gamma(s)}{W_\infty}\right\}^2 \tag{2.29}$$

The lift coefficient, through (2.4) and (2.3), now becomes

$$C_L = \frac{2\Gamma}{W_\infty l} = \frac{2}{l}\left\{\cos \alpha_\infty \sum_{n=1}^{M} \gamma_u(s_n) \Delta s_n + \sin \alpha_\infty \sum_{n=1}^{M} \gamma_v(s_n) \Delta s_n\right\} \tag{2.30}$$

where l is taken as the chord length of the aerofoil.

67

2.5.2 Specification of aerofoil geometry

Although the above analysis can be applied to aerofoils of any shape whatsoever, in engineering practice systematic procedures are desirable for profile design resulting in progressive series of easily reproducible aerofoils. American and British practices are alike in constructing aerofoil profiles by the superposition of a standard profile thickness shape y_t normal to a camber line y_c, Fig. 2.8. Fortunately this method proves equally ideal in surface vorticity analysis for generating profile data points, since, as we shall illustrate later, the surface vorticity method yields maximum accuracy if the pivotal points on the upper and lower surfaces lie directly opposite to one another in pairs. This was of course true for the ellipse considered in the previous section for which profile data

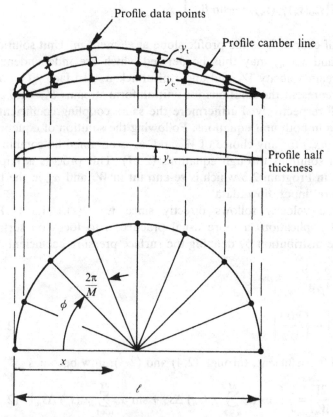

Fig. 2.8. Aerofoil profile construction from camber line and base profile, for prescribing profile data points.

generation was exactly as prescribed in Fig. 2.8. A brief description will be helpful.

To generate M surface elements (where M must be even), the circle of radius $a = \ell/2$ is constructed and divided into $M/2$ equal segments $\Delta\phi = 2\pi/M$. Camber-line x coordinates are then given by

$$\frac{x_c}{\ell} = \tfrac{1}{2}(1 - \cos \phi) \tag{2.31}$$

From tabulated data the profile half-thickness y_t and camber-line coordinates y_c are obtained by interpolation and the camber-line slopes θ_c are evaluated. Data points a and b then follow from

$$\left.\begin{aligned} X &= x_c \pm y_t \sin \theta_c \\ Y &= y_c \pm y_t \cos \theta_c \end{aligned}\right\} \tag{2.32}$$

This process yields the two-fold benefit of normally opposed data points and a concentration of elements close to both leading and trailing edge regions. Surface velocities vary most rapidly in the vicinity of an aerofoil's leading edge. Although trailing edge flows are usually smooth this is only due to cancellation of opposing recirculatory flows due to W_∞ and the bound vorticity Γ to satisfy the trailing edge Kutta condition. Furthermore, concentration of pivotal points in these regions ensures a fairly uniform distribution of the parameter $\Delta s_n/2y_t$ which we identified in Section 2.3.1 as having a critical influence upon the accuracy of numerical modelling associated with opposite pivotal points.

2.5.3   Comparison with Joukowski aerofoils

The Joukowski transformation (2.10) may be used to generate a family of aerofoils for which we may derive the exact solution for the potential flow. This may be achieved by transforming the circle C of radius r_0 with off-set centre P at $(-\varepsilon 1, \varepsilon 2)$, in the z plane, Fig. 2.9, into the Joukowski aerofoil C' in the ζ plane. The one restriction is that the singular points of the transformation A and B located at $\pm a$ must lie either inside or on the circle C whose equation is

$$z = r_0 \mathrm{e}^{\mathrm{i}\phi} - \varepsilon 1 + \mathrm{i}\varepsilon 2 \tag{2.33}$$

Following a similar analysis to that in Section 2.3, the coordinates

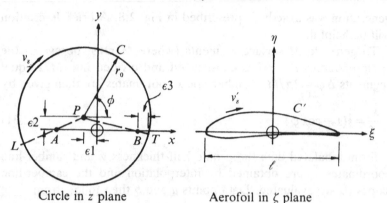

Circle in *z* plane   Aerofoil in *ζ* plane

(*a*) Geometrical transformation $\zeta = z + a^2/z$

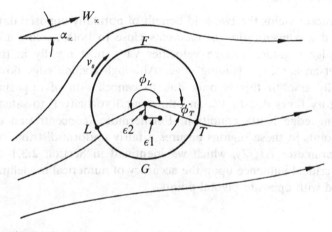

(*b*) *Trailing edge Kutta condition*

Fig. 2.9. Joukowski's transformation.

of the aerofoil in the *ζ* plane become

$$\left.\begin{aligned}
\xi &= \left(1 + \left(\frac{a}{r}\right)^2\right)(r_0 \cos \phi - \varepsilon 1) \\
\eta &= \left(1 - \left(\frac{a}{r}\right)^2\right)(r_0 \sin \phi + \varepsilon 2)
\end{aligned}\right\} \tag{2.34}$$

where the polar coordinate (r, θ) of the circle in the *z* plane may be expressed in terms of the convenient prescriptive variables r_0, ϕ, $\varepsilon 1$

70

and $\varepsilon 2$ as follows

$$r = \sqrt{[\{(r_0 \cos \phi - \varepsilon 1)^2 + (r_0 \sin \phi + \varepsilon 2)^2\}]}$$
$$\theta = \arctan\left\{\frac{r_0 \sin \phi + \varepsilon 2}{r_0 \cos \phi - \varepsilon 1}\right\} \qquad (2.35)$$

By specifying a, r_0, $\varepsilon 1$ and $\varepsilon 2$ it is a simple matter to evaluate the profile coordinates (ξ, η), enabling us to generate a family of related aerofoil profiles. Several categories can be identified as follows.

Case 1. Ellipse and flat plate
If both offsets $\varepsilon 1$ and $\varepsilon 2$ are set to zero we obtain the family of ellipses already considered provided $r_0 > a$. The limiting case of $r_0 = a$ corresponds to the flat plate aerofoil.

Case 2. Aerofoils with a cusped trailing edge
The simplest type of Joukowski aerofoil may be obtained if curve C is arranged to pass through the singular point B located at $(a, 0)$ in the z plane. From Fig. (2.9) we observe that for this case

$$r_0 = \sqrt{[(a + \varepsilon 1)^2 + \varepsilon 2^2]} \qquad (2.36)$$

Unfortunately however such aerofoils have a cusped trailing edge and are not representative of practical aerofoils.

Case 3. Aerofoils with a rounded trailing edge
Glauert (1926) presented a more realistic series of aerofoils with sharp trailing edges by introducing a further transformation. On the other hand, as shown by Lewis (1984b), the Joukowski transformation alone may be used to progress directly to aerofoils with a rounded trailing edge and with profile characteristics similar to those used for turbomachine blades. This can be achieved as illustrated in Fig. (2.9) provided that r_0 exceeds the value given by (2.36). To ensure this and gain fine control over profile generation, we may introduce the circle C trailing edge T offset $\varepsilon 3$, whereupon

$$r_0 = \sqrt{[(a + \varepsilon 1 + \varepsilon 3)^2 + \varepsilon 2^2]} \qquad (2.37)$$

Sample profiles for various offsets are shown in Fig. (2.10).

There remains indeterminacy regarding the location of the leading edge L and trailing edge T and an arbitrary choice must be made. Reasonable grounds for this choice would be to locate L and T on the circle radius vectors PL and PT which pass through the

Two-dimensional aerofoils and cascades

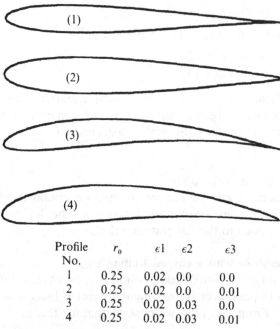

Profile No.	r_0	$\epsilon 1$	$\epsilon 2$	$\epsilon 3$
1	0.25	0.02	0.0	0.0
2	0.25	0.02	0.0	0.01
3	0.25	0.02	0.03	0.0
4	0.25	0.02	0.03	0.01

Fig. 2.10. Selection of typical Joukowski profiles.

singular points A and B of the transformation. The angles ϕ_L and ϕ_T thus defined are then

$$
\left.
\begin{aligned}
\phi_L &= \pi + \arctan\left(\frac{\epsilon 2}{a - \epsilon 1}\right) \\
\phi_T &= -\arctan\left(\frac{\epsilon 2}{a + \epsilon 1}\right)
\end{aligned}
\right\}
\tag{2.38}
$$

Repeating the analysis of Section 2.3 the surface velocity on the aerofoil reduces to

$$
v_s = \frac{2W_\infty \sin(\phi - \alpha_\infty) + \Gamma/2\pi r_0}{\sqrt{\left[\left(1 - \left(\frac{a}{r}\right)^2 \cos 2\phi\right)^2 + \left(\frac{a}{r}\right)^4 \sin^2 2\phi\right]}}
\tag{2.39}
$$

where the bound vortex strength Γ required to make T a stagnation point is given by

$$
\Gamma = 4\pi r_0 W_\infty \sin(\alpha_\infty - \phi_T)
\tag{2.40}
$$

72

resulting in the lift coefficient

$$C_L = \frac{8\pi r_0}{\ell} \sin(\alpha_\infty - \phi_T) \qquad (2.41)$$

We are now able to evaluate profile coordinates and related surface velocity for given W_∞ and α_∞ by inserting successive values of ϕ into the above equation to move around the perimeter of the circle C. However further thought is needed to select ϕ values which generate coordinates suitable for comparison with the surface vorticity analysis. Suppose we first define the upper and lower surface quadrant angles in Fig. 2.9(b) as

$$\Delta\phi_u = \widehat{TFL} = \phi_L + \phi_T, \qquad \Delta\phi_l = \widehat{LGT} = 2\pi - \Delta\phi_u \qquad (2.42)$$

The simplest procedure for generating M surface elements would be to divide these angles each into $M/2$ equal sub-divisions. However, as illustrated by case 1 of Fig. 2.11, this produces an unsuitable distribution of opposite facing elements. More important, the two trailing edge elements are the most badly affected, being of different lengths. To overcome this difficulty, Lewis (1984b) defines alternative angular locations α proceeding from trailing edge to leading edge on the upper and lower surfaces by

$$\alpha_u = \left(\frac{n}{M/2}\right)\Delta\phi_u$$
$$\qquad\qquad\qquad \text{where} \quad n = 1\ldots M/2 \qquad (2.43)$$
$$\alpha_l = \left(\frac{n}{M/2}\right)^P \Delta\phi_l$$

If we now enforce equal element angles on both surfaces at the trailing edge ($n = 1$), the required exponent in (2.43) becomes

$$p = 1 + \frac{\ln(\Delta\phi_l/\Delta\phi_u)}{\ln(M/2)} \qquad (2.44)$$

Although the trailing edge elements are now equal, ensuring an accurate Kutta condition by Wilkinson's method, the distribution of opposite pivotal points elsewhere on the profile is made rather worse, Fig. 2.11. However, the back diagonal correction implemented in Program 2.3 is able to minimise these errors too, resulting in a remarkably good prediction of surface pressure compared with exact theory. The prediction without the trailing edge point correction, ($p = 1$), has an unacceptable margin of error by comparison. To emphasise the relationship of opposite profile points a reduced number of elements $M = 20$ are shown on the

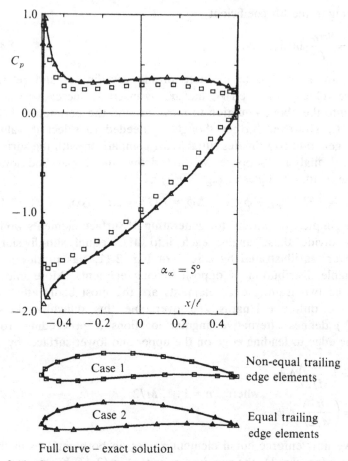

Fig. 2.11. Comparison of exact solution for flow past a Joukowski aerofoil with surface vorticity method. (Profile 4 of Fig. 2.10 with $\alpha_\infty = 5°$)

aerofoil profiles in Fig. 2.11, although 40 elements were used for the surface vorticity calculation, resulting in conservative values of profile thickness/element length. Predicted lift coefficients were as shown in Table 2.4.

It is quite evident that careful specification of the trailing edge Kutta–Joukowski condition is crucial, requiring equal length trailing edge elements. With this proviso and care to maintain $\Delta s_n / \Delta T_m <$ 1.0, extremely accurate predictions may be obtained.

74

Table 2.4. *Predicted lift coefficients*

Method	C_L
Exact solution	1.145 616
Case 1. Non-equal trailing edge elements	1.105 920
Case 2. Equal trailing edge elements	1.139 354

2.6 Turbomachine Linear Cascades

With suitable modifications, the foregoing analysis for single aerofoils may readily be extended to deal with turbomachinery blade cascades. Consider the infinite rectilinear array of aerofoils set at equal pitch intervals t parallel to the y axis, Fig. 2.12. One possible approach would be to treat this problem as an assembly of separate but mutually interfering bodies each with its own initially unknown surface vorticity distribution. Limiting the array to say 50 aerofoils might yield an adequate representation for the centre five or six aerofoils but would result in $50M$ elements with massive computing requirements. Because the flow is periodic in the y direction, such an approach is in fact quite unnecessary since the nth element on all aerofoils will have the same vortex strength $\gamma(s_n)\,\Delta s_n$. Consequently the velocity induced at element s_m becomes that induced by an infinite array of vortex elements each of strength $\gamma(s_n)\,\Delta s_n$. All that is necessary is to derive a modified coupling coefficient $K(s_m, s_n)$ to express this.

2.6.1 Cascade coupling coefficient

A suitable starting point is to consider the velocity induced by an infinite array of point vortices, say of strength Γ and pitch t, Fig. 2.13(a) located along the y axis between $y = \pm\infty$. The flow field induced by this array in the z plane may be transformed into the equivalent flow in the Z plane due to a vortex of strength Γ located at $(1, 0)$ and a second vortex at the origin of strength $-\Gamma/2$, by the conformal transformation

$$z = \ln Z \tag{2.45}$$

The point P, $(Z = \mu e^{i\phi})$, in the Z plane and its equivalent point

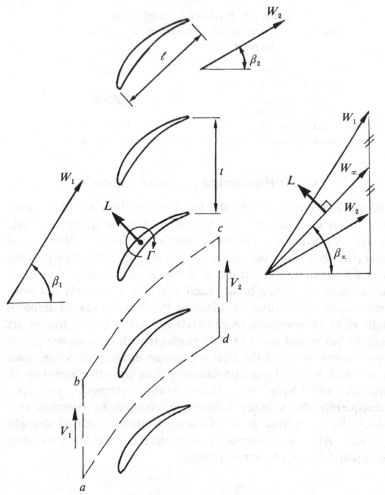

Fig. 2.12. Cascade geometry and velocity triangles.

p, $(z = x + iy)$, in the z plane are then related through

$$x + iy = \ln \mu + i\phi$$

so that

$$\left.\begin{array}{l} x = \ln \mu \\ y = \phi \end{array}\right\} \qquad (2.46)$$

Thus the circle $P'P'$ of radius μ is the transform of the line $p'p'$ parallel to the y axis, along which the flow is periodic. From (2.46b)

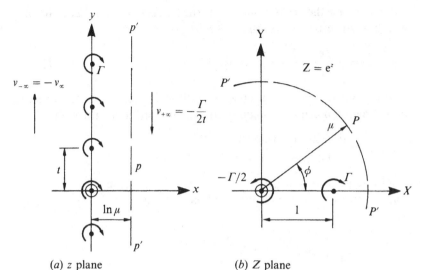

(a) z plane   (b) Z plane

Fig. 2.13. Transformation of vortex array in z plane to vortex pair in Z plane.

we thus have the relationship

$$t = 2\pi \qquad (2.47)$$

The singularity of the transformation at $\mu = 0$ in the Z plane transforms to $x = -\infty$, where the vortex $-\Gamma/2$ is transformed into a vertical uniform stream $v_{-\infty}$ in the z plane. We may now write down the complex potential from the Z plane, namely

$$\omega = \frac{i\Gamma}{2\pi} \ln(Z - 1) - \frac{i\Gamma}{4\pi} \ln Z \qquad (2.48)$$

from which the velocity components in the Z plane may be obtained

$$U - iV = \frac{d\omega}{dZ} = \frac{i\Gamma}{4\pi Z}\left(\frac{Z+1}{Z-1}\right) \qquad (2.49)$$

The velocity components induced by the vortex array in the z plane then become

$$u - iv = \frac{d\omega}{dz} = \frac{d\omega}{dZ}\frac{dZ}{dz}$$

$$= \frac{i\Gamma}{2t}\left(\frac{Z+1}{Z-1}\right)$$

77

Introducing the transformation formula (2.45) this reduces to the well known result, W. Traupel (1945)

$$u - iv = \frac{i\Gamma}{2t}\left(\frac{e^z + 1}{e^z - 1}\right) = \frac{i\Gamma}{2t}\cosh\left(\frac{z}{2}\right) \qquad (2.50)$$

After further reduction, (Dwight (1963) formula 665.4), this may be expressed more conveniently in (x, y) coordinates through

$$u - iv = \frac{i\Gamma}{2t}\left\{\frac{\sinh x - i \sin y}{\cosh x - \cos y}\right\}$$

Although it is not strictly necessary, most cascade treatments transform this to normalised coordinates by use of (2.47), resulting in

$$u - iv = \frac{i\Gamma}{2t}\left\{\frac{\sinh\dfrac{2\pi x}{t} - i\sin\dfrac{2\pi y}{t}}{\cosh\dfrac{2\pi x}{t} - \cos\dfrac{2\pi y}{t}}\right\} \qquad (2.51)$$

From (2.50) we observe that as $z \to \pm\infty$,

$$u_\infty = 0, \qquad v_\infty = \pm\frac{\Gamma}{2t} \qquad (2.52)$$

As expected, u vanishes at infinity. On the other hand by arguments of symmetry about the y axis we can deduce that $v_\infty = -v_{-\infty}$, which may be confirmed then by taking the circulation about one vortex following a contour such as *abcd* shown on the blade cascade, Fig. 2.12.

Returning to the cascade then, Γ may be replaced by $\gamma(s_n)\,\Delta s_n$, resulting in the modified cascade coupling coefficient as follows

$$K(s_m, s_n) = u_{mn}\cos\beta_m + v_{mn}\sin\beta_m$$

$$= \frac{\Delta s_n}{2t}\left\{\frac{\sin\dfrac{2\pi}{t}(y_m - y_n)\cos\beta_m - \sinh\dfrac{2\pi}{t}(x_m - x_n)\sin\beta_m}{\cosh\dfrac{2\pi}{t}(x_m - x_n) - \cos\dfrac{2\pi}{t}(y_m - y_n)}\right\}$$

for $m \ne n$ \qquad (2.53)

The same linear equations (1.21) as those used for the single aerofoil, may then be used to state the surface vorticity model for cascade flow, the modified coupling coefficient being the only essential change. The self inducing coupling coefficient $K(s_m, s_m)$ needs brief consideration in view of the velocities at s_m on one aerofoil induced by the remaining vortices in the $\gamma(s_m)\,\Delta s_m$ array. Referring to Fig. 2.13(a) however, we observe that the drift velocities of the vortex Γ located at $(0, 0)$ due to each pair of vortices at $y = \pm nt$ (where $n = 1, 2, 3 \ldots \infty$), cancel due to anti-symmetry about the x axis. The drift velocity of the array of vortices is thus zero. Applying this to the surface vorticity model it follows that the self-inducing coupling coefficients for a cascade are identical to those for a single aerofoil, namely

$$K(s_m, s_m) = -\tfrac{1}{2} - \frac{\Delta \beta_m}{4\pi} \tag{1.31}$$

as derived in Chapter 1.

2.6.2 Cascade dynamics and parameters

The function of a cascade is to produce fluid deflection from a uniform velocity W_1 at $-\infty$ to W_2 at $+\infty$, Fig. 2.12. Alternatively, for the purpose of fluid dynamic analysis, we may regard the complete flow as the superposition of a vortex array and a single uniform stream W_∞. As we have already shown (2.52) if the blade bound circulation is Γ, the equivalent vortex array induces equal and opposite veloties $v_{-\infty}$ and $v_{+\infty}$ of strength $\pm\Gamma/2t$ at a long distance upstream and downstream of the cascade. Consequently for the cascade, W_∞ is the vector mean of W_1 and W_2. By taking the circulation about path *abcd* for one blade pitch, Γ may be related to t, W_∞ and the flow angles through

$$\Gamma = t(V_1 - V_2) = t(\tan \beta_1 - \tan \beta_2)W_\infty \cos \beta_\infty \tag{2.54}$$

From the velocity triangles, an additional important relationship may be obtained, linking β_∞ to β_1 and β_2, namely

$$\tan \beta_\infty = \tfrac{1}{2}(\tan \beta_1 + \tan \beta_2) \tag{2.55}$$

To obtain information about the lift force $L = L_x + iL_y$ for an aerofoil cascade, following Dixon (1975) and Horlock (1958) we may write momentum equations in the x and y directions, namely (for inviscid flow)

$$\left.\begin{aligned} L_x &= (p_2 - p_1)t = \tfrac{1}{2}\rho(W_1^2 - W_2^2)t = \rho V_\infty \Gamma \\ L_y &= \rho U_\infty t(V_1 - V_2) = \rho U_\infty \Gamma \end{aligned}\right\} \tag{2.56}$$

where use has been made of (2.54) and velocity triangle relationships from Fig. 2.12. We observe that the Magnus law applies individually to the components U_∞ and V_∞ of the vector mean velocity. Consequently the net lift force L is given by

$$L = L_x + iL_y = \rho W_\infty \Gamma$$

and is normal to W_∞. The lift coefficient based on vector mean velocity, as previously defined for single aerofoils (2.30) then becomes

$$C_{L_\infty} = \frac{L}{\frac{1}{2}\rho W_\infty^2 \ell} = 2\left(\frac{t}{\ell}\right)(\tan \beta_1 - \tan \beta_2) \cos \beta_\infty \qquad (2.57)$$

and is thus a function of the pitch/chord ratio t/ℓ and the flow angles β_1, β_2 and β_∞

As presented above and following the strategy for analysis of single aerofoils, it would seem necessary for surface vorticity modelling to prescribe W_∞ and β_∞ as initial input data. In engineering practice on the other hand the usual problem posed is the prediction of outlet conditions (β_2, W_2) for a range of prescribed inlet flows (β_1, W_1) for a cascade of given geometry, vector mean quantities being of little value. Further analysis is required to accomplish this as follows.

Following the procedure described in Section 2.5.1, separate solutions $\gamma_u(s)$ and $\gamma_v(s)$ may be derived for unit velocities $U_\infty = 1$ and $V_\infty = 1$ respectively. The corresponding unit bound circulations are then given by

$$\left.\begin{aligned} \Gamma_u &= \sum_{n=1}^{M} \gamma_u(s_n)\,\Delta s_n \\ \Gamma_v &= \sum_{n=1}^{M} \gamma_v(s_n)\,\Delta s_n \end{aligned}\right\} \qquad (2.58)$$

Implementing (2.52b), the velocities in the y direction at $x = \pm\infty$, introducing the actual velocity components U_∞, V_∞, may be written

$$V_1 = V_\infty + \frac{\Gamma_{u_{U_\infty}}}{2t} + \frac{\Gamma_{v_{V_\infty}}}{2t} \quad \text{at} \quad x = -\infty$$

$$V_2 = V_\infty - \frac{\Gamma_{u_{U_\infty}}}{2t} - \frac{\Gamma_{v_{V_\infty}}}{2t} \quad \text{at} \quad x = +\infty$$

Since $U_1 = U_2 = U_\infty$, dividing throughout by U_∞ we have

$$\left.\begin{aligned}
\tan \beta_1 &= \tan \beta_\infty + \frac{\Gamma_u}{2t} + \frac{\Gamma_v}{2t} \tan \beta_\infty \\
\tan \beta_2 &= \tan \beta_\infty - \frac{\Gamma_u}{2t} - \frac{\Gamma_v}{2t} \tan \beta_\infty
\end{aligned}\right\} \tag{2.59}$$

After subtracting these equations and rearranging terms, β_2 may be expressed as a function of the unit circulations Γ_u and Γ_v and β_1 though

$$\beta_2 = \text{arc tan}\left\{\left(\frac{1 - \Gamma_v/2t}{1 + \Gamma_v/2t}\right) \tan \beta_1 - \left(\frac{2}{1 + \Gamma_v/2t}\right) \frac{\Gamma_u}{2t}\right\} \tag{2.60}$$

We note also that the sum of equations (2.59) agrees with (2.55), the definition of vector mean flow angle β_∞.

A surface pressure coefficient normalised by vector mean dynamic head may be defined as before according to (2.29). However, since in cascade design and testing, inlet conditions are usually specified, surface pressure coefficient may also be defined according to axial compressor practice,

$$C_{p1} = \frac{p - p_1}{\frac{1}{2}\rho W_1^2} = C_{p\infty}\left(\frac{\cos \beta_1}{\cos \beta_\infty}\right)^2 \tag{2.61}$$

2.6.3 Program bladerow.pas and sample calculations

All the necessary equations are now available for the solution of cascade flows by the surface vorticity model and these have been embodied in the pascal program Bladerow.pas which is included in the Appendix as Program 2.4. In many respects this program is similar to that for the single aerofoil, from which it has in fact been evolved. The essential differences are attributable to the cascade coupling coefficients and velocity triangle relationships developed in the preceding two sections. From the user's viewpoint these are reflected mainly in the input and output data referred to in boxes 1 and 10 of the flow diagram for Program 2.4. All of the derivations down to box 8, the procedure for the unit solutions, are calculated once and for all. Following this the program is re-entrant at procedures 'input flow data" and 'solutions' to permit repetition for a range of inlet conditions W_1 and β_1. The computational sequence is as follows.

Two-dimensional aerofoils and cascades

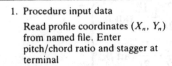

Flow diagram for bladerow program 2.4

1. Procedure input data

   Read profile coordinates (X_n, Y_n)
   from named file. Enter
   pitch/chord ratio and stagger at
   terminal

2. Procedure data-preparation

   Prepare pivotal points (x_n, y_n),
   element lengths ds_n and profile
   slopes β_n

3. Procedure coupling-coefficients
   Cascade coefficients if $t/l < 30$
   otherwise single aerofoil
   coefficients

4. Procedure right-hand-sides

   Unit right hand sides for $U_\infty = 1.0$
   and $V_\infty = 1.0$

5. Procedure back-diagonal-
   correction

6. Procedure Kutta-condition

7. Procedure invert-matrix

8. Procedure unit solutions

9. Procedure input-flow-data

   Enter W_1 and β_1 at terminal

10. Procedure solution

    Solve for β_2, Γ, C_L and pressure
    distribution C_{p1}

new
case?

yes

no

Spurred on by the rapid growth of computation facilities in the 1960s, there was a concentration of research effort into the development of methods for predicting the incompressible flow through turbomachine cascades. Linearised singularity theories such as those of Ackeret (1942) and Schlichting (1955b), being just within the limits of electrical desk calculators, attracted a good deal of interest at that time, together with more precise conformal transformation solutions such as those of Merchant & Collar (1941), Garrick (1944) and Howell (1948). In view of the errors inherent in the more numerically adaptable linearised singularity theories, Gostelow (1964) recognised the need for an absolute standard against which to check their validity. In response to this he set about providing several standard cascade solutions by the exact conformal transformation theory of Merchant & Collar, two of which we will now consider. Others, such as Czibere (1962), (1963) and Fuzy (1970), followed the alternative route of removing the linearisation of the singularity theory by distributing the source/vortex singularities along a *carrier line* within the blade profile. In parallel with this a range of other techniques for cascade calculation have been explored which were ably reviewed by Gostelow (1984). A full exposition of Merchant & Collar's theory has been given by Gostelow to which the reader is referred for further analytical details.

Case 1. Profile 10C4/70C50

By experimentation with the conformal transformation parameters, Gostelow was able to produce a blade profile quite similar to the then current U.K. compressor practice. As illustrated by Fig. 2.14 the analytical profile compared well with a C4 base profile distributed upon a 70° circular arc camber line. Flow predictions have been given by Gostelow for a pitch/chord ratio of 0.900 364 and zero stagger angle, with inlet angles β_1 of ±35°. These are compared with the output of Program 2.4 for a 50-element representation, in Fig. 2.14. Surface pressure distributions are in excellent agreement with the exact theory, bearing in mind the slight differences which one would expect due to profile mismatch. The run with $\beta_1 = -35°$ provides a particularly exacting test of the surface vorticity theory which is also vindicated by the predictions of outlet angle tabulated in Table 2.5.

The profile coordinates for C4/70/C50 used in this test calculation

Two-dimensional aerofoils and cascades

Table 2.5. *Comparison of Martensen's method with exact cascade theory –*
Case 1

C4/70C50 profile with $\lambda = 0$, $t/\ell = 0.900\,364$

Method	$\beta_1 = 35°$	$\beta_1 = -35°$
Exact solution, Gostelow (1984), $\beta_2 =$	23.80	24.84
Surface vorticity theory, Program 2.4, $\beta_2 =$	23.85	25.04
Schlichting linearised theory, (1955b), $\beta_2 =$	20.28	22.57

were generated by means of the author's profile design software
PROMOD.PAS, and are given in Table 2.6.

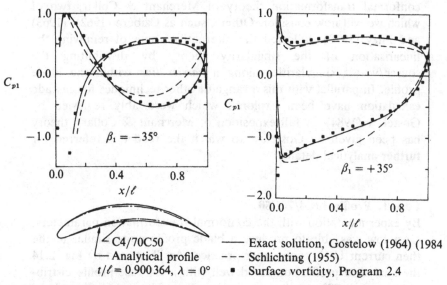

Fig. 2.14. 70° camber cascade.

Case 2. Highly cambered impulse cascade

To expose Schlichting's linearised theory to a more extreme test,
Gostelow developed the highly cambered cascade profile ($\theta = 112°$)
shown in Fig. 2.15. Although linearised theory coped reasonably
well with the 70° cambered cascade, it proved unable to handle
more highly cambered profiles, resulting in added pressure to
develop more advanced theories. The immediate outcome was the
surface vorticity analysis of Wilkinson (1967a). The output from
Program 2.4 compares well with Gostelow's conformal transforma-

84

Table 2.6. *Input data coordinates for C4/70/50*

x upper	y upper	x lower	y lower
0.000 000	0.000 000	0.995 474	0.001 902
−0.001 331	0.010 364	0.982 130	0.007 511
0.005 112	0.026 605	0.960 629	0.016 584
0.020 890	0.045 899	0.932 161	0.029 227
0.044 910	0.068 663	0.897 301	0.044 355
0.077 124	0.093 177	0.856 346	0.060 111
0.116 978	0.118 066	0.810 038	0.075 216
0.163 905	0.141 551	0.759 238	0.088 598
0.216 900	0.162 479	0.704 900	0.099 530
0.274 882	0.179 811	0.647 952	0.107 441
0.336 656	0.192 905	0.589 296	0.111 950
0.401 069	0.201 078	0.529 800	0.112 830
0.466 915	0.203 984	0.470 295	0.110 183
0.532 990	0.201 338	0.411 550	0.104 122
0.598 085	0.193 250	0.354 327	0.094 790
0.661 065	0.180 253	0.299 339	0.082 697
0.720 879	0.162 978	0.247 273	0.068 438
0.776 589	0.142 319	0.198 671	0.053 045
0.827 386	0.119 380	0.154 053	0.037 519
0.872 623	0.095 474	0.113 859	0.023 052
0.911 716	0.071 873	0.078 783	0.010 403
0.944 146	0.049 839	0.049 333	0.000 783
0.969 148	0.030 097	0.026 305	−0.005 114
0.986 453	0.013 980	0.009 216	−0.004 875
0.996 640	0.003 587	0.000 000	0.000 000
1.000 000	0.000 000		

tion theory except in the trailing edge region, Fig. 2.15. Differences here may well be attributable to arbitrariness in selection of the trailing edge stagnation point in Gostelow's analysis. The surface vorticity analysis, which implements Wilkinson's Kutta–Joukowski condition, predicts greater loading in the trailing edge region associated with a larger outlet angle as shown by Table 2.7. Turning angles $(\beta_1 + \beta_2)$ on the other hand agree to within 2.5%.

Profile coordinates for the exact solution have been given by Gostelow (1984). Unfortunately these are unsuitable for direct entry into the Martensen analysis, so that replotting was required for data input following the approach outlined in Section 2.5.2. The profile coordinate data finally used by the author are recorded in Table 2.8.

Two-dimensional aerofoils and cascades

Table 2.7. *112° Camber Impulse Cascade, Gostelow
(1964) (1984)*

$\lambda = 0°$, $t/\ell = 0.5\,899\,644$, $\beta_1 = 50°$

Method	β_2
Exact solution, Gostelow	51.17
Surface vorticity, program (2.4)	53.45
Linearised theory, Schlichting	46.32

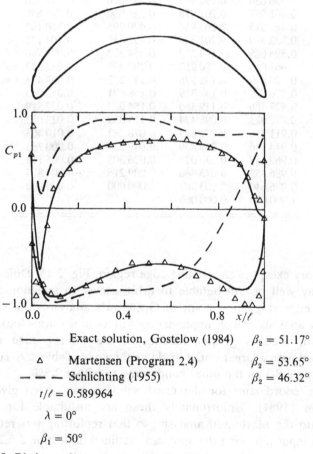

——————— Exact solution, Gostelow (1984) $\beta_2 = 51.17°$

△   Martensen (Program 2.4) $\beta_2 = 53.65°$

— — — Schlichting (1955) $\beta_2 = 46.32°$

$t/\ell = 0.589\,964$

$\lambda = 0°$

$\beta_1 = 50°$

Fig. 2.15. Blade profile and pressure distribution for a highly cambered impulse cascade ($\theta = 112°$)

86

Table 2.8. *112° Camber cascade (Gostelow 1984)*

x upper	y upper	x lower	y lower
0.010 001	−0.008 145	1.000 001	−0.000 145
0.000 964	0.003 977	0.990 702	0.001 593
−0.004 460	0.040 981	0.969 807	0.015 477
0.006 080	0.087 637	0.943 525	0.042 463
0.031 732	0.132 531	0.912 663	0.073 745
0.063 568	0.176 405	0.876 988	0.106 475
0.103 440	0.220 826	0.835 644	0.137 768
0.151 083	0.262 324	0.790 353	0.168 518
0.205 373	0.299 794	0.741 314	0.196 554
0.265 191	0.332 860	0.689 960	0.220 017
0.329 619	0.358 132	0.637 629	0.237 244
0.397 197	0.374 532	0.584 516	0.246 056
0.466 578	0.382 811	0.530 322	0.247 722
0.536 691	0.381 963	0.476 116	0.245 250
0.605 958	0.370 344	0.422 269	0.237 935
0.673 499	0.351 063	0.369 654	0.225 353
0.737 148	0.321 596	0.318 817	0.209 033
0.795 226	0.285 914	0.270 126	0.186 900
0.847 322	0.246 208	0.224 351	0.159 730
0.893 261	0.203 404	0.181 935	0.129 985
0.931 782	0.158 886	0.143 156	0.099 836
0.963 355	0.114 070	0.108 958	0.071 139
0.986 035	0.070 701	0.082 252	0.044 027
0.998 691	0.033 607	0.054 993	0.013 715
1.001 409	0.008 560	0.026 741	−0.007 420
		0.010 001	−0.008 145

Case 3. Inlet nozzle guide vane

In view of the above slight discrepancies between Gostelow's published results and the present surface vorticity theory, a third case has been considered by the author based upon a simple extension of the conformal transformation theory developed in Section 2.5.3. The uniform stream W_∞ is replaced by a source m and vortex Γ located at point $P_1(r_1, \phi_1)$ in the z plane, Fig. 2.16. The potential flow past the circle in the z plane may then be modelled by introducing singularities $+m$ and $-\Gamma$ at the inverse point P_2 and singularities $-m$, and Γ at the centre of the circle. The coordinates of these locations are given by

$$\left.\begin{array}{ll} x_1 = r_1 \cos \phi_1 - \varepsilon 1, & y_1 = r_1 \sin \phi_1 + \varepsilon 2, \\ x_2 = \dfrac{r_0^2}{r_1} \cos \phi_1 - \varepsilon 1, & y_2 = \dfrac{r_0^2}{r_1} \sin \phi_1 + \varepsilon 2, \\ x_3 = -\varepsilon 1, & y_3 = \varepsilon 2 \end{array}\right\} \qquad (2.62)$$

Two-dimensional aerofoils and cascades

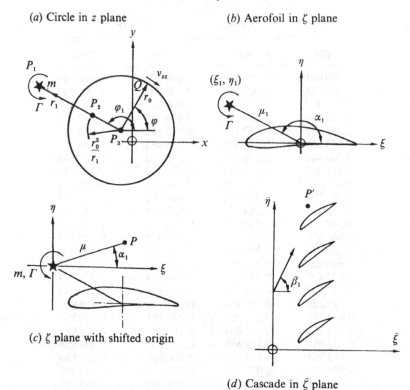

(a) Circle in z plane (b) Aerofoil in ζ plane

(c) ζ plane with shifted origin

(d) Cascade in $\bar{\zeta}$ plane

Fig. 2.16. Transformation of a circle into a cascade.

The velocity components at Q on the cylinder due to the singularities are then given by

$$
\left.\begin{aligned}
u &= \frac{1}{2\pi} \sum_{n=1}^{3} \frac{m_n(x_Q - x_n) - \Gamma_n(y_Q - y_n)}{\sqrt{[(x_Q - x_n)^2 + (y_Q - y_n)^2]}} \\
v &= \frac{1}{2\pi} \sum_{n=1}^{3} \frac{m_n(y_Q - y_n) + \Gamma_n(x_Q - x_n)}{\sqrt{[(x_Q - x_n)^2 + (y_Q - y_n)^2]}}
\end{aligned}\right\}
\tag{2.63}
$$

from which we may obtain the velocity parallel to the surface through

$$
v_{sz} = u \sin \phi_Q - v \cos \phi_Q
\tag{2.64}
$$

The coordinates of the Joukowski aerofoil in the ζ plane are already given by equations (2.34). The source/vortex transforms

88

into a source/vortex of equal strength located in the ζ plane at

$$\left.\begin{aligned}
\xi_1 &= x_1\left(1 + \left(\frac{a}{r}\right)^2\right) \\
\eta_1 &= y_1\left(1 - \left(\frac{a}{r}\right)^2\right)
\end{aligned}\right\}$$ (2.65)

It is now helpful to move the coordinate origin in the ζ plane to the location of the source/vortex by subtracting (ξ_1, η_1) from all dimensions. The coordinates of the Joukowski aerofoil then become

$$\left.\begin{aligned}
\xi &= \left(1 + \left(\frac{a}{r}\right)^2\right)(r_0 \cos \phi_Q - \varepsilon 1 - x_1) \\
\eta &= \left(1 - \left(\frac{a}{r}\right)^2\right)(r_0 \sin \phi_Q + \varepsilon 2 - y_1)
\end{aligned}\right\}$$ (2.66)

Now Fisher and Lewis (1971) have shown that a single aerofoil in the ζ plane may be transformed into a cascade in the $\bar{\xi}$ plane through the same transformation used in Section 2.6 to generate an array of point vortices, namely

$$\bar{\zeta} = \ln \zeta$$ (2.67)

from which the cascade coordinates become

$$\left.\begin{aligned}
\bar{\xi} &= \ln \mu = \tfrac{1}{2}\ln(\xi^2 + \eta^2) \\
\bar{\eta} &= \alpha = \arctan(\eta/\xi)
\end{aligned}\right\}$$ (2.68)

Working through the transformations the surface velocity on the cascade is given finally by

$$v_{s\bar{\xi}} = v_{sz}\left|\frac{dz}{d\bar{\zeta}}\right| = v_{sz}\sqrt{\left[\frac{\xi^2 + \eta^2}{\left(1 - \left(\frac{a}{r_0}\right)^2\right)\cos 2\phi + \left(\frac{a}{r_0}\right)^4 \sin 2\phi}\right]}$$ (2.69)

The source/vortex transforms into the uniform stream W_1 upstream of the cascade at $-\infty$, where

$$\beta_1 = \arctan(\Gamma/m)$$

and (2.70)

$$W_1 = \frac{1}{2\pi}\sqrt{(m^2 + \Gamma^2)}$$

The blade chord and stagger may be evaluated from the profile coordinates in the ζ plane. From the transformation equation (2.66) for the $\bar{\eta}$ direction, the pitch is given by

$$t = 2\pi \tag{2.71}$$

The trailing edge Kutta–Joukowski condition must be satisfied by introducing a bound vortex Γ_B at the centre of the circle of strength

$$\Gamma_B = -2\pi r_0 v_{szB} \tag{2.72}$$

where use is made of (2.63) to evaluate v_{szB}. A final correction to these equations must then be made to replace Γ_3 by $\Gamma_3 + \Gamma_B$, thereby including the effect of bound circulation upon the surface velocity distribution.

Unfortunately this transformation scheme produces considerable distortions making it difficult to obtain low pitch/chord ratio cascades of realistic profiles. Furthermore coordinates generated with even ϕ intervals in the z plane may produce trailing edge coordinates which are badly conditioned for surface vorticity analysis. After experimentation the profile shown in Fig. 2.17 was

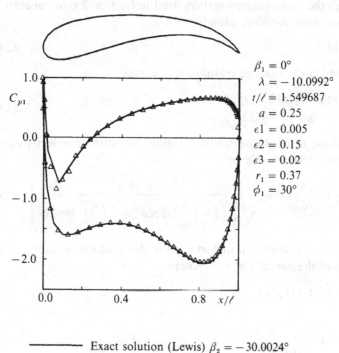

$\beta_1 = 0°$
$\lambda = -10.0992°$
$t/\ell = 1.549687$
$a = 0.25$
$\epsilon 1 = 0.005$
$\epsilon 2 = 0.15$
$\epsilon 3 = 0.02$
$r_1 = 0.37$
$\phi_1 = 30°$

——— Exact solution (Lewis) $\beta_2 = -30.0024°$

△   Martensen (Program 2.4) $\beta_2 = -29.7948°$

Fig. 2.17. Inlet guide vane profile and surface pressure distribution.

Table 2.9. *Coordinates of inlet guide vane – Case 3*

x upper	y upper	x lower	y lower
0.000 000	0.000 000	0.995 161	0.001 851
−0.003 295	0.007 914	0.993 372	0.003 026
−0.005 915	0.025 654	0.991 147	0.004 561
−0.000 437	0.051 383	0.988 414	0.006 500
0.018 802	0.081 428	0.985 083	0.008 889
0.053 079	0.110 407	0.981 053	0.011 770
0.099 151	0.134 333	0.976 202	0.015 181
0.152 118	0.152 101	0.970 391	0,019 155
0.207 824	0.164 498	0.963 459	0.023 710
0.263 516	0.172 826	0.955 222	0.028 849
0.317 601	0.178 223	0.945 474	0.034 549
0.369 249	0.181 496	0.933 988	0.040 758
0.418 098	0.183 168	0.920 523	0.047 384
0.464 051	0.183 569	0.904 832	0.054 291
0.507 159	0.182 903	0.886 670	0.061 293
0.547 549	0.181 300	0.865 807	0.068 158
0.585 382	0.178 846	0.842 042	0.074 614
0.620 831	0.175 604	0.815 208	0.080 356
0.654 064	0.171 619	0.785 174	0.085 063
0.685 239	0.166 930	0.751 847	0.088 414
0.714 501	0.161 574	0.715 152	0.090 098
0.741 979	0.155 583	0.675 022	0.089 820
0.767 788	0.148 989	0.631 380	0.087 306
0.792 028	0.141 827	0.584 125	0.082 293
0.814 787	0.134 131	0.533 124	0.074 530
0.836 137	0.125 937	0.478 230	0.063 778
0.856 141	0.117 285	0.419 313	0.049 849
0.874 850	0.108 219	0.356 345	0.032 717
0.892 305	0.098 789	0.289 549	0.012 790
0.908 537	0.089 049	0.255 000	0.002 000
0.923 566	0.079 067	0.219 653	−0.008 457
0.937 402	0.068 921	0.185 000	−0.018 000
0.950 044	0.058 706	0.148 344	−0.027 282
0.961 476	0.048 542	0.115 000	−0.034 000
0.971 665	0.038 579	0.079 398	−0.035 951
0.980 554	0.029 012	0.050 000	−0.034 500
0.988 050	0.020 097	0.021 947	−0.022 737
0.994 003	0.012 174	0.009 500	−0.013 000
0.998 171	0.005 705	0.000 000	0.000 000
1.000 000	0.000 000		

obtained typical of an inlet guide vane. Very much better agreement was obtained with surface vorticity theory for this cascade but only after some replotting of data notably in the trailing and leading edge regions. The actual data used for these calculations are recorded in Table 2.9.

Two-dimensional aerofoils and cascades

2.7 Multiple bodies and aerofoils with slots and flaps

So far we have considered only single bodies or periodic cascades of aerofoils for which the body shapes and surface velocilties are identical. As shown by Jacob and Riegels (1963) and Wilkinson (1967a), the flow past an assembly of P mutually interacting bodies may be represented by a simple adaptation of equations (1.21) to read

$$\sum_{q=1}^{P}\sum_{n=1}^{M_q} K_{mn}{}^{pq}(s_{qn}) = -U_\infty \cos \beta_{pm} - V_\infty \sin \beta_{pm} \tag{2.73}$$

where also

$$p = 1, 2 \dots P \quad \text{and} \quad m = 1, 2 \dots M_p$$

Equation (2.73) states the Dirichlet boundary condition at surface element m of body p. The left hand side includes the contributions from all elements ($n = 1, 2 \dots M_q$) on each of the bodies ($q = 1, 2 \dots P$). The number of elements M_q chosen for each body may of course differ depending upon the individual geometrical requirements.

The coupling coefficient representing the induced velocity at pivotal point m of body p due to element n of body q is then given by

$$K_{mn}{}^{pq} = \frac{\Delta s_{qn}}{2\pi} \left\{ \frac{(y_{pm} - y_{qn})\cos\beta_{pm} - (x_{pm} - x_{qn})\sin\beta_{pm}}{(x_{pm} - x_{qn})^2 + (y_{pm} - y_{qn})^2} \right\} \tag{2.74}$$

In order to extend this model to deal with mutually interfering or slotted cascades, Wilkinson introduced the suitably modified cascade coupling coefficient, namely

$$K_{mn}{}^{pq}$$

$$= \frac{\Delta s_{qn}}{2t} \left\{ \frac{\cos\beta_{pm}\sin\dfrac{2\pi}{t}(y_{pm} - y_{qn}) - \sin\beta_{pm}\sinh\dfrac{2\pi}{t}(x_{pm} - x_{qn})}{\cosh\dfrac{2\pi}{t}(x_{pm} - x_{qn}) - \cos\dfrac{2\pi}{t}(y_{pm} - y_{qn})} \right\} \tag{2.75}$$

In both cases the self-inducing coupling coefficients (when $p = q$) and $m = n$) are given as before for single aerofoils or cascades (1.31). To illustrate this model for the three mutually interfering

92

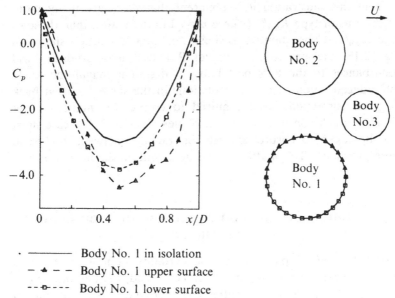

- ———— Body No. 1 in isolation
- − ▲ − − Body No. 1 upper surface
- − − ▫ − − − Body No. 1 lower surface

Fig. 2.18. Flow past three cylinders in close proximity.

cylinders shown in Fig. 2.18 the equations can be expressed in matrix form as follows.

$$
\begin{pmatrix}
\begin{array}{ccc}
A^{11} & A^{12} & A^{13} \\
A^{21} & A^{22} & A^{23} \\
A^{31} & A^{32} & A^{33}
\end{array}
\end{pmatrix}
\begin{pmatrix} \gamma(s) \end{pmatrix}
=
\begin{pmatrix} \text{rhs} \end{pmatrix}
\tag{2.76}
$$

where the coupling coefficient matrix has been partitioned to clarify the various body interactions. Thus typically, the sub-matrix A^{23} contains all of the coefficients accounting for the interference experienced by body 2 due to body 3. The sub-matrices forming the back diagonal A^{11}, A^{22} etc, account for the influence of each body upon itself and are thus identical to those obtained for each body considered in isolation.

93

From this illustration it is apparent that there are many more coefficients of type K_{mn}^{pq} (where $p \neq q$) in the matrix than those of type K_{mn}^{pp}. Consequently, as illustrated by the solution shown in Fig. 2.18, the presence of bodies 2 and 3 may produce local disturbances to the flow past body 1 of similar magnitude to its self-induced potential flow in isolation. In the same way that back diagonal corrections were required to correct for numerical net internal circulation errors for thin bodies, Section 2.3.3, care must also be taken to ensure accurate interactions between bodies in close proximity. This matter will be dealt with in the next section.

2.7.1 Internal circulation correction for bodies in close proximity

Wilkinson (1967a) pointed out the certainty of numerical errors should the gap between points pm and qn on adjacent bodies be less than the local element lengths Δs_{pm} or Δs_{qn}. The circulation induced around the perimeter of body p due to a unit vortex placed at the centre of element n of body q is then given by

$$\Delta\Gamma = \oint k_{mn}^{pq}\, ds_{pm}$$

or in the present numerical form

$$\Delta\Gamma = \frac{1}{\Delta s_{qn}} \sum_{m=1}^{M_p} K_{mn}^{pq}\, \Delta s_{pm}$$

which involves only the coupling coefficients in column n of sub-matrix A^{pq}. To enforce zero net circulation $\Delta\Gamma$, we must replace, say, the ith coupling coefficient in column n by the value

$$K_{in}^{pq} = -\frac{1}{\Delta s_{pi}} \sum_{\substack{m=1 \\ m \neq i}}^{M_p} K_{mn}^{pq}\, \Delta s_{pm} \qquad (2.77)$$

The ith element of body p should be the one in closest proximity to element n of body q, Fig. 2.19. Since this element will normally receive the greatest induced velocity due to element $\gamma(s_{qn})\, \Delta s_{qn}$, the best and fastest computational procedure to determine i is to assume that K_{in}^{pq} is the coupling coefficient in column n of sub-matrix A^{pq} having the largest absolute value.

94

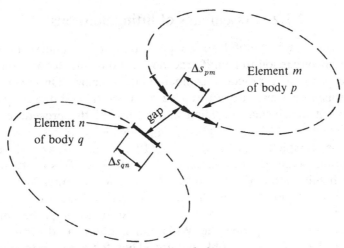

Fig. 2.19. Effect of elements in close proximity.

Unfortunately, as pointed out also in Section 2.3.3 for single non-lifting bodies, the outcome of this procedure is a singular matrix, since the sum of all equations, each being first multiplied in turn by the constant Δs_{pm}, is now zero. A method for dealing with this, introduced in Section 2.4.1 for single lifting bodies with prescribed bound circulation Γ, involved the addition to all equations of the matrix of the extra circulation equation

$$\sum_{n=1}^{M} \gamma(s_m)\, \Delta s_n = \Gamma \qquad\qquad [2.8]$$

The indeterminacy implied by the singular matrix stems from the fact that unless Γ is thus enforced an infinity of solutions is possible. Applying this principle to the present instance, one such equation must be added to all equations of each leading diagonal sub-matrix A^{pp} with the value $\Gamma_p = 0$ to enforce the condition of zero circulation individually upon each body. It is insufficient to apply this to all coefficients of the matrix, thus specifying zero circulation of the system taken as a whole, since one body could then assume arbitrary positive circulation compensated for by negative circulations of other bodies in the assembly. It is absolutely essential to modify the A^{pp} sub-matrices individually, for which the equation

$$\sum_{m=1}^{M_p} \gamma(s_{mp})\, \Delta s_{mp} = 0 \qquad\qquad (2.78)$$

must be added to each row of sub-matrix A_{pp}.

2.7.2 Assemblies of lifting aerofoils

Although it was helpful to adopt a partitioned matrix in the foregoing discussion to identify the various body interactions, this is unnecessary in a practical computational scheme. Once surface elements, slopes and curvatures have been determined, the coupling coefficient matrix may be filled making use of the procedure for single bodies. Back diagonal and opposite point corrections may then be completed for each sub-matrix as just described. For non-lifting bodies the zero circulation equation (2.77) is next added to each sub-matrix A^{pp}. For lifting bodies, on the other hand, we must instead introduce the trailing edge Kutta condition to each individual aerofoil of the assembly. This can be accomplished most satisfactorily by applying the Wilkinson method, as described in Section 2.4.3, to each profile in turn. Thus for body p, following (2.26), we must subtract column te$p$ + 1 from column te$p$, since the trailing edge vortices are equal and opposite,

$$\gamma(\text{te}p) = -\gamma(\text{te}p + 1) \qquad (2.79)$$

Since the coupling coefficient matrix is now reduced by P columns, one for each lifting body, it must be restored to square before inversion, by elimination of P rows. This may best be achieved, as proposed in Section 2.4.3, by subtracting equation te$p$ + 1 from equation te$p$ for each body p. In all other respects the method of analysis is identical to that for single body problems.

This technique was applied to aerofoils with slots and flaps by Jacob and Riegels (1963) and by Wilkinson (1967b). Obviously the purpose of such aerodynamic control devices is to obtain high lift by taking deliberate advantage of the mutual interference between the elements comprising the multiple aerofoil. Application of the author's program 'polyfoil' to the NACA 65_3–118 aerofoil with a double slotted flap is shown in Fig. 2.20 in comparison with experimental data published by Abbott, Von Doenhoff & Stivers (1945) and Abbott & Von Doenhoff (1959). For flap angles in the range $0 < \delta < 45°$ the two flaps pivot together about the coordinate position (0.806, −0.212) for a unit chord aerofoil. For $\delta > 45°$ the small vane remains fixed and the larger flap rotates about the pivot position (0.875, −0.046). Predicted potential flow values are, as expected, in excess of the experimental lift coefficients and of course are unable to predict the onset of stall. Despite this, both the

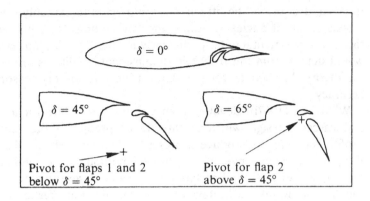

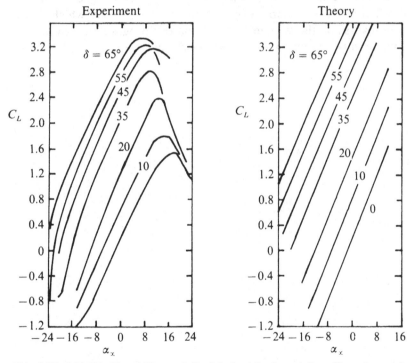

Fig. 2.20. NACA $65_3 - 118$ aerofoil with double-slotted flap. Experimental results by courtesy, Abbott & Von Doenhoff (1959). *Theory of Wing Sections*, Dover Publications.

trends and slopes of the lift curve are predicted remarkably well for a wide range of angles of attack and flap settings. It is also likely that the experimental data, particularly for the high lift flap settings, would depart from the truly two-dimensional conditions implicit in the theory. Predictions for low flap settings are of very reasonable accuracy.

Wilkinson (1967b) developed an iterative procedure for aerofoil and cascade design which is capable of predicting the required profile geometry to produce a prescribed surface velocity distribution, the so-called 'inverse' or 'design' method. He extended this technique also to slotted or tandem cascades or aerofoils to enable the designer to take advantage of aerodynamic interference in order to obtain stable cascades capable of high lift with low drag. More detailed reference to this will be made in Section 7.5 which deals with inverse methods. The author's analysis program tandem.pas is compared favourably with the Wilkinson method in Fig. 7.17.

CHAPTER 3

Mixed-flow and radial cascades

3.1 Introduction

As early in the history of gas turbines and internal aerodynamics as 1952, C. H. Wu recognised the truly three-dimensional nature of the flow in turbomachines and proposed a remarkably sophisticated scheme for numerical analysis illustrated by Fig. 3.1. The fully three-dimensional flow was treated by the superposition of a number of two-dimensional flows which were of two types located on the so-called S-1 and S-2 stream surfaces. S-2 surfaces follow the primary fluid deflection caused by the blade profile curvature and its associated aerodynamic loading. Due to the blade-to-blade variation in static pressure the curvature of each S-2 stream surface will differ, calling for several surfaces for adequate modelling of the flow. S-1 surfaces account for consequent twist in the so-called 'through-flow' or 'meridional flow' which comprises a family of stream surfaces which approach axisymmetry close to the hub and casing and exhibit maximum departure from axisymmetry at the blade passage mid height. By solution of the flows on this mesh for successively improved estimates of the S-1 and S-2 surfaces, allowing for fluid dynamic coupling between them, an iterative approach to the fully three-dimensional flow was fairly comprehensively laid out by Wu in a paper which was truly twenty years ahead of its time.

Until relatively recently such calculation procedures have been ruled out by lack of suitable computing facilities. It was in 1966 that Marsh gave a strong impetus to computer application of Wu's method by developing the well known matrix through-flow analysis. However this was restricted to a simpler model for meridional flows based upon the assumption of axisymmetric or circumferentially averaged S-2 flow. Only very recently have solutions been derived to the fully three-dimensional flow such as the time marching methods published by Denton (1974), (1976), (1982) for high Mach number axial turbomachine blade rows and by Potts (1987) with

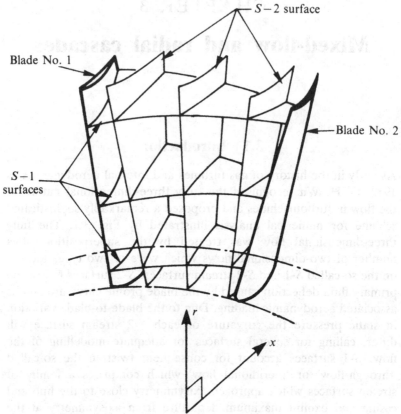

Fig. 3.1. *S*-1 and *S*-2 stream surfaces (after Wu (1952)).

applications to strongly swept blade rows for which *S*-1 flow becomes especially important. For such applications involving high Mach number and possibly transonic or supersonic three-dimensional flows, channel or mesh methods of one sort or another seem certain to offer the way ahead as the route to accommodating the dominant interactions between the *S*-1 and *S*-2 surfaces.

For many other applications the traditional design technique of superimposing two-dimensional blade-to-blade flows on an assumed axisymmetric meridional flow remains perfectly adequate and geometrically convenient for both mechanical design and fluid dynamic analysis. This alternative to the use of *S*-1, *S*-2 surfaces is illustrated for the Francis turbine runner depicted in Fig. 3.2 and has been described more fully elsewhere by the author (Lewis (1964a)). The streamlines shown diagrammatically are obtained by

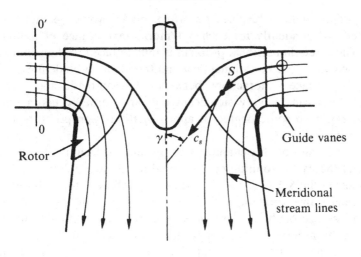

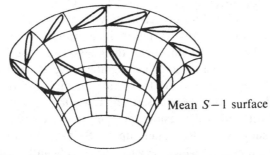

Mean $S-1$ surface

Fig. 3.2. Meridional flow through a Francis Turbine. (Reproduced from the Proceedings of the Institution of Mechanical Engineers by permission of the Council of the Institution.)

circumferential projection of actual streamlines starting from $00'$ onto the meridional (x, r) plane of the page. The meridional streamline pattern thus generated will vary periodically if the starting line $00'$ is rotated about the axis relative to the blades. Since this variation is often small, a reasonable assumption is to take the circumferential average and to regard the mean meridional flow as axisymmetric. Defined this way, which is equivalent to the circumferential S-2 average of Marsh (1966), the meridional flow axisymmetric S-1 surfaces form a useful structure upon which to design the blade system. The machine is now broken down into a series of elementary turbines for which blade profile geometry is selected independently for each S-1 meridional surface of revolution. The three-dimensional problem is thus reduced to the solution of the axisymmetric meridional flow and the blade-to-blade flows on each S-1 surface. Interactions between blade-to-blade and meridional flows can be very significant and are normally taken into consideration to some degree of approximation appropriate to the application.

The main aim of this chapter is to present extensions of surface vorticity theory for solution of the flow through mixed-flow rotating cascades on the S-1 surfaces. This is usually achieved through initial conformal transformation of the geometry to a straight cascade to take advantage of the analysis presented in Chapter 2 and is to be dealt with in Section 3.2. Application to radial guide vanes is presented in Sections 3.3 and 3.4 and to rotating radial or mixed-flow cascades in Section 3.5 followed by the derivation of datum exact solutions in Section 3.6, obtained by conformal transformation techniques. As already mentioned interactions between the meridional and cascade flows can be important. In axial compressors for example variation in meridional velocity due to change in the S-1 stream sheet thickness can modify deflection properties of the blade aerofoils significantly, Pollard & Horlock (1963). These so-called AVR (axial velocity ratio) effects are introduced into the Martensen analysis in Section 3.5.3 and compared with Pollard's results for compressor cascades in Section 3.7. The subject is also taken up in Section 3.6 in relation to mixed-flow fans for which AVR can significantly influence the interactions between the blade-to-blade and meridional flows.

3.2 Transformation of a mixed-flow cascade into a straight cascade

In order to simplify the blade to blade problem it is possible to transform each S-1 surface of revolution into a plane containing an infinite rectilinear cascade, Fig. 3.3. For the transformation to be conformal we require that

$$\frac{d\xi}{d\eta} = \frac{ds}{r \, d\theta} = \frac{1}{r \sin \gamma} \frac{dr}{d\theta} \tag{3.1}$$

where γ is the local cone angle of the meridional streamline and s is the distance measured along the streamline, Fig. 3.2.

Transformation of a mixed-flow cascade into a straight cascade

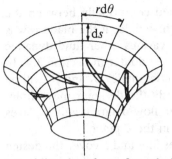

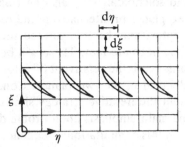

$S-1$ meridional surface of revolution in $z(s, \theta)$ plane with local cone angle γ

Rotor blade section transformed into straight cascade in ζ plane

$$d\xi = \frac{1}{\sin \gamma} \frac{dr}{r}, \quad d\eta = d\theta$$

(*a*) Transformation of $S-1$ surface into a straight cascade

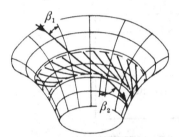

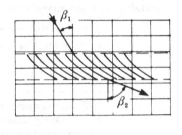

(*b*) Actuator disc model of circumferential average $S-2$ surfaces

Fig. 3.3. Transformation of S-1 and S-2 surfaces into rectilinear cascade plane. (Reproduced from the Proceedings of the Institution of Mechanical Engineers by permission of the Council of the Institution.)

Following Young (1958), this may be achieved by the coordinate transformations

$$\left. \begin{aligned} d\xi &= \frac{ds}{r} = \frac{1}{r \sin \gamma} dr \\ d\eta &= d\theta \end{aligned} \right\} \tag{3.2}$$

The conical surface of revolution containing say N blades in the z plane transforms into a cartesian coordinate plane system (ξ, η) in the ζ plane containing an equivalent infinite cascade parallel to the η axis of pitch

$$t = 2\pi/N \tag{3.3}$$

Furthermore for a stator the absolute flow field, being irrotational and solenoidal, may also be transformed conformally between the two planes or alternatively the relative flow field, if the blade row is a rotor, may be transformed provided various other fluid dynamic conditions are met. These will be considered later in Section 3.4. In either case the relative fluid inlet and outlet angles β_1 and β_2 will be identical in the two planes. Consequently the designer may implement the iterative strategy shown in the flow diagram which places the main emphasis upon profile design in the ζ plane.

In this computational sequence, boxes 3, 4 and 5 cover the design and analysis of the flow on each S-1 surface from hub to casing while boxes 2 and 6 permit iterative interactions with the meridional flow. As illustrated by the lower two diagrams of Fig. 3.3, the

Overall design and analysis sequence for mixed-flow turbomachines

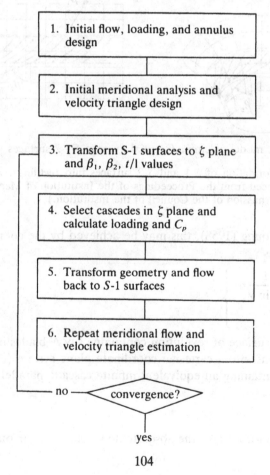

104

circumferentially averaged effect of the S-1 or blade-to-blade flow is frequently used to account for the effect of blade loading upon the meridional or S-2 flow. The difficult but important part of any quasi-three-dimensional scheme of this type is evaluating and transferring this coupling data as the output from box 5 into box 6. Marsh (1966) presented fairly full equations and Lewis & Mughal (1986) have presented a simplified technique which treats the blade row as a mixed flow actuator disc. Horlock (1978) extensively reviewed actuator disc theory which can in some circumstances provide useful simplified models for solving the meridional flow. Although the related classical solutions are limiting, the basic actuator disc model of an equivalent infinite number of tightly packed blades of infinitesimal thickness, Fig. 3.3, is entirely relevant to more flexible numerical schemes for reduction of the meridional flow.

3.2.1 Axial and radial blade rows

Axial and radial blade rows are special cases for which the transformations may be integrated directly. Thus for an axial blade row, the S-1 surface of constant radius r_1 transforms through

$$\xi = x/r_1, \qquad \eta = (r\theta)/r_1 \tag{3.4}$$

where (x, r_1, θ) are coordinates on the cylinder.

This is equivalent to simply unwrapping the cylindrical intersection through the blade row reproducing similar geometry in the ζ plane and is thus a justification for the use of straight cascade test modelling for axial turbomachines.

For the radial cascade shown in Fig. 3.4, the transformation equations reduce to

$$d\xi = \frac{dr}{r}, \quad d\eta = d\theta \tag{3.5}$$

and upon integration

$$\left.\begin{array}{l} \xi = \ln r \\ \eta = \theta \end{array}\right\} \tag{3.5a}$$

which corresponds to the well known transformation

$$\zeta = \ln z \tag{3.6}$$

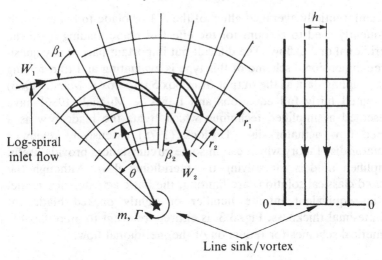

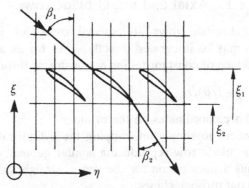

Fig. 3.4. Conformal transformation of radial guide vanes in $z(r, \theta)$ plane into rectilinear cascade in $\zeta(\xi, \eta)$ plane. (Reproduced from the Proceedings of the Institution of Mechanical Engineers by permission of the Council of the Institution.)

The polar coordinate system $z = re^{i\theta}$ and radial guide vane cascade in the z plane transform into cartesian coordinates $\zeta = \xi + i\eta$ and an equivalent rectilinear cascade in the ζ plane. As illustrated in Fig. 3.4, for stator guide vanes the entry and exit flows some radial distance from the blades tend towards logarithmic spirals of constant angle equivalent to flow due to sink/vortex at the origin, provided the meridional streamline thickness h is constant. These entry and exit streamlines transform to uniform streams W_1

106

Transformation of a mixed-flow cascade into a straight cascade

and W_2 at the identical swirl angles β_1 and β_2 in the ζ plane. The velocity $q_z = u_z + iv_z$ in the z plane may be obtained from the cascade solution in the ζ plane through the transformation

$$u_z - iv_z = (u_\zeta - iv_\zeta)\frac{\mathrm{d}\zeta}{\mathrm{d}z} \tag{3.7}$$

which in this case reduces to

$$q_z = q_\zeta/r \tag{3.8}$$

The surface pressure coefficient for the radial cascade, following the previous definition (2.61), is then

$$C_{p1} = 1 - \left(\frac{v_s}{W_1}\right)^2 = 1 - \left(\frac{v_{s\zeta}}{rW_1}\right)^2 \tag{3.9}$$

Another advantage of this transformation process is the possibility to categorise mixed-flow cascades in terms of the usual straight cascade geometrical parameters such as pitch/chord ratio, stagger and camber. Thus the blade chord ℓ, Fig. 3.5, follows from (3.2a), where suffix 1 and 2 denote leading edge and trailing edge

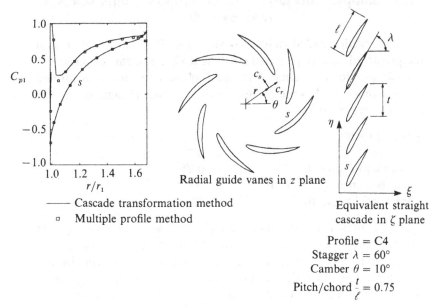

——— Cascade transformation method
□   Multiple profile method

Radial guide vanes in z plane

Equivalent straight cascade in ζ plane

Profile = C4
Stagger $\lambda = 60°$
Camber $\theta = 10°$
Pitch/chord $\dfrac{t}{\ell} = 0.75$

Fig. 3.5. Predicted surface pressure distribution for radial diffuser guide vanes designed in the ζ plane.

107

locations, namely

$$\ell = \frac{\xi_2 - \xi_1}{\cos \lambda}$$

$$\left.\begin{array}{ll} = \displaystyle\int_1^2 \frac{1}{r \sin \gamma \cos \lambda}\, dr & \text{for mixed-flow cascades} \\[2mm] = \ln(r_2/r_1)/\cos \lambda & \text{for radial cascades} \end{array}\right\} \qquad (3.10)$$

The stagger angle is given by

$$\lambda = \arctan\left(\frac{\theta_2 - \theta_1}{\xi_2 - \xi_1}\right) \qquad (3.11)$$

and the pitch/chord ratio for our radial cascade, making use of (3.3), becomes

$$\frac{t}{\ell} = \frac{2\pi \cos \lambda}{N \ln(r_2/r_1)} \qquad (3.12)$$

3.3 Sample calculation for an outflow radial diffuser vane cascade

An eight-bladed radial diffuser is shown in Fig. 3.5 together with the predicted surface pressure coefficient C_{p1}. For this example the cascade profile was designed in the ζ plane and then transformed into the radial guide vane diffuser in the ζ plane. Design data in the ζ plane were as follows

Base profile	C4
Stagger λ	60°
Pitch chord ratio t/ℓ	0.75
Camber angle (circular arc)	10°
Inlet swirl angle β_1	70°

It is of interest to note that the modest positive camber angle of 10°, introduced to produce a reduction in swirl angle, cannot be visualised in the z plane due to the distortion introduced by the transformation. Indeed, the guide vanes in the z plane appear to have strong negative camber and it is hard to realise that the concave surface labelled s is actually the 'suction surface'. The predicted surface pressure rises throughout the blade passage on both surfaces mainly due to the increase in area but more rapidly on the concave surface due to the camber-induced blade loading, required to reduce the swirl angle β_2.

3.3.1 Surface pressure distribution

Two methods of analysis are compared in Fig. 3.5. The cascade method as outlined in Section 2.6 may be applied in the ζ plane provided the blades are identical, with truly periodic flow, offering the maximum economy since M pivotal points only need be specified for a single blade. On the other hand the multiple aerofoil method of Section 2.7 may also be applied to this problem, although of course (x, y) coordinates must then be specified for all eight blades resulting in $8M$ pivotal points. This is obviously necessary if there are variations in individual blade geometry but otherwise involves massive redundancy. The surface pressure distributions predicted by these two methods with eight identical guide vanes are shown in Fig. 3.5 employing 50 elements for the cascade transformation and 22 elements per blade for the multiple aerofoil method. Results are shown for blade number 1 only in the latter case but solutions for all eight blades were of course the same and in close agreement with the cascade transformation method.

3.3.2 Inlet and outlet angles

In applying the multiple aerofoil method to a radial cascade the primary flow is no longer the uniform stream W_∞ required for plane flows, Section 2.7. Instead we may introduce a source m and vortex Γ at the origin resulting in a prewhirl inlet flow of angle and velocity

$$\beta_1 = \text{arc } \tan(\Gamma/m) \tag{3.13}$$

$$W_1 = \sqrt{(\Gamma^2 + m^2)}/2\pi r_1 \tag{3.14}$$

The boundary integral equation for this problem, c.f. (2.72), then reduces to

$$\sum_{q=1}^{P} \sum_{n=1}^{M} K_{mn}{}^{pq}\gamma(s_{qn}) = c_{rm}\cos(\beta_m - \theta_m) + c_{\theta m}\sin(\beta_m - \theta_m) \tag{3.15}$$

where c_{rm} and $c_{\theta m}$ are velocity components in the (r, θ) directions at element m of value $m/2\pi r_m$ and $\Gamma/2\pi r_m$ respectively. For greater computational convenience the right hand side may be expressed instead in cartesian coordinates through

$$\text{rhs} = \frac{m}{2\pi r_m}(x_m \cos \beta_m + y_m \sin \beta_m) + \frac{\Gamma}{2\pi r_m}(x_m \sin \beta_m - y_m \cos \beta_m)$$

$$(3.16)$$

Unit source and vortex solutions may then be derived by inserting successively $(m = 1, \Gamma = 0)$ and $(m = 0, \Gamma = 1)$, the principle outlined in Section 2.5.1. Combining these for any prescribed β_1, W_1 inlet flow, having also built in the Kutta–Joukowski condition, will result in a bound circulation for the nth blade of

$$\Gamma_n = \sum_{p=1}^{M} \gamma(s_p) \, \Delta s_p \qquad (3.17)$$

By taking the circulation around the outer radius of the guide vane row, the average fluid efflux angle then follows from

$$\tan \beta_2 = \bar{c}_{\theta 2}/\bar{c}_{r2}$$

$$= \tan \beta_1 + \frac{1}{m} \sum_{n=1}^{N} \Gamma_n \qquad (3.18)$$

In the case of the cascade radial to straight transformation method on the other hand, β_2 is obtained directly from the standard cascade analysis in the ζ plane. For the present example the predicted outlet angles were 61.05° from the multiple aerofoil model with 22 elements per blade and 60.31° from the cascade transformation method with 50 elements. Although the discrepancy seems rather large the predicted overall pressure rise is in extremely close agreement.

3.4 Rotor/stator interference in centrifugal compressors

Inoue (1980) undertook extensive experimental investigations of the interactions between a centrifugal compressor rotor and its surrounding diffuser vanes, Fig. 3.6. The rotor comprised 26 blades with axial inducer inlet and purely radial blade geometry at exit with a tip radius r_t of 147 mm and rotational speed 2900 rpm. Various radial diffusers were constructed with either 10 or 20

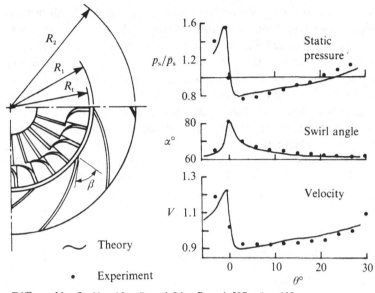

Diffuser No. 3 $N = 10$, $R_1 = 1.04$, $R_2 = 1.587$, $\beta = 60°$

Fig. 3.6. Inlet traverse of radial diffuser vanes, Fisher & Inoue (1981) (Reproduced from the Proceedings of the Institution of Mechanical Engineers by permission of the Council of the Institution.)

blades permitting variation in the entry vane angle β and the inlet and outlet dimensionless radii r_1/r_t and r_2/r_t. Circumferential traverses of swirl angle and velocity at entry to the diffuser vanes were undertaken together with wall static pressure measurements to record the blade to blade variation of these quantities. At a given (r, θ) measuring location the hot wire anemometer in such circumstances will register a periodic or fluctuating signal due to the regular sweep past of the rotor passage non-uniform exit flow. To eradicate this additional difficulty of data interpretation an ensemble averaging technique was adopted to estimate the average velocity at each (r, θ) location of the diffuser entry traverse.

Fisher & Inoue (1981) and Inoue (1980) and Fisher (1980) have published comparisons between these results and surface vorticity predictions by the above cascade transformation method for 17 diffuser configurations of which one sample is shown in Fig. 3.6 for diffuser No. 3. The surface vorticity method is extremely powerful for dealing with such problems as this where the flow variations in the vaneless space between r_t and r_1 are clearly dominated by the diffuser potential flow. Thus the excellent agreement between

111

experiment and theory shown here was typical of nearly all diffuser configurations and mass flow rates investigated except for situations when the diffuser leading edge stalled due to excessive angles of attack. The study revealed that the diffuser blade to blade potential flow variations in the vaneless space are virtually unaffected by the superimposed periodic flow due to the fairly strong wake jet profile emerging from the rotor blade passages. It is sufficient to adopt the circumferential average of this when designing diffuser exit vanes. Furthermore the surface vorticity method can provide large amounts of reliable design and research data at minimal cost. The study by Fisher and Inoue is of considerable importance in establishing this credibility for radial cascade analysis.

3.5 Mixed-flow and radial rotor blade rows

As early as 1928 Busemann published his classic paper on the flow through centrifugal pump rotors with logarithmic spiral blades, using conformal transformation theory. It was already fully realised that the flow viewed relative to a centrifugal pump or fan is strongly influenced by the so-called 'relative eddy' introduced when transforming coordinates from a stationary system to one which rotates with the rotor. For example, suppose that the blade row previously considered in Fig. 3.5 is made to rotate with angular velocity Ω. Since the absolute flow is irrotational, the vorticity ω (also defined as anticlockwise positive) may be expressed

$$\omega = \frac{\partial c_\theta}{\partial r} + \frac{c_\theta}{r} - \frac{1}{r}\frac{\partial c_r}{\partial \theta} = 0 \tag{3.19}$$

The fluid velocity components (w_r, w_θ) relative to coordinates which rotate with the rotor are related to those in stationary coordinates (c_r, c_θ), through

$$\left.\begin{array}{l} c_r = w_r \\ c_\theta = w_\theta + r\Omega \end{array}\right\} \tag{3.20}$$

Transforming to the rotating coordinates we then have relative vorticity ω_{rel} throughout the entire flow field of strength

$$\omega_{\text{rel}} = \frac{\partial w_\theta}{\partial r} + \frac{w_\theta}{r} - \frac{1}{r}\frac{\partial w_r}{\partial \theta} = -2\Omega \tag{3.21}$$

The effect of this relative vorticity is to produce a flow rotation within the blade passages opposite in direction to the blade rotation and known as the 'relative eddy' or 'slip flow'. Its effect generally is to reduce blade loading. Busemann predicted 'slip factors' for a wide range of log-spiral blade geometries, which have proved extremely valuable as design aids for estimating efflux angles and overall pressure rise for this restricted class of machines.

Another important feature of radial or mixed-flow rotors is the presence of Coriolis accelerations, which produce very significant contributions to power output independent of blade shape. For example, as shown by Lewis *et al.* (1972), the Euler pump equation for the stagnation pressure rise through our centrifugal pump, may be expressed

$$\frac{1}{\rho}(p_{02} - p_{01}) = \Omega(r_2 c_{\theta 2} - r_1 c_{\theta 1})$$

$$= \Omega(r_2 w_{\theta 2} - r_1 w_{\theta 1}) + \Omega^2(r_2^2 - r_1^2)$$

$$= \begin{pmatrix} \text{Energy input due} \\ \text{to fluid deflection} \end{pmatrix} + \begin{pmatrix} \text{Energy input due} \\ \text{to Coriolis forces} \end{pmatrix} \quad (3.22)$$

The first term, involving relative inlet and outlet swirl velocities $w_{\theta 1}$ and $w_{\theta 2}$, is linked directly to the deflection or 'aerodynamic' properties of the blades. The second term however is dependent only upon Ω and the radial limits r_1 and r_2 of the rotor. While wind tunnel cascade tests are perfectly appropriate for estimating the performance of axial turbomachine blade profile characteristics, they are clearly totally unable to model correctly either radial or mixed-flow rotors. For these turbomachines the second term in (3.22) is usually significant in magnitude, in some cases accounting for most of the blade loading. Accurate methods for theoretical analysis are thus crucial for all such applications and have proved invaluable as design aids for fans, pumps and hydraulic turbines.

Following a review of these and other problems by Lewis (1964a), Pollard (1965) was the first author to publish a numerical method for mixed-flow and radial rotor cascades, applied to Francis turbine design followed by Railly (1967), Railly, Houlton & Murugesan (1969). His method was based on the linearised singularity cascade analysis of Schlichting (1955b) with corrections for the added influence of the relative eddy. The immediate success of this work and its impact upon design methodology soon led Wilkinson (1967b) to extend the more accurate surface vorticity theory to deal with this problem following similar guide-lines concerning transformation to and from a straight cascade. The basis of this will be presented in the next sub-section, followed by an outline of work by Fisher (1975), (1986) who succeeded in developing a more precise solution for the flow in the ζ plane to provide a benchmark method. More recently Lewis & Mughal (1986) have reported the combination of a Wilkinson type blade to blade analysis with a form of mixed-flow actuator disc theory for desktop microcomputer analysis of the quasi three-dimensional flow of mixed-flow fans as described earlier in Section 3.2.

3.5.1 Transformation of the 'relative eddy' to the straight cascade plane

The previous analysis may be generalised to mixed flow machines with clockwise rotation Ω, Fig. 3.7, if we observe that the relative eddy may be resolved into two components $\omega_z = 2\Omega \sin \gamma$ normal to the z plane and $\omega_s = 2\Omega \cos \gamma$ in the s direction. Although the second of these components can cause strong departures from axisymmetry of the meridional flow as shown by Nyiri (1970), Lewis & Fairbairn (1980) and Fairbairn & Lewis (1982), it is the first component which influences the blade-to-blade flow and is thus of interest here. The relative vorticity distribution ω_z in the z plane may be transformed to equivalent vorticity $\omega(\xi)$ in the cascade plane by considering the circulations about equivalent area elements, Fig. 3.3, whereupon

$$\omega(\xi) \, d\xi \, d\eta = \omega_z \, ds \, rd\theta$$

or, from (3.2)

$$\left.\begin{aligned} \omega(\xi) &= \omega_z r^2 \\ &= 2\Omega r^2 \sin \gamma \end{aligned}\right\} \tag{3.23}$$

We observe that the ζ plane is then filled with a vorticity distribution which is a function of r (and therefore ξ) and local cone angle γ. When undertaking cascade analysis in the ζ plane we must account for the influence of the relative vorticity which lies between the leading and trailing edges of the blade row, ξ_1 and ξ_2. Thus the undisturbed streamlines in the absence of the blades would have the curved appearance illustrated in Fig. 3.7 for the case of a mixed-

114

Mixed-flow and radial rotor blade rows

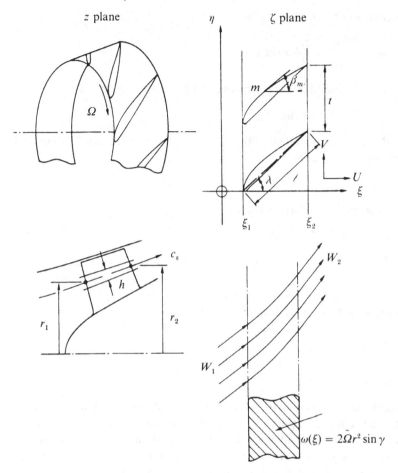

Fig. 3.7. Transformation of a mixed-flow fan rotor into a straight cascade with relative eddy $\omega(\xi)$.

flow fan, sometimes called the displacement flow. It is the interaction of the blade cascade with this rotational mainstream flow which we wish to calculate. To achieve this we may begin by defining the velocity components of the displacement flow in the ζ plane through

$$\left.\begin{aligned} U &= U_\infty \\ V &= V_\infty + v_\Omega \end{aligned}\right\} \tag{3.24}$$

where v_Ω is the disturbance due to $\omega(\xi)$ and is given by the

115

definition of vorticity

$$\omega(\xi) = \frac{dv_\Omega}{d\xi} = \frac{dv_\Omega}{dr} r \sin \gamma$$

Introducing (3.23b) we thus have upon integration

$$v_\Omega = \Omega r^2 + \text{constant}$$

By arguments of symmetry we can show that for the leading and trailing edge planes

$$v_{\Omega 2} = -v_{\Omega 1}$$

so that finally (3.24b) becomes

$$V = V_\infty + \Omega\{r^2 - \tfrac{1}{2}(r_1^2 + r_2^2)\} \tag{3.25}$$

To apply this in the ζ plane r must be expressed as a function of ξ through the transformation (3.2a). These equations, which link blade profile geometry between the z and ζ planes, may be integrated to provide the more useful form

$$\left. \begin{array}{l} \xi - \xi_1 = \displaystyle\int_{s_1}^{s} \frac{1}{r}\, ds = \int_{r_1}^{r} \frac{1}{r} \sin \gamma\, dr \\[2mm] \eta - \eta_1 = \theta - \theta_1 \end{array} \right\} \tag{3.26}$$

where suffix 1 refers to some inlet flow datum such as the blade leading edge location. Since γ is a known function of s (and therefore of r) along the prescribed meridional surface of revolution, (3.26a) may be evaluated by the trapezium rule to provide the functional relationship $\xi(r)$ in tabular form. This is usually completed as the first step of a computational scheme to facilitate the transformation of blade profile geometry from z plane to ζ plane or vice versa. $\eta(\theta)$ follows directly from (3.26b) whereas $\xi(r)$ is achieved by interpolation of the tabulated transformation. The same table may then be used to replace ξ by its equivalent r value in the displacement flow equation (3.25).

Introducing these modifications into the surface vorticity model for cascades of Section 2.6, the modified Martensen equation for element m becomes

$$\sum_{n=1}^{M} K_{mn}\gamma(s_n) = -U_\infty \cos \beta_m - (V_\infty + \Omega\{r^2 - \tfrac{1}{2}(r_1^2 + r_2^2)\}) \sin \beta_m \tag{3.27}$$

116

As expressed here the equations are similar to those for a single aerofoil, equations (1.21), applicable to a cascade with modified coupling coefficients given by equations (2.53), differing in one respect only. The additional disturbance due to rotation Ω is included in the right hand side. At first sight it would seem that extension of Martensen's method to mixed-flow rotating cascades is almost a trivial matter. Unfortunately the above formulation overlooks one important assumption of Martensen's method, that the fluid inside the blade profile region is irrotational and indeed motionless. We will now deal with this crucial matter.

3.5.2 Correction for irrotationality of the inner blade profile region

Contributors in this field have followed the two possible alternative approaches of considering the absolute or the relative flow. Thus Nyiri (1970), (1972) developed a Martensen-type solution to the mixed-flow rotor problem for hydraulic turbo-machines based upon the absolute and therefore irrotational flow in stationary coordinates. Wilkinson (1967b), (1969), Fisher (1975), (1986), Lewis *et al.* (1972) and Lewis & Mughal (1986) on the other hand solved the same problem in relative rotating coordinates as presented here. One advantage of the first approach is that the flow is irrotational throughout including the inner blade region. However an unexpected problem arises when working in stationary coordinates regarding the transformation of blade speed to the straight cascade. From the velocity transformation equation (3.8) we observe that a point on the blade surface with blade speed $U = r\Omega$ in the z plane has a transformed velocity in the ζ plane parallel to the η direction

$$v_\zeta = r^2\Omega \tag{3.28}$$

Thus the transformed absolute flow in the ζ plane generates a blade profile which is no longer rigid. For the fan shown in Fig. 3.7, elements of the blade surface near to the trailing edge will translate in the η direction more rapidly than those close to the leading edge resulting in a shearing motion. Analyses of the absolute flow must therefore account for this by including the correct individual translational velocity of each surface element and an appropriate Dirichlet-type boundary condition which accounts for the absolute translational motion of the blade surface parallel to its local direction. The analyses developed by Nyiri account for this but involve fairly elaborate arguments springing from detailed consideration of the related boundary integral theorems.

Analyses of the present type, based upon the relative flow, in general assume rigid blades in the ζ plane immersed in the curved rotational displacement flow, which we have just considered, Fig. 3.7. Use of the principle of superposition then eliminates the need to refer further to boundary integral theorems provided the individual superimposed flows are correctly chosen. With regard to this our rotational displacement flow, based upon the relative vorticity with the blades removed, is incorrect in one important respect. The blade profiles, once inserted, must be modelled in the ζ plane by motionless fluid to satisfy the ground rules of Martensen's method*. Analyses which ignore this observation lead to serious errors and (3.27) must be modified as follows

$$\sum_{n=1}^{M} K_{mn}\gamma(s_n) = -U_\infty \cos \beta_m$$
$$- (V_\infty + \Omega\{r^2 - \tfrac{1}{2}(r_1{}^2 + r_2{}^2)\}) \sin \beta_m + \Omega c_{\Omega m} \quad (3.29)$$

The principle involved in evaluating the correction term $c_{\Omega m}$ is illustrated in Fig. 3.8. Making use of the principle of superposition, the desired flow model, permitting relative vorticity $\omega(\xi)$ within the inter-blade space only, may be constructed from the full displacement flow already considered in Section 3.4.1 minus the vorticity distribution within the blade profile which it implies. This idea is attributable to Wilkinson (1967b) and the method of solution published by Lewis & Mughal (1986) akin to this is the one we shall recommend here. Because of its simplifying assumptions Fisher (1975), (1986) developed a more precise analysis generally regarded as a datum and we will refer to this in Section 3.5.5.

The vorticity within the blade profile which we wish to extract from the flow field may be modelled quite conveniently by an equivalent distribution of line vorticity, along the camber line, Fig. 3.8. Indeed, for numerical simplicity, a set of $M/2$ discrete vortices will suffice, each to represent the effect of the trapezia formed by joining the ends of opposite surface elements. If the camber line coordinates of one such element are (ξ_c, η_c), then the velocities

---

* See Section 1.4 where it is shown that the Dirichlet boundary condition is adequate only if there are no internal vortex or source singularity distributions.

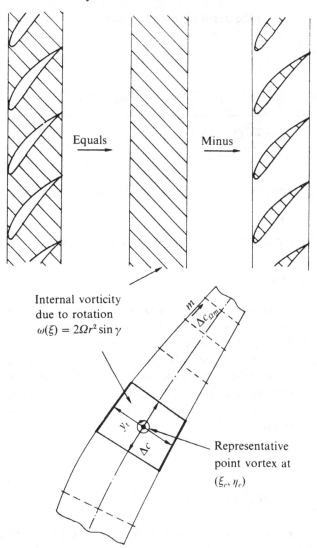

Fig. 3.8. Representation of profile interior vorticity by a camber line vortex sheet.

induced parallel to the aerofoil profile at element m are given by

$$\Delta c_{\Omega m} = \frac{1}{2t}$$

$$\times \left\{ \frac{-\sinh\frac{2\pi}{t}(\xi_m - \xi_c)\sin\beta_m + \sin\frac{2\pi}{t}(\eta_m - \eta_c)\cos\beta_m}{\cosh\frac{2\pi}{t}(\xi_m - \xi_c) - \cos\frac{2\pi}{t}(\eta_m - \eta_c)} \right\} \Delta A_c \omega(\xi_c)$$

(3.30)

where the blade element area approximates to

$$\Delta A_c = \Delta c \cdot y_t \tag{3.31}$$

Summing for all internal singularities we have finally

$$c_{\Omega m} = \sum_{c=1}^{M/2} \Delta c_{\Omega m} \tag{3.32}$$

Since the spatial vorticity $\omega(\xi)$ within the profile has been treated as if concentrated on the camber line in the form of point vortices, (3.32) will be subject to some error, especially of course due to implied net circulation around the profile. A numerical procedure which has proved satisfactory is to check the net circulation around the profile perimeter due to the camber line vorticity model against that due to the undisturbed displacement flow. The ratio of these may be expressed

$$\chi = \frac{\sum\limits_{m=1}^{M} c_{\Omega m}\Delta s_m}{\sum\limits_{m=1}^{M} \{r_m^2 - \frac{1}{2}(r_1^2 + r_2^2)\}\sin\beta_m\,\Delta s_m} \tag{3.33}$$

To enforce the correct condition of zero net profile internal circulation we must now scale each $c_{\Omega m}$ value according to

$$c_{\Omega m} := \chi c_{\Omega m} \tag{3.34}$$

If the above camber line vorticity distribution model were perfect then χ would be unity. In practice χ is subject to errors which increase with profile thickness. Application of this analysis to the radial cascade previously considered in Fig. 3.5, but treated as a centrifugal rotor with clockwise rotation and zero prewhirl, resulted in the values of χ for decreasing profile thickness given in Table 3.1.

Table 3.1

Profile thickness scaling factor	χ
1.0 (as Fig. 3.5)	0.645 981
0.5	0.754 589
0.25	0.858 129
0.125	0.930 002
0.05	0.980 971

The predicted surface pressure distribution for the normal C4 profile as illustrated in Fig. 3.5 is shown in Fig. 3.9. Other input and output data were as follows:

$r_1 = 1.0$ m
$r_2 = 1.68809$ m
$c_{r1} = 1.0$ m/s
$\Omega = 18$ r.p.m.
Profile = C4
Camber = $10°$
Stagger = $60°$ Specified in the ζ plane
$t/\ell = 0.75$
$\alpha_1 = 0°$
$\beta_1 = 62.0530°$
Predicted data
$\beta_2 \quad = 66.849°$
$\alpha_2 \quad = 71.751°$
Slip factor = 0.43533

Slip factor is defined here as

$$\mu = \frac{\text{Euler (frictionless) head rise}}{\text{Euler head rise for radial outflow}}$$

$$= 1 - \frac{c_{r2} \tan \beta_2}{r_2 \Omega} \tag{3.35}$$

It is of interest to note that despite the poor value of χ for this case, namely 0.645 981, a good solution was obtained. For comparison a second solution is also shown in Fig. 3.9 with χ increased artificially by 1% to 0.652 441 in order to bring out the disastrous effects of residual 'numerical' circulation, which are to vie with the

121

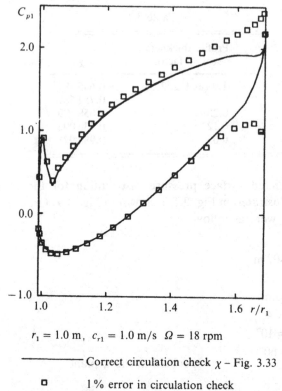

$r_1 = 1.0$ m, $c_{r1} = 1.0$ m/s $\Omega = 18$ rpm

——————— Correct circulation check χ – Fig. 3.33

□ 1% error in circulation check

Fig. 3.9. Pressure distribution for 8-bladed centrifugal rotor (Fig. 3.5) and effect of 1% error in circulation when removing profile internal vorticity due to relative eddy.

imposed trailing edge Kutta condition, resulting in excessive loading in the trailing edge region. Indeed, a good if subjective check of an acceptable solution, not subject to internal net numerical circulation or leakage flux, is a smooth predicted surface pressure and unloading approaching the trailing edge. Fig. 3.9 confirms that the simplified model and circulation check put forward here are successful in handling this problem for typical turbomachine profiles.

3.5.3 Influence of meridional streamline thickness (AVR)

So far we have considered only blade rows for which the gap h between adjacent meridional streamlines, Fig. 3.4, is constant. For mixed-flow turbomachines this is in general not normally the case as may be seen from the Francis turbine meridional flow illustrated in Fig. 3.2. The annulus itself may expand or contract. In addition to this, streamline adjustments to accommodate the meridional and blade-to-blade flow interactions may result in considerable variation of h with s, exercising a strong influence upon the blade-to-blade flow. The correct strategy for dealing with this matter, when transforming mixed-flow cascades to the ζ plane, is to introduce the same meridional stream sheet thickness $h(\xi)$ into the cascade plane, Fisher (1975). The problem is then similar to that of axial velocity ratio effects (AVR) in axial cascades, which have been shown to have substantial influence upon outlet angle and blade loading, Pollard & Horlock (1963), Gostelow (1984). Thus we begin by defining the equivalent overall AVR for mixed-flow cascades as the meridional stream sheet thickness ratio h_1/h_2. From mass flow continuity through the stream sheet we have

$$2\pi r_1 h_1 c_{s1} = 2\pi r_2 h_2 c_{s2}$$

Hence, making use of (3.8) for velocity transformation,

$$\text{AVR} = \frac{h_1}{h_2} = \frac{r_2 c_{s2}}{r_1 c_{s1}} = \frac{U_2}{U_1} \tag{3.36}$$

which confirms that transfer of h directly to the ζ plane satisfied the continuity equation for the through flow velocity U,

$$Uh = U_1 h_1 = U_2 h_2 = \text{constant} \tag{3.37}$$

Although $U(\xi)$ is now known in terms of the prescribed stream sheet thickness

$$U(\xi) = U_1 h_1 / h(\xi) \tag{3.38}$$

a simple approach, Lewis & Mughal (1986), is to assume linear variation between leading and trailing edges.

$$U = r_1 c_{s1} \left\{ 1 + \left(\frac{\xi - \xi_1}{\xi_2 - \xi_1} \right) (\text{AVR} - 1) \right\} \tag{3.38a}$$

Martensen's equation (3.29) for mixed-flow rotors may now be modified to include also a correction for 'AVR' as follows

$$\sum_{n=1}^{M} K_{mn} \gamma(s_n) = -r_1 c_{s1} \left\{ 1 + \left(\frac{\xi - \xi_1}{\xi_2 - \xi_1} \right) (\text{AVR} - 1) \right\} \cos \beta_m$$
$$+ r_1 c_{s1} c_{\sigma m} - V_\infty \sin \beta_m$$
$$- \Omega \{ [r^2 - \tfrac{1}{2}(r_1^2 + r_2^2)] \sin \beta_m - c_{\Omega m} \} \tag{3.39}$$

123

As shown by Wilkinson (1967b), this variation of through flow velocity $U(\xi)$ implies an equivalent source distribution $\sigma(\xi)$ throughout the ζ plane given

$$\sigma(\xi) = \frac{dU(\xi)}{d\xi} \tag{3.40}$$

and for linear variation of $U(\xi)$

$$\sigma(\xi) = r_1 c_{s1}(\text{AVR} - 1)/(\ell \cos \lambda) \tag{3.40a}$$

In other words $\text{div}(U(\xi))$ is non-zero. Since neither source nor vortex singularities can be allowed inside the blade profile a procedure analogous to that just described in Section 3.4.2 for internal vorticity must be adopted to remove the implied internal fluid divergence. The term $c_{\sigma m}$ in Martensen's equations (3.39) accommodates this correction by direct analogy if we prescribe a camber line distribution of equivalent line sources. The velocity at boundary element m due to the camber line source element at (ξ_c, η_c) then becomes

$$\Delta c_{\sigma m} = \frac{1}{2t}$$

$$\times \left\{ \frac{\sinh\dfrac{2\pi}{t}(\xi_m - \xi_c)\cos\beta_m + \sin\dfrac{2\pi}{t}(\eta_m - \eta_c)\sin\beta_m}{\cosh\dfrac{2\pi}{t}(\xi_m - \xi_c) - \cos\dfrac{2\pi}{t}(\eta_m - \eta_c)} \right\} \Delta A_c \sigma(\xi_c)$$

$$\tag{3.41}$$

Summing for all camber line source elements we then have finally

$$c_{\sigma m} = \sum_{c=1}^{M/2} \Delta c_{\sigma m} \tag{3.42}$$

3.5.4 Unit solutions for mixed-flow cascades and prediction of flow angles

The best strategy for solution of (3.39) is to define three unit components of the surface vorticity $\gamma(s)$ linked to the meridional velocity through $U_1 = r_1 c_{s1}$, the vector mean transverse velocity V_∞ and the rotation Ω, namely

$$\gamma(s) = r_1 c_{s1} \gamma_u(s) + V_\infty \gamma_v(s) + \Omega \gamma_\Omega(s) \tag{3.43}$$

124

Mixed-flow and radial rotor blade rows

Introduction of this into (3.39) permits its reduction to three independent sets of equations, namely

$$\sum_{n=1}^{M} \gamma_u(s_n) K_{mn} = -\left\{1 + \left(\frac{\xi - \xi_1}{\xi_2 - \xi_1}\right)(\text{AVR} - 1)\right\}\cos\beta_m + c_{om}$$

$$\sum_{n=1}^{M} \gamma_v(s_n) K_{mn} = -\sin\beta_m \qquad\qquad (3.44)$$

$$\sum_{n=1}^{M} \gamma_\Omega(s_n) K_{mn} = -\{r^2 - \tfrac{1}{2}(r_1^2 + r_2^2)\}\sin\beta_m + c_{\Omega m}$$

The first two equations provide solutions due to unit strength velocities normal ($r_1 c_{s1} = U_1 = 1$) and parallel ($V_\infty = 1$) to the cascade respectively and the third for unit angular velocity ($\Omega = 1$). The same coupling coefficient matrix K_{mn} applies to all three and the Kutta condition must also be applied as described in Section 2.6.2. For each unit solution there is then an associated bound circulation given by

$$\Gamma_u = \sum_{n=1}^{M} \gamma_u(s_n)\,\Delta s_n, \qquad \Gamma_v = \sum_{n=1}^{M} \gamma_v(s_n)\,\Delta s_n$$

$$\Gamma_\Omega = \sum_{n=1}^{M} \gamma_\Omega(s_n)\,\Delta s_n \qquad\qquad (3.45)$$

If we now scale the unit solutions by U_1, V_∞ and Ω respectively, their associated induced velocities (U, V) upstream and downstream of the cascade in the ζ plane are as illustrated in Fig. 3.10. If these three flows are then recombined we have inlet and outlet transverse velocities

$$V_1 = \frac{\Gamma_u U_1}{2t} + V_\infty\left(1 + \frac{\Gamma_v}{2t}\right) + \Omega\left\{\frac{\Gamma_\Omega}{2t} + \tfrac{1}{2}(r_1^2 - r_2^2)\right\}$$

$$V_2 = -\frac{\Gamma_u U_1}{2t} + V_\infty\left(1 - \frac{\Gamma_v}{2t}\right) - \Omega\left\{\frac{\Gamma_\Omega}{2t} + \tfrac{1}{2}(r_1^2 - r_2^2)\right\}$$

From these results we may derive the vector mean and outlet flow angles as follows

$$\tan\beta_\infty = \frac{\tan\beta_1 - \dfrac{\Gamma_u}{2t} - \dfrac{\Omega}{r_1 c_{s1}}\left\{\dfrac{\Gamma_\Omega}{2t} + \tfrac{1}{2}(r_1^2 - r_2^2)\right\}}{1 + \dfrac{\Gamma_v}{2t}} \qquad (3.46)$$

$$\tan\beta_2 = 2\tan\beta_\infty - \tan\beta_1 \qquad\qquad (3.47)$$

125

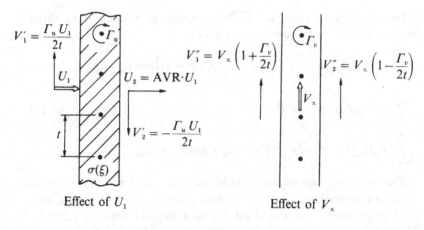

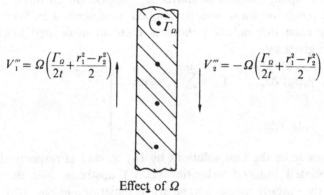

Fig. 3.10. Effect of bound vortex strength upon (U, V) velocities for the three unit solutions.

The fluid velocity on the blade surface likewise may then be calculated for any combination of blade rotation Ω or through flow velocity c_{s1}. In non-dimensional form we have,

$$\frac{v_s}{c_{s1}} = \left\{ \gamma_u(s) + \tan \beta_\infty \gamma_v(s) + \frac{\Omega}{r_1 c_{s1}} \gamma_\Omega(s) \right\} \frac{r_1}{r} \qquad (3.48)$$

3.5.5 More precise method for removal of profile internal vorticity

As an alternative to the camber line singularity distribution approximation which has just been described, Fisher (1975), (1986)

126

developed precise formulations for the profile internal vorticity correction. In view of the complexity of his analysis a brief review only will be given here to indicate the underlying principles of this valuable piece of work.

The source and vorticity distributions within the blade profiles, due to AVR and relative rotation respectively, may be represented by a finite number of trapezia each containing a singularity distribution of constant strength, Fig. 3.11. The influence of each cascade of trapezia may be deduced from the periodic line singularity solution of Ackeret (1942) as follows. We begin by considering the flow due to a line vorticity $\gamma(y)$ located at $x = x_0$ of square wave strength

$$\gamma(y) = \gamma \quad \text{for} \quad y_0 - \tfrac{1}{2}\alpha t \leqslant y \leqslant y_0 + \tfrac{1}{2}\alpha t$$
$$= 0 \quad \text{for} \quad y_0 + \tfrac{1}{2}\alpha t \leqslant y \leqslant y_0 + t - \tfrac{1}{2}\alpha t$$

periodic over the blade pitch t, Fig. 3.12(a).

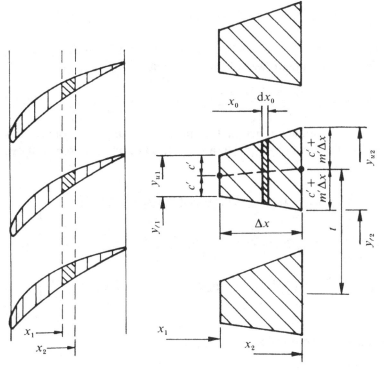

Fig. 3.11. Trapezia block representation of blade profile interior vorticity and source distributions.

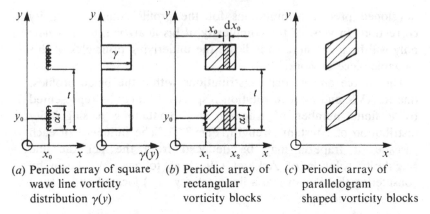

(a) Periodic array of square wave line vorticity distribution $\gamma(y)$ (b) Periodic array of rectangular vorticity blocks   (c) Periodic array of parallelogram shaped vorticity blocks

Fig. 3.12. Periodic arrays of vorticity distributions to model relative eddy correction for blade profile interior in mixed-flow turbomachines.

Following Ackeret we may expand $\gamma(y)$ as a Fourier series

$$\gamma(y) = \gamma_0 + \sum_{n=1}^{\infty} \gamma_n \cos p_n(y - y_0)$$

with $p_n = 2\pi n/t$. The coefficients follow from Fourier analysis, which, for the square wave form results in

$$\gamma(y) = \gamma\alpha + \frac{4\gamma}{t} \sum_{n=1}^{\infty} \frac{1}{p_n} \sin\left(\frac{p_n \alpha t}{2}\right) \cos p_n(y - y_0) \tag{3.49}$$

Since the stream function for the surrounding irrotational flow satisfies the equation $\nabla^2 \psi = 0$, a suitable solution may be postulated of the form

$$
\left.
\begin{aligned}
\psi &= (\pm)A_0(x - x_0) + \sum_{n=1}^{\infty} A_n \cos p_n(y - y_0)e^{(\pm)p_n(x-x_0)} \\
c_x &= \frac{\partial \psi}{\partial y} = -\sum_{n=1}^{\infty} p_n A_n \sin p_n(y - y_0)e^{(\pm)p_n(x-x_0)} \\
c_y &= -\frac{\partial \psi}{\partial x} = -(\pm)\left\{A_0 + \sum_{n=1}^{\infty} p_n A_n \cos p_n(y - y_0)e^{(\pm)p_n(x-x_0)}\right\}
\end{aligned}
\right\}
$$

$$\tag{3.50}$$

with the sign convention

   $(+)$ for $x < x_0$

   $(-)$ for $x > x_0$

Mixed-flow and radial rotor blade rows

The coefficients $A_0, A_1 \ldots A_n$ follow directly by matching c_y to (3.49) at $x = x_0$, where $c_y = (\pm)\frac{1}{2}\gamma(y)$, whereupon

$$A_0 = -\tfrac{1}{2}\gamma\alpha, \qquad A_n = -\frac{2\gamma}{t p_n^{~2}} \sin\left(\frac{p_n \alpha t}{2}\right) \tag{3.51}$$

Finally we have the solution

$$\left.\begin{aligned}
\psi &= -\gamma\left\{(\pm)\frac{\alpha}{2}(x - x_0)\right. \\
&\quad \left.+\frac{2}{t}\sum_{n=1}^{\infty}\frac{1}{p_n^{~2}}\sin\left(\frac{p_n \alpha t}{2}\right)\cos p_n(y - y_0)e^{(\pm)p_n(x-x_0)}\right\} \\
c_x &= \frac{2\gamma}{t}\sum_{n=1}^{\infty}\frac{1}{p_n}\sin\left(\frac{p_n \alpha t}{2}\right)\sin p_n(y - y_0)e^{(\pm)p_n(x-x_0)} \\
c_y &= (\pm)\gamma\left\{\frac{\alpha}{2}+\frac{2}{t}\sum_{n=1}^{\infty}\frac{1}{p_n}\sin\left(\frac{p_n \alpha t}{2}\right)\cos p_n(y - y_0)e^{(\pm)p_n(x-x_0)}\right\}
\end{aligned}\right\} \tag{3.52}$$

To illustrate the technique used by Fisher to apply this result to trapezium-shaped cascades of distributed vorticity let us first consider the simpler problem of an array of rectangular blocks of constant vorticity ω bounded by x_1 and x_2, Fig. 3.12(b). Equations (3.52) may now be applied to the elementary vortex strip at x_0 of width dx_0, noting that $\gamma = \omega\, dx_0$. The stream function for this strip is thus

$$d\psi = -\omega\left\{(\pm)\frac{\alpha}{2}(x - x_0)\right.$$
$$\left.+\frac{2}{t}\sum_{n=1}^{\infty}\frac{1}{p_n^{~2}}\sin\left(\frac{p_n \alpha t}{2}\right)\cos p_n(y - y_0)e^{(\pm)p_n(x-x_0)}\right\}dx_0$$

which may be integrated between $x_0 = x_1$ to $x_0 = x_2$ to give the solution for the array of rectangular vorticity blocks,

$$\psi = -\omega\left(I_0 + \sum_{n=1}^{\infty} I_n\right)$$

129

The integrals I_0 and I_n may be evaluated as follows.

$$
\left.
\begin{aligned}
I_0 &= (\pm)\frac{\alpha}{2}\int_{x_1}^{x_2}(x-x_0)\,dx_0 = (\pm)\frac{\alpha}{2}\Delta x(x-x_m) \\[2mm]
I_n &= \frac{2}{tp_n^2}\int_{x_1}^{x_2}\left\{\sin\!\left(\frac{p_n\alpha t}{2}\right)\cos p_n(y-y_0)e^{(\pm)p_n(x-x_0)}\right\}dx_0 \\[2mm]
&= \frac{4}{tp_n^3}\sinh\!\left(\frac{p_n\,\Delta x}{2}\right)e^{(\pm)p_n(x-x_m)}
\end{aligned}
\right\}
\qquad (3.53)
$$

where $\Delta x = (x_2 - x_1)$ is the block width and $x_m = \frac{1}{2}(x_1 + x_2)$ is its central x location. Finally the flow field for the array of rectangular vorticity blocks becomes

$$
\left.
\begin{aligned}
\psi &= -\omega\Biggl\{(\pm)\frac{\alpha\Delta x}{2}(x-x_m)+\frac{4}{t}\sum_{n=1}^{\infty}\frac{1}{p_n^3}\sin\!\left(\frac{p_n\alpha t}{2}\right) \\[2mm]
&\quad \times \sinh\!\left(\frac{p_n\Delta x}{2}\right)\cos p_n(y-y_0)e^{(\pm)p_n(x-x_m)}\Biggr\} \\[2mm]
c_x &= \frac{4\omega}{t}\sum_{n=1}^{\infty}\frac{1}{p_n^2}\sin\!\left(\frac{p_n\alpha t}{2}\right)\sinh\!\left(\frac{p_n\Delta x}{2}\right) \\[2mm]
&\quad \times \sin p_n(y-y_0)e^{(\pm)p_n(x-x_m)} \\[2mm]
c_y &= (\pm)\omega\Biggl\{\frac{\alpha\Delta x}{2}+\frac{4}{t}\sum_{n=1}^{\infty}\frac{1}{p_n^2}\sin\!\left(\frac{p_n\alpha t}{2}\right)\sinh\!\left(\frac{p_n\Delta x}{2}\right) \\[2mm]
&\quad \times \cos p_n(y-y_0)e^{(\pm)p_n(x-x_m)}\Biggr\}
\end{aligned}
\right\}
\qquad (3.54)
$$

Fisher (1975), (1986) extended this analysis first to arrays of parallelograms for which y_0 varies linearly with x_0, Fig. 3.12(c) and then to trapezia for which the tangential blade thickness $t\alpha$ also varies linearly with x_0, Fig. 3.11(b). Evaluation of the integral I_n (3.53b), in closed form is still possible though extremely complex. Details are given by Fisher based on the notation shown in Fig. 3.11(b) where the mid-line EF is defined by the equation

$$
y^* = m^*x^* + c^* \qquad (3.55)
$$

His final solution for the velocity components is as follows.

$$c_{x\omega} = \frac{\omega}{t} \sum_{n=1}^{\infty} \frac{1}{p_n^2}$$

$$\times \left[\frac{1}{1+(m'+m^{-*})^2} \left\{ \begin{array}{l} -[(\pm)\cos p_n(y-y_{l2}) - (m'-m^*) \\ \qquad \sin p_n(y-y_{l2})]e^{(\pm)p_n(x-x_2)} \\ +[(\pm)\cos p_n(y-y_{l1}) - (m'-m^*) \\ \qquad \sin p_n(y-y_{l1})]e^{(\pm)p_n(x-x_1)} \end{array} \right\} \right.$$

$$\left. + \frac{1}{1+(m'+m^*)^2} \left\{ \begin{array}{l} [(\pm)\cos p_n(y_{u2}-y) - (m'+m^*) \\ \qquad \sin p_n(y_{u2}-y)]e^{(\pm)p_n(x-x_2)} \\ -[(\pm)\cos p_n(y_{u1}-y) - (m'+m^*) \\ \qquad \sin p_n(y_{u1}-y)]e^{(\pm)p_n(x-x_1)} \end{array} \right\} \right]$$

$$c_{y\omega} = (\pm)\frac{\omega}{t} \left[\frac{m'(x_2^2 - x_1^2)}{2} + (c' - m'x_1)(x_2 - x_1) \right]$$

$$(\pm)\frac{\omega}{t} \sum_{n=1}^{\infty} \frac{1}{p_n^2}$$

$$\times \left[\frac{1}{1+(m'-m^*)^2} \left\{ \begin{array}{l} -[(\pm)\sin p_n(y-y_{l2}) + (m'-m^*) \\ \qquad \cos p_n(y-y_{l2})]e^{(\pm)p_n(x-x_2)} \\ +[(\pm)\sin p_n(y-y_{l1}) + (m'-m^*) \\ \qquad \cos p_n(y-y_{l1})]e^{(\pm)p_n(x-x_1)} \end{array} \right\} \right.$$

$$\left. + \frac{1}{1+(m'+m^*)^2} \left\{ \begin{array}{l} -[(\pm)\sin p_n(y_{u2}-y) + (m'+m^*) \\ \qquad \cos p_n(y_{u2}-y)]e^{(\pm)p_n(x-x_2)} \\ +[(\pm)\sin p_n(y_{u1}-y) + (m'+m^*) \\ \qquad \cos p_n(y_{u1}-y)]e^{(\pm)p_n(x-x_1)} \end{array} \right\} \right]$$

$$(3.56)$$

where c' and m', as defined by Fisher in Fig. 3.11(b), are related to our previous variable for tangential blade thickness α through

$$\left. \begin{array}{l} c' = \tfrac{1}{2}\alpha_1 t \\ m' = \dfrac{(\alpha_2 - \alpha_1)t}{2\,\Delta x} \end{array} \right\} \qquad (3.57)$$

A similar analysis for arrays of trapezia filled with source distributions of constant strength σ leads to an orthogonal flow field given by

$$c_{x\sigma} = \frac{\sigma}{\omega} c_{y\omega}, \qquad c_{y\sigma} = -\frac{\sigma}{\omega} c_{x\omega} \tag{3.58}$$

In performing the integrals I_0 and I_n the implicit assumption was made that the above expressions apply only to x locations outside the vortex or source arrays with the corresponding sign convention

$$(+) \text{ for } x < x_1, \qquad (-) \text{ for } x > x_2 \tag{3.59}$$

For locations within the region $x_1 < x < x_2$ an appropriate strategy is to subdivide the vorticity or source strength into two trapezia over the regions (x_1 to x) and (x to x_2) for which equations (3.56) and (3.57) are then applicable.

Inspection of equations (3.56) shows that it is necessary to sum series of the form

$$S_1 = \sum_{n=1}^{\infty} \frac{\cos p\phi}{n^2} e^{-nD}, \qquad S_2 = \sum_{n=1}^{\infty} \frac{\sin n\phi}{n^2} e^{-nD}$$

Fortunately for finite values of D, which is true for all surface elements under the influence of a given vorticity block except those defining the block itself, these series converge and mostly rapidly. However, Fisher (1975) (Appendix IV) developed further powerful reductions which speed convergence for small values of D. Of particular importance are those formulae applicable to the effect of a vortex trapezium upon itself which, with sub-division into two trapezia as mentioned above, involve $D = 0$. Clearly, over 1000 terms might then be required for convergence to six-figure accuracy. For this special case S_1 reduces surprisingly to the following simple closed form, Bromwich (1908).

$$\sum_{n=1}^{\infty} \frac{\cos n\phi}{n^2} = \frac{\phi^2}{4} + \frac{\pi^2}{6} - \frac{\pi\phi}{2} \tag{3.60}$$

and S_2 reduces to

$$\sum_{n=1}^{\infty} \frac{\sin n\phi}{n^2} = S_m + \sin\frac{\phi}{2} - \phi \ln\left(2 \sin\frac{\phi}{2}\right) \tag{3.61}$$

where the rapidly convergent series S_m is given by

$$S_m = 2 \sum_{j=1}^{\infty} \left\{ \frac{1 \cdot 3 \cdot 5 \ldots (2j-1)(\sin \phi/2)}{2 \cdot 4 \ldots (2j)(2j+1)^2} \right\} \tag{3.62}$$

132

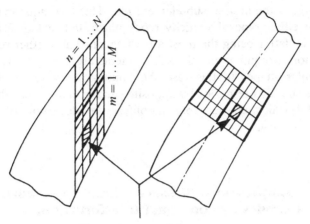

Elementary area ΔA_{mn}

(*a*) Fisher type configuration   (*b*) Wilkinson type configuration

Fig. 3.13. Use of sub-elements to model blade profile interior vortex/source singularity corrections for mixed-flow turbomachines.

This analysis was a natural development from Ackeret's linearised aerofoil theory which was well suited to first generation computers and extended to mixed-flow turbomachines in the late 60s by Railly (1967) and Railly *et al.* (1969). Most Martensen mixed-flow cascade programs now implement either the camberline model of Wilkinson or Fisher's exact solution since they are well proven and known to yield excellent results. However a more convenient numerical approach to the problem of the removal of the blade profile internal relative eddy or AVR source distribution is to break down each block into sub-elements, Fig. 3.13. Making use of the expressions derived in Chapter 2 for a point vortex array, equations (2.51), the induced velocity components at (x, y) are then

$$
\left.
\begin{aligned}
c_{x\omega} &= \frac{\omega}{2t} \sum_{n=1}^{N} \sum_{m=1}^{M} \frac{\Delta A_{mn} \sin \frac{2\pi}{t}(y - y_{mn})}{\cosh \frac{2\pi}{t}(x - x_{mn}) - \cos \frac{2\pi}{t}(y - y_{mn})} \\
c_{y\omega} &= -\frac{\omega}{2t} \sum_{n=1}^{N} \sum_{m=1}^{M} \frac{\Delta A_{mn} \sinh \frac{2\pi}{t}(x - x_{mn})}{\cosh \frac{2\pi}{t}(x - x_{mn}) - \cos \frac{2\pi}{t}(y - y_{mn})}
\end{aligned}
\right\}
\quad (3.63)
$$

where ΔA_{mn} is the area of sub-element *mn*. This technique may be applied to either type of vorticity block illustrated in Fig. 3.12(*a*) and (*b*), the latter being the most attractive since no further profile interpolation is required and the sub-elements are normally close to a rectangular shape. In this case the boundaries of the vorticity blocks are already prescribed by opposite surface elements as shown in Fig. 3.8 which considerably simplifies the definition of sub-element geometry.

3.6 Comparison with exact solutions for radial cascades by conformal transformation

Exact solutions were derived for a family of conical mixed-flow rotors by Fisher & Lewis (1971), for the express purpose of providing a datum check for numerical methods. Their conformal transformation technique, which produces a series of cambered Joukowski type aerofoils located upon straight conical meridional surfaces of constant thickness, leads to intricate analysis given in full by these authors in NEL Report Nos. 498 and 524. A brief description only is possible here limited to radial centrifugal rotors.

Consider the radial cascade with N blades in the Z plane, Fig. 3.14. This may be transformed into a single blade shape in the ζ

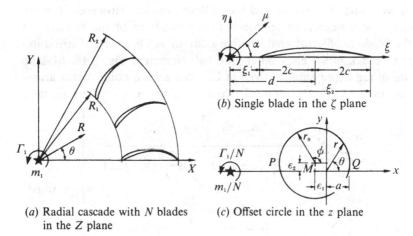

(*a*) Radial cascade with N blades in the Z plane

(*b*) Single blade in the ζ plane

(*c*) Offset circle in the z plane

Fig. 3.14. Conformal transformation of unit circle in the z plane to a radial cascade in the Z plane.

plane through

$$\zeta = Z^N \tag{3.64}$$

Now the means for generating a typical blade profile shape in the ζ plane through the Joukowski transformation has already been explored in Section 2.3. For a Joukowski aerofoil offset by d and scaled by c, the transformation equation (2.10) becomes

$$\zeta = d + c(z + a^2/z) \tag{3.65}$$

Consider the circle of radius r_0 in the z plane with its centre offset by ε_1 and ε_2. Its equation is given by

$$z = re^{i\theta} = r_0 e^{i\phi} - \varepsilon_1 + i\varepsilon_2$$

Substituting into (3.65) we then have the (ξ, η) coordinates of the Joukowski aerofoil in the ζ plane.

$$\left.\begin{aligned}
\xi &= d + c\left(1 + \left[\frac{a}{r}\right]^2\right)(r_0 \cos\phi - \varepsilon_1) \\
\eta &= c\left(1 - \left[\frac{a}{r}\right]^2\right)(r_0 \sin\phi + \varepsilon_2)
\end{aligned}\right\} \tag{3.66}$$

where (r, θ) may be expressed in terms of the prescribed values of r_0, ε_1 and ε_2 through

$$\left.\begin{aligned}
r &= \sqrt{[(r_0 \cos\phi - \varepsilon_1)^2 + (r_0 \sin\phi + \varepsilon_2)^2]} \\
\theta &= \arctan\{(r_0 \sin\phi + \varepsilon_2)/(r_0 \cos\phi - \varepsilon_1)\}
\end{aligned}\right\} \tag{3.67}$$

The constants c and d may be determined from the inlet and outlet radii of the blade row R_1 and R_2, Fig. 3.14. If the intersections P and Q of the circle in the z plane with the x axis, for which $\theta = \pi$ and 0 respectively, are assumed to transform to R_1 and R_2 in the Z plane, then the equivalent coordinates ξ_1 and ξ_2 in the ζ plane reduce to

$$\xi_1 = R_1^N = d - ac\left\{1 + \frac{2\varepsilon_1}{a} + \frac{1}{1 + \frac{2\varepsilon_1}{a}}\right\}$$

$$\xi_2 = R_2^N = d + ac$$

from which we may obtain c and d

$$
\left.
\begin{aligned}
c &= \frac{\left(1 + \dfrac{2\varepsilon_1}{a}\right)}{4a\left(1 + \dfrac{\varepsilon_1}{a}\right)^2}(R_2{}^N - R_1{}^N) \\[2em]
d &= R_2{}^N - 2ac
\end{aligned}
\right\}
\tag{3.68}
$$

By varying r_0, ε_1 and ε_2 a wide range of profiles was considered by Fisher & Lewis (1971) from which a selection is shown in Fig. 3.15. Profile thickness is controlled primarily by ε_1, for which values of 0.02 and 0.04 were chosen here. Camber is controlled by ε_2 for which four values were selected, 0.0, 0.1, 0.3, 0.5. As it stands this analysis was restricted to profiles with cusped trailing edges and with zero stagger. Both limitations could be removed by further development of the theory along the lines of the transformations discussed in Section 2.5.3. Since Fisher's objective was to provide sample exact solutions as datum checks for the mixed-flow Martensen method, the introduction of stagger with its added complications was not a necessity. Furthermore Martensen's method is not well adapted to thin profiles, so that the retention of a cusped trailing edge presents a particularly severe test.

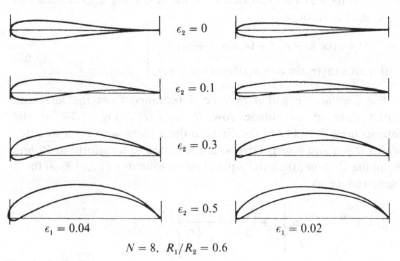

Fig. 3.15. Radial cascade profiles obtained by conformal transformation. Influence of ε_1 upon profile thickness and ε_2 upon camber.

3.6.1   Flow analysis of the transformation

The velocity due to the through-flow and prewhirl are modelled by introducing a source m_1 and vortex Γ_1 at the origin of the Z plane. For a rotor with N blades, these transform into a source m_1/N and vortex Γ_1/N in the ζ and z planes. Flow past the circle in the z plane due to the source/vortex may then be modelled by introducing the appropriate mirror image singularities at the inverse point and circle centre, following the method already prescribed for radial stators in Section 2.6.3. Since we are now dealing with a rotor however, the velocities in the z plane must also include a component to represent blade motion. The velocity components may thus be expressed by

$$u_z = u_z{'} + u_z{''}, \qquad v_z = v_z{'} + v_z{''}$$

where the through-flow components $(u_z{'}, v_z{'})$ follow directly from equations (2.63) for the stator case. As shown by Fisher & Lewis, the $(u_2{''}, v_2{''})$ components for blade rotation may be found by first postulating a complex potential for the absolute irrotational flow of series form

$$\omega{''} = \sum_{n=1}^{\infty} \frac{a_n + i b_n}{z_M{}^n} \tag{3.69}$$

z_M is the complex coordinate of a point on the circle measured from its centre M, Fig. 3.14, and thus given by

$$z_M = r_0 e^{i\phi}$$

Differentiation leads to the velocity components $(u_z{''}, v_z{''})$ which may then be combined to give expressions for the fluid velocity components normal and parallel to the cylinder body surface in the z plane, namely

$$
\left.
\begin{aligned}
u_z{''} \cos\phi_s + v_z{''} \sin\phi_s &= \sum_{n=1}^{\infty} (A_n \cos n\phi_s + B_n \sin n\phi_s) \quad \text{normal} \\
-u_z{''} \sin\phi_s + v_z{''} \cos\phi_s &= \sum_{n=1}^{\infty} (A_n \sin n\phi_s - B_n \cos n\phi_s) \quad \text{parallel}
\end{aligned}
\right\}
$$
$$\tag{3.70}$$

where

$$
\left.
\begin{aligned}
A_n &= \frac{n a_n}{\{(1 + \varepsilon_1)^2 + \varepsilon_2{}^2\}^{(n+1)/2}}, \\[2mm]
B_n &= \frac{n b_n}{\{(1 + \varepsilon_1)^2 + \varepsilon_2{}^2\}^{(n+1)/2}}
\end{aligned}
\right\}
\tag{3.71}
$$

The coefficients (A_n, B_n) may be found by applying equation (3.70a) as boundary condition for the rotating blade surface. If the velocity of the circle r_0 normal to itself is $f(\phi_s)$ we then have

$$\sum_{n=1}^{\infty} \{A_n \cos n\phi_s + B_n \sin n\phi_s\} = f(\phi_s)$$

from which the coefficients (A_n, B_n) follow by Fourier analysis

$$\left.\begin{aligned} A_n &= \frac{2}{\pi} \int_0^{\pi} f(\phi_s) \cos n\phi_s \, \mathrm{d}\phi_s \\[2mm] B_n &= \frac{2}{\pi} \int_0^{\pi} f(\phi_s) \sin n\phi_s \, \mathrm{d}\phi_s \end{aligned}\right\} \tag{3.72}$$

$f(\phi_s)$ is directly calculable from the known blade velocity $R\Omega$ in the Z plane through the two stages of conformal transformation, equations (3.64) and (3.65), resulting finally in

$$f(\phi_s) = \frac{c\Omega}{N} \mu^{2(1/N-1)} \{-(\xi_s E + \eta_s F) \cos \phi_s + (-\eta_s F + \xi_s E) \sin \phi_s\} \tag{3.73}$$

where E and F are the real and imaginary parts of $\mathrm{d}z/\mathrm{d}\zeta$, namely

$$\left.\begin{aligned} E &= c \left\{ \frac{1 - \left(\dfrac{a}{r}\right)^2 \cos 2\theta}{1 + \left(\dfrac{a}{r}\right)^4 - 2\left(\dfrac{a}{r}\right)^2 \cos 2\theta} \right\} \\[4mm] F &= \left\{ \frac{c \left(\dfrac{a}{r}\right)^2 \sin 2\theta}{1 + \left(\dfrac{a}{r}\right)^4 - 2\left(\dfrac{a}{r}\right)^2 \cos 2\theta} \right\} \end{aligned}\right\} \tag{3.74}$$

and the coordinates of the blade surface in the ζ plane, (ξ_s, η_s), are given by equations (3.65). For computation it is convenient to specify the circle radius r_0 in the z plane at, say, even intervals of ϕ to define the blade profile. Equivalent (r, θ) values then follow from equations (3.67)

Having evaluated the series coefficients (A_n, B_n), (3.70b) may be used to calculate the velocity parallel to the cylinder due to rotation and hence the Kutta Joukowski trailing edge condition and bound

138

circulation Γ. The slip factor, as defined by (3.35), is then

$$\mu = \frac{N\Gamma}{2\pi \left(\Omega R_2{}^2 - \dfrac{\Gamma_1}{2\pi} \right)} \tag{3.75}$$

The reader is referred to Fisher & Lewis (1971) for a fuller treatment including techniques for simplifying some of the series to deal with convergence problems.

3.6.2 Sample solutions

A large number of test cases have been presented by Fisher & Lewis (1971) for a wide range of centrifugal rotor geometries. Two of these solutions are shown in Fig. 3.16 for an eight-bladed rotor with highly cambered blades of radius ratio $R_1/R_2 = 0.6$ and for rotational speeds of ± 750 rpm. The sign convention for Ω and the absolute swirl angles α_1 and α_2 is anti-clockwise positive. Appropriate prewhirl angles α_1 were chosen to produce a fairly highly loaded leading edge. Surface velocity v_s was expressed as a fraction of the inlet radial velocity V_{R1}.

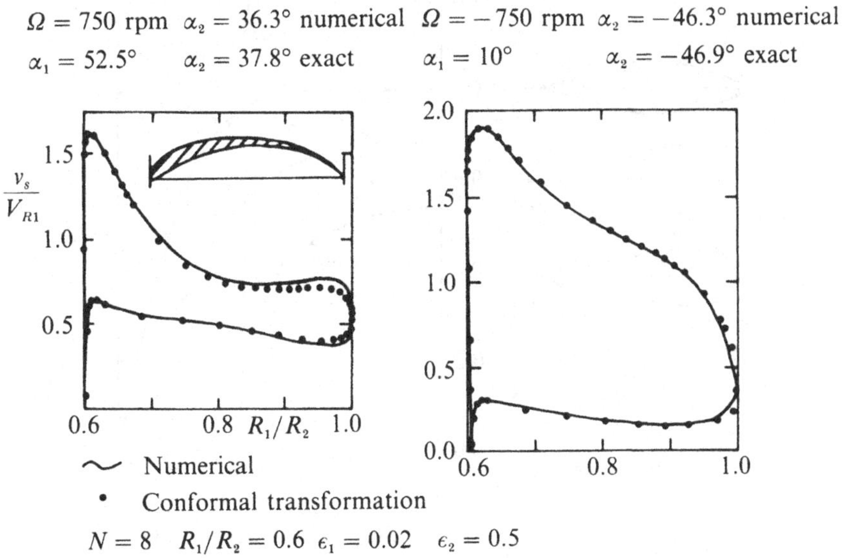

$\Omega = 750$ rpm $\quad \alpha_2 = 36.3°$ numerical $\quad \Omega = -750$ rpm $\quad \alpha_2 = -46.3°$ numerical

$\alpha_1 = 52.5° \quad\quad \alpha_2 = 37.8°$ exact $\quad\quad\quad \alpha_1 = 10° \quad\quad\quad \alpha_2 = -46.9°$ exact

$\sim$ Numerical

• Conformal transformation

$N = 8 \quad R_1/R_2 = 0.6 \quad \epsilon_1 = 0.02 \quad \epsilon_2 = 0.5$

Fig. 3.16. Surface velocity distributions for a centrifugal rotor by Fisher's surface vorticity method compared with conformal transformation solution.

Mixed-flow and radial cascades

Surface velocity distributions predicted by the surface vorticity method agreed extremely well with exact conformal transformation theory in both cases, the main error occurring in the trailing edge region. Outlet angles α_2 generally agreed to within 0.5° of conformal transformation predictions although larger discrepancies occurred with highly cambered blades such as the first case shown here. These errors, linked to poor trailing edge predictions of v_s/V_{R1}, were largely due to difficulties in setting up suitably distributed pivotal points from the transformation theory. Accurate re-plotting proves difficult for the highly cambered blades in the region of the cusped trailing edge. Despite these problems the mixed-flow Martensen method agreed well with exact solutions for a wide range of geometries and operating variables Ω and α_1.

Slip factors μ for uncambered radial rotors with 2, 4 or 8 blades are compared with the mixed-flow numerical method in Fig. 3.17 for a range of radius ratios. Blade profiles of course differ for each $(N, R_1/R_2)$ combination and samples are shown for $N = 8$ only. Agreement was within plotting accuracy in all cases. Also shown on Fig. 3.17 are contours of constant pitch/chord ratio t/l as defined by (3.12). It will be observed that the well formed passages for

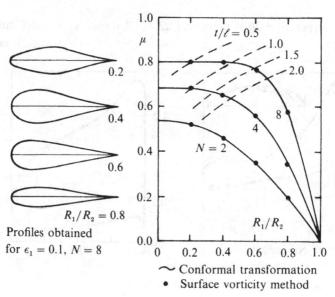

Fig. 3.17. Slip factors μ for radial bladed centrifugal rotors obtained by conformal transformation.

pitch/chord ratios less than 1·0 lead to a constant value of slip factor for rotors with 4 or 8 blades, since the exit angle is uninfluenced by further reduction in R_1/R_2. This illustrates the usefulness of defining t/l in the transformed straight cascade as an indicator of the actual radial cascade behaviour.

3.6.3   Comparisons with experimental test

The mixed-flow fan research rig illustrated in Fig. 3.18(a) was designed for free/vortex loading using the mixed-flow Martensen method as a design tool. To maintain constant average meridional velocity the annulus was reduced in width progressively resulting in significant AVR effects. The correction for this outlined in Section 3.4.3 was built into the surface vorticity design analysis assuming equal AVR for all blade sections. A sample surface velocity distribution measured at the mean section was compared with the design prediction by Fisher (1986), showing excellent agreement, Fig. 3.18(b).

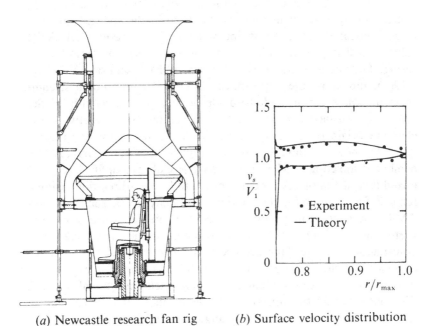

(a) Newcastle research fan rig   (b) Surface velocity distribution

Fig. 3.18. Comparison of predicted surface velocity for mean section of the Newcastle University mixed-flow research fan.

141

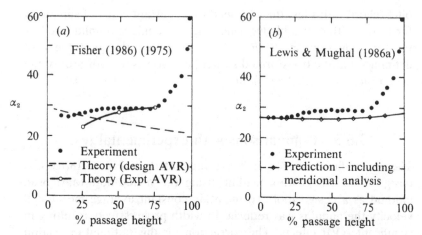

Fig. 3.19. Exit swirl angle distribution for design point of the Newcastle research fan rig.

On the other hand a downstream traverse of outlet angle α_2 at the design point revealed considerable departure from the free vortex design. As a first check upon the capability of the mixed-flow cascade analysis to deal with the blade-to-blade flow adequately, Fisher repeated his analysis introducing the experimental AVR values which in practice differed for each blade section from hub to casing. The outcome was a considerable improvement of AVR, Fig. 3.19(a), indicating the importance of AVR, or meridional stream-sheet thickness, upon blade-to-blade flow and the capability of the mixed-flow surface vorticity method to handle this correctly. However the origin of these variations in AVR lies in the meridional flow. In view of this, Lewis & Mughal (1986), Lewis (1987a) combined mixed-flow Martensen analysis with a simplified form of mixed-flow actuator disc theory into a full quasi-three-dimensional scheme for computer aided design and analysis. In such schemes AVR is able to adjust naturally, resulting in departure from free vortex flow in this case but with appropriate meridional flow adjustments. Of equal importance in such schemes is the influence of losses, which, in the author's simplified actuator disc model, are included as total-to-total efficiencies for each meridional section. The outcome of this quasi-three-dimensional analysis is a considerable improvement in outlet angle prediction as shown in Fig. 3.19(b).

It is clear from these studies that the merits of a blade-to-blade analysis cannot be judged in isolation from meridional effects. Both losses and AVR can play a significant part in the interactions between the blade-to-blade and meridional flows.

3.7 Effects of AVR in compressor cascades

Cascade experiments by Rhoden (1956), Montgomery (1958), Pollard (1964) and many others in the early days of axial compressors, revealed large contractions due to sidewall boundary layer growth and corner stall on the suction surface, Gostelow (1984), resulting in increased axial velocities up to 30% or more. The consequence of AVR was found to be a significant change in blade circulation and fluid outlet angle when compared with true two-dimensional flow (for which AVR = 1.0). To counter this difficulty NACA introduced side wall suction into cascade testing at at early stage to enforce constant axial velocity ratio, Herrig *et al.* (1951); Mellor (1956), a strategy adopted by other later experimenters such as Pollard & Gostelow (1967). In parallel with this, theoretical adaptations of the linearised cascade method by Schlichting (1955b) were undertaken by several authors including Pollard & Horlock (1963), Shaalan & Horlock (1966), Mani and Acosta (1968) and Soundranayagam (1971). These methods in general follow the technique outlined in Section 3.4.3 whereby the change in meridional velocity caused by three-dimensional effects is modelled by introducing source strips into a two-dimensional cascade flow. On the other hand Shaalan & Horlock (1966) and Montgomery (1959) also approached the problem by consideration of the three-dimensional solenoidal flow.

Without entering further into these analyses, surface vorticity predictions following Section 3.4.3 are compared in Fig. 3.20 with the method of Pollard & Horlock (1963) for their compressor cascade 10C4 – 30C50 with $t/\ell = 1.0$, stagger = 36° and $\beta_1 = 52.8°$. The deviation angle δ is defined as the difference between the blade outlet angle and the fluid outlet angle β_2, which for a circular arc camber θ is given by

$$\delta = \beta_2 - \lambda + \theta/2 \qquad (3.76)$$

Although the values of δ disagree, both indicate a decrease in the deviation angle as AVR increases, which corresponds to a decrease in β_2 of significant proportions. For example an increase in AVR

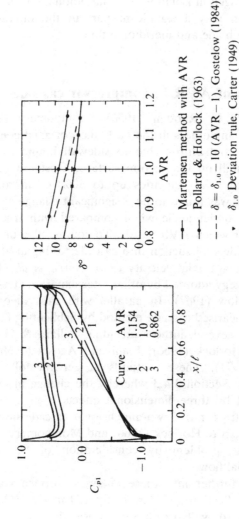

Fig. 3.20. Surface pressure distribution for compressor cascade 10C4 – 30C50 with varying axial velocity ratio.

from 1.0 to 1.2 would produce roughly 1° change in outlet angle. Also shown in Fig. 3.20 is an empirical estimate based on Carter's deviation rule for unity AVR and tests on a NACA cascade reported by Gostelow for which the following approximation held,

$$\delta = \delta_{1.0} - 10(\text{AVR} - 1) \qquad\qquad (3.77)$$

There is reasonable agreement bearing in mind the importance of viscous effects in diffusing cascades, which tend to increase the deviation angle. Also shown in Fig. 3.20 are the surface pressure distributions predicted by surface vorticity theory. Apart from the leading edge region these are in fair agreement with Pollard & Horlock's linearised cascade theory which has therefore not been shown here. The general effect of AVR is to reduce the rate of static pressure rise and in particular on the less stable suction surface.

Although these studies were completed to examine errors in supposed two-dimensional cascade testing, changes in meridional velocity arise also from other sources in real machines, such as variations in annulus height and curvature and the effects of fluid radial equilibrium, i.e. the interference between $S1$ and $S2$ flows, as we have already discussed at the beginning of this chapter. These datum checks linked to the mixed-flow fan traverses referred to in the previous section, are thus extremely useful for validation of the quasi-three-dimensional analysis of mixed-flow turbomachines presented in this chapter.

CHAPTER 4

Bodies of revolution, ducts and annuli

4.1 Introduction

Over the next three chapters we shall develop analyses to deal with progressively more complex problems in the fields of ducted propellers or fans and turbomachine meridional flows. As illustrated in Chapter 3, a design strategy frequently adopted for such devices involves representation of the fully three-dimensional flow as a series of superimposed and connected two-dimensional flows. These are of two main types, blade-to-blade and meridional flow. Having dealt with the first of these, we now turn our attention to the second principal turbomachine problem, calculation of the meridional flow. Turbomachine annuli, Fig. 4.1, are of many different configurations but are usually axisymmetric. For design purposes meridional through-flows are likewise often assumed to be axisymmetric. In general it is important to build into meridional analysis the interactions of the blade-to-blade flow which results in vortex shedding and stagnation pressure or enthalpy gradients. These matters will be dealt with in Chapters 5 and 6, including extension to a consideration of some three-dimensional flows which have been studied by surface vorticity modelling. In the present chapter the foundations will simply be laid for the analysis of axisymmetric potential flows by the surface vorticity method with applications to bodies of revolution, engine or ducted propeller cowls, wind tunnel contractions and turbomachinery annuli.

Axisymmetric flows are in fact two-dimensional in the mathematical sense, even in the presence of circumferentially uniform swirling velocities. By direct analogy with the use of rectilinear vortex elements in plane two-dimensional flows, we may therefore adopt ring surface vortex elements to model the flow past axisymmetric bodies or ducts. Many of the preceding principles may then be applied to this situation also. We shall begin by re-stating the governing integral equation to suit axisymmetric flow. Solutions for the flow field induced by a ring vortex element will be expressed in

146

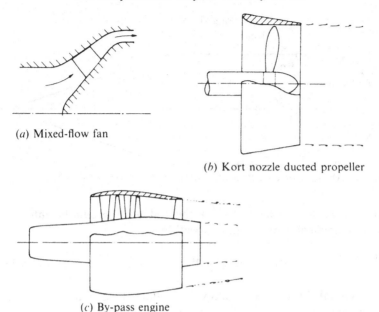

(*a*) Mixed-flow fan

(*b*) Kort nozzle ducted propeller

(*c*) By-pass engine

Fig. 4.1. Turbomachine annulus configurations.

terms of elliptic integrals. Special attention will be paid to problems inherent in 'smoke ring' vortex modelling which do not occur in plane flows. Numerical schemes will be proposed for bodies of revolution, ducts of finite length (i.e. engine cowls), wind tunnel contractions and turbomachine annuli, including the development of the semi-infinite vortex tube to model inlet and outlet duct flows. Several computer programs to cover these situations are included in the Appendix.

4.2 The axisymmetric surface vorticity model

Consider the flow past a body of revolution placed in a uniform stream W parallel to the x axis of a cylindrical coordinate system (x, r, θ), Fig. 4.2. Following the arguments of Chapter 1, we see that the potential flow past the body is bounded by a sheet of ring vorticity adjacent to the body surface of local strength $\gamma(s) = v_s$, where v_s is the fluid velocity close to the surface. The Dirichlet boundary condition of zero velocity actually on the body surface

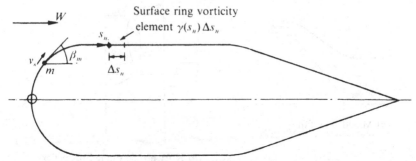

Fig. 4.2. Axisymmetric surface vorticity model for body of revolution.

and parallel to it is described by the following Fredholm integral equation applicable at any body location s_m.

$$\oint \bar{K}(s_m, s_n)\gamma(s_n) \, \mathrm{d}s_n - \tfrac{1}{2}\gamma(s_m) + W \cos \beta_m = 0 \qquad (4.1)$$

In general form this equation is identical to that for plane two-dimensional flows (1.19). The principal difference lies in the coupling coefficient $\bar{K}(s_m, s_n)$, which, in this case, represents the velocity parallel to the body surface at s_m induced by a ring vortex of unit strength located at s_n. Expressions for $\bar{K}(s_m, s_n)$ convenient for computation and involving elliptic integrals will shortly be given. First we shall follow a more basic approach, originally developed by Kuchemann & Weber (1953), which begins with application of the Biot–Savart Law (1.8), to a short line vortex element, Fig. 4.3. If the vortex element is of strength Γ and length $\mathrm{d}s$ with vector direction $\mathbf{t}$, then the velocity which it induces at a vectorial distance away of $\mathbf{R}$, is given by

$$\mathrm{d}\mathbf{q} = \frac{\Gamma \, \mathrm{d}s \, \mathbf{t} X \mathbf{R}}{4\pi \, |R|^3} \qquad (4.2)$$

Let us now apply this to the element $\mathrm{d}s = r_n \, \mathrm{d}\theta'$ of a ring vortex, where $\mathbf{t}$ is a unit vector parallel to the element. First we select the meridional (x, r) datum plane $\theta' = 0$ and define unit vectors $\mathbf{i}, \mathbf{j}$ in the x, r directions lying in this plane with $\mathbf{k}$ normal to the plane. The unit vector $\mathbf{t}$ and radial vector $\mathbf{R}$ may then be expressed

$$\left.\begin{array}{l} \mathbf{t} = -\sin \theta' \mathbf{j} + \cos \theta' \mathbf{k} \\[4pt] \mathbf{R} = (x_m - x_n)\mathbf{i} + (r_m \cos \theta - r_n \cos \theta')\mathbf{j} \\[4pt] \quad + (r_m \sin \theta - r_n \sin \theta')\mathbf{k} \end{array}\right\} \qquad (4.3)$$

148

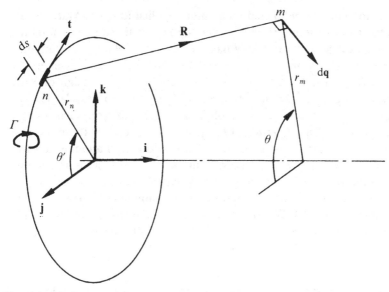

Fig. 4.3. Modelling of ring vortex by the Biot–Savart law.

Introduction of these expressions into (4.2) and evaluation of the cross product results in

$$d\mathbf{q} = \frac{\Gamma \, ds}{4\pi R^3} \{(-r_m \cos(\theta - \theta') + r_n)\mathbf{i} + (x_m - x_n) \cos \theta' \mathbf{j}$$

$$+ (x_m - x_n)\sin \theta' \mathbf{k}\} \qquad (4.4)$$

where

$$R = \sqrt{[(x_m - x_n)^2 + r_m^2 + r_n^2 - 2r_m r_n \cos(\theta - \theta')]} \qquad (4.5)$$

If (4.4) is integrated over the range $\theta' = 0$ to 2π, the $\mathbf{k}$ component vanishes and we obtain the induced velocity components at m due to the complete unit ring vortex at n of strength $\Gamma = 1.0$.

$$u_{mn} = \frac{1}{4\pi} \int_0^{2\pi} \left\{ \frac{r_n - r_m \cos(\theta - \theta')}{[(x_m - x_n)^2 + r_m^2 + r_n^2 - 2r_m r_n \cos(\theta - \theta')]^{\frac{3}{2}}} \right\} d\theta'$$

$$v_{mn} = \frac{1}{4\pi} \int_0^{2\pi} \left\{ \frac{(x_m - x_n) \cos \theta'}{[(x_m - x_n)^2 + r_m^2 + r_n^2 - 2r_m r_n \cos (\theta - \theta')]^{\frac{3}{2}}} \right\} d\theta'$$

$$(4.6)$$

The coupling coefficient then becomes

$$\bar{K}(s_m, s_n) = u_{mn} \cos \beta_m + v_{mn} \sin \beta_m \qquad (4.7)$$

where the body profile slope is $\beta_m = \arctan(dr/dx)$.

149

Ryan (1970) published the first application of Martensen's method to axisymmetric flows, making use of these expressions for analysis of the potential flow past bodies of revolution and annular aerofoils. This work was quickly extended to ducted propellers by Ryan & Glover (1972) for direct application to the design of Kort nozzle propulsors for supertankers and large cargo vessels. Earlier works based on linearised aerofoil singularity theory, such as that published by Bagley *et al.* (1961), also made use of these expressions and analogous ones for ring sources to model annular aerofoil or duct flows. However, other formulations more suitable for fast computation were subsequently developed making use of elliptic integrals of the first and second kinds, recommended even earlier by Riegels (1949) (1952). Gibson (1972) finally established the simple expressions now commonly used for the unit ring vortex induced velocities, which are as follows.

$$
\left.
\begin{aligned}
u_{mn} &= -\frac{1}{2\pi r_n \sqrt{[x^2 + (r+1)^2]}} \left\{ K(k) - \left[1 + \frac{2(r-1)}{x^2 + (r-1)^2}\right] E(k) \right\} \\
v_{mn} &= \frac{x/r}{2\pi r_n \sqrt{[x^2 + (r+1)^2]}} \left\{ K(k) - \left[1 + \frac{2r}{x^2 + (r-1)^2}\right] E(k) \right\}
\end{aligned}
\right\}
$$

$$(4.8)$$

$K(k)$ and $E(k)$ are complete elliptic integrals of the first and second kind respectively and we define

$$
k = \sqrt{\left[\frac{4r}{x^2 + (r+1)^2}\right]} = \sin \phi
\qquad (4.9)
$$

where x and r become the dimensionless coordinates

$$
x = \frac{x_m - x_n}{r_n}, \qquad r = \frac{r_m}{r_n}
\qquad (4.10)
$$

4.2.1 Evaluation of complete elliptic integrals. Use of look-up tables

For calculation of the coupling coefficients, $K(k)$ and $E(k)$ may be evaluated by quadrature from the well known formulae given by

The axisymmetric surface vorticity model

Dwight (1963),

$$K(k) = \int_0^{\pi/2} \frac{1}{\sqrt{[\{1 - k^2 \sin^2 \alpha\}]}} \, d\alpha \\ E(k) = \int_0^{\pi/2} \sqrt{[\{1 - k^2 \sin^2 \alpha\}]} \, d\alpha \Bigg\}$$

$$(4.11)$$

A short Pascal code, Program No. 4.1, is given in the Appendix to evaluate these expressions, together with tabulated results for the range $\phi = 0(0.5)89.5$ degrees, making use of the trapezium integration rule. Unless $K(k)$ and $E(k)$ are available as functions on a given computer facility, it would at first sight appear necessary to evaluate equations (4.11) directly whenever elliptic integrals are required in a Martensen scheme. To avoid this unnecessarily time-consuming process, an alternative approach is advocated using a table 'look-up' procedure, Lewis (1984a). The recommended sequence is as follows.

Sequence for use of look-up tables for elliptic integrals

(i) Evaluate ϕ for the given coordinate input (x_m, r_m), (x_n, r_n) from (4.10) and (4.9).

(ii) Read the tabulated elliptic integrals from file into real array variables $K[i]$, $E[i]$ with $i = 1 \ldots m$. The last values on file are $fi[m]$, $K[m]$ and $E[m]$.

(iii) Calculate the nearest position in the table from

$$i = \text{round}(\phi/fi[m] * (m - 1) + 1) \qquad (4.12)†$$

(iv) $K[i]$, $E[i]$ are then the nearest values appropriate to the given value of ϕ. For further accuracy linear or higher order interpolation may be undertaken.

Since $K(k)$ is singular for $\phi \to 90°$, the table in Appendix I, at $5°$ intervals, has been terminated at $\phi = 89.5°$. For $\phi > 89.5°$ asymptotic expressions are available, Dwight (1963), Lewis (1984a),

$$K(k) \to \ln(4/\cos \phi) \\ E(k) \to 1 + \tfrac{1}{2}(K(k) - 1/1.2) \cos^2 \phi \Bigg\}$$

$$(4.13)$$

On the other hand we observe from (4.9) that the limiting case of $\phi \to 90°$ corresponds to

$$a = x^2 + (r + 1)^2 \to 0$$

---

† In Pascal round(x) produces the integer nearest in value to a real number x.

which refers to locations (x_m, r_m) close to the ring vortex. An alternative approach therefore could be to redefine u_{mn} and v_{mn} for this special case by adopting the solution for a rectilinear vortex, namely

$$\left. \begin{aligned} u_{mn} &= \frac{r-1}{2\pi\sqrt{[x^2+(r-1)^2]}} \\[2ex] v_{mn} &= -\frac{x}{2\pi\sqrt{[x^2+(r-1)^2]}} \end{aligned} \right\} \quad \begin{aligned} &\text{when (say)} \\ &\sqrt{[x^2+(r-1)^2]} < 0.01 \end{aligned} \qquad (4.14)$$

Although this works quite well, the asymptotic expressions are used in the computer programs in Appendix I. Let us now illustrate this process with an example.

Example

Let us consider the following data:

Coordinates of ring vortex $x_n = 0.0$, $r_n = 1.0$
Coordinates of point m $x_m = 0.2$, $r_m = 1.2$
Thus $\phi = 82.6438°$
Nearest location in table is line i where
$i = \text{round } (82.6438/895 * (180 - 1) + 1) = 166$

Extract from tables for elliptic integrals, Appendix

i	ϕ	$K(k)$	$E(k)$
.	.	.	.
.	.	.	.
.	.	.	.
165	82.0	3.369 868	1.027 844
166	82.5	3.432 887	1.025 024
167	83.0	3.500 422	1.022 313
.	.	.	.
.	.	.	.
.	.	.	.

Comparing the 'look-up' values with the actual values we have

	$K(k)$	$E(k)$
Actual value from (4.11)	3.451 821	1.024 233
Look-up table	3.432 887	1.025 024
Look-up table with linear interpolation	3.452 310	1.024 244

Thus reasonable accuracy was obtained by the direct look-up procedure with considerable improvement using linear interpolation between items $i = 166$ and 167.

4.2.2 Numerical representation of the integral equation for axisymmetric flow

Representing the body surface by M discrete ring vortex elements $\gamma(s_n)\Delta s_n$ in exactly the same manner as that already adopted in Chapter 1 for plane flows, the governing integral equation (4.1), using trapezoidal integration, may be expressed

$$\sum_{n=1}^{M} \bar{K}(s_m, s_n)\gamma(s_n)\,\Delta s_n = -W\cos\beta_m \tag{4.15}$$

where the term $-\frac{1}{2}\gamma(s_m)$ has been absorbed into the self-inducing coupling coefficient, i.e.

$$\bar{K}(s_m, s_n)\,\Delta s_m = \bar{K}(s_m, s_n)\,\Delta s_m - \tfrac{1}{2} \quad \text{for} \quad n = m \tag{4.16}$$

For computational purposes it is convenient also to absorb the element length Δs_n into the coupling coefficient, whereupon the set of linear simultaneous equations (4.15) takes the form analogous to equations (1.25)

$$\begin{pmatrix} K_{11} & K_{12} & K_{13} & \cdots & K_{1M} \\ K_{21} & K_{22} & K_{23} & \cdots & K_{2M} \\ \cdot & & & & \\ \cdot & & & & \\ K_{M1} & K_{M2} & K_{M3} & \cdots & K_{MM} \end{pmatrix} \begin{pmatrix} \gamma_1 \\ \gamma_2 \\ \cdot \\ \cdot \\ \gamma_M \end{pmatrix} = \begin{pmatrix} \text{rhs}_1 \\ \text{rhs}_2 \\ \cdot \\ \cdot \\ \text{rhs}_M \end{pmatrix} \tag{4.17}$$

where the coupling coefficients are now re-defined from the unit vortex coupling coefficients (4.7), through

$$
\begin{aligned}
K(s_m, s_n) &= \bar{K}(s_m, s_n)\Delta s_n \\
&= (u_{mn}\cos\beta_m + v_{mn}\sin\beta_m)\Delta s_n \quad \text{for} \quad m \neq n \\
K(s_m, s_m) &= (u_{mm}\cos\beta_m + v_{mm}\sin\beta_m)\Delta s_m - \tfrac{1}{2} \quad \text{for} \quad m = n
\end{aligned} \right\} \quad (4.18)
$$

The induced velocity components u_{mn}, v_{mn} due to a unit ring vortex, equations (4.8), are then directly applicable provided $n \neq m$. On the other hand we observe that for $n = m$ the self-inducing coupling coefficient $K(s_m, s_m)$ is singular. Thus for this special case the dimensionless coordinates become $x = 0$, $r = 1$, so that $\phi = 90°$ and the elliptic integral $K(k)$ is infinite. Furthermore the coefficients of $E(k)$ in equations (4.8) are also infinite. In other words the self-inducing velocity (u_{mm}, v_{mm}) of a concentrated ring vortex is infinite. In the case of plane flows modelled by rectilinear vortex elements this problem did not arise. By arguments of symmetry it is easy to deduce that the self-induced velocity of a concentrated line vortex is zero.

On the other hand each coupling coefficient $K(s_m, s_n)$ really represents the influence of the vorticity sheet $\gamma(s_n)$ spread over the element length Δs_n. The use of central pivotal points to model this is simply equivalent to locating concentrated ring vortices of strength $\gamma(s_n)\,\Delta s_n$ at each pivotal point. This model is quite adequate to handle the influence between different elements for $m \neq n$, for which equations (4.8) provide a sufficiently accurate representation. For the case $m = n$ on the other hand we would expect the self-induced velocity of vortex element $\gamma(s_m)\,\Delta s_m$ to be finite. Following the strategy laid out in Section 1.7 for plane flows, the present model and formulation for evaluating the self-inducing coupling coefficients $K(s_m, s_m)$ must be abandoned. As shown there for rectilinear two-dimensional aerofoils, the surface curvature in the (x, y) plane actually gives rise to a self-induced velocity which may be of significant proportions. In the case of an axisymmetric body flow, the surface vorticity sheet exhibits double curvature resulting in two contributions to its self-inducing velocity. This problem merits the special consideration which it will be given in the next sub-section.

4.2.3 Self-induced velocity of a ring vorticity element

Fig. 4.4 illustrates the two curvatures possessed by a ring vorticity element in the (x, r) and (r, θ) planes, which can be expressed in

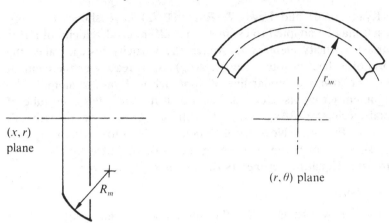

Fig. 4.4. Double curvature of a ring vortex surface element.

terms of the two surface radii of curvature R_m and r_m respectively. As already stated, both curvatures contribute to the self-induced velocity of the element. For simplicity we shall assume that they act independently and can be estimated separately.

For body profile curvature in the (x, r) plane we will further assume that (1.31), which expresses the self-induced velocity of a rectilinear surface vorticity element, will suffice, a reasonable assumption provided $\Delta s_m/r_m$ is small. This will usually be the case except in the stagnation point regions of a body of revolution where the vorticity strength itself is in any case small. The contribution due to curvature in the (x, r) plane is thus

$$\delta q_1 = \frac{\gamma(s_m)\, \Delta s_m}{4\pi R_m} \tag{4.19}$$

Regarding curvature in the (r, θ) plane, a useful starting point in our considerations is the approximate expression given by Lamb (1945) for the self-induced velocity of a smoke ring vortex with a core of radius a containing unit strength vorticity, namely

$$\delta q_2 = -\frac{1}{4\pi r_m}\left\{\ln\frac{8r_m}{a} - \frac{1}{4}\right\} \tag{4.20}$$

It is well known that a smoke ring vortex will propel itself parallel to the x direction without change of radius r_m and it is clear from Lamb's solution that the velocity δq_2 will depend upon the dimensionless core radius a/r_m.

155

Ryan (1970) and Lewis & Ryan (1972) have shown how this result may be adapted to estimate the self-induced velocity of a ring surface vorticity element for which the vorticity instead takes the form of a sheet of total strength $\gamma(s_m)\,\Delta s_m$ spread over the element Δs_m. Following a similar line of approach to Ryan, to simplify the argument let us first treat the sheet as if it were a flattened tube of total perimeter $2\Delta s_m$ covered with sheet vorticity of strength $\frac{1}{2}\gamma(s_m)$, Fig. 4.5. Now let us open up the tube into a circular cross-section of radius a, keeping the same perimeter and vorticity strength. Equating perimeters the radius s is thus given by

$$a = \Delta s_m/\pi \qquad\qquad (4.21)$$

Ryan now assumes that the self-induced velocity of the ring vortex tube remains unchanged and furthermore that it is equal to

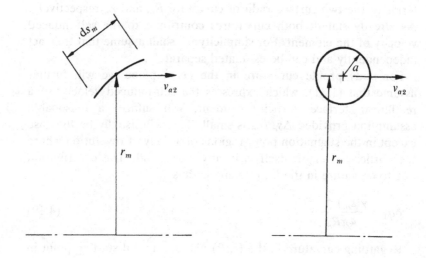

Fig. 4.5. Ring vortex element Δs_m with smoke-ring vortex equivalent.

that of a Lamb type smoke-ring vortex with uniform core vorticity. As will be shown in Chapter 5 when considering flows with distributed vorticity in the mainstream, Section 5.7.2, this is a quite reasonable assumption and results in the propagation velocity contribution for unit total vortex strength

$$\delta q_2 = -\frac{1}{4\pi r_m}\left\{\ln\frac{8\pi r_m}{\Delta s_m} - \frac{1}{4}\right\} \tag{4.20a}$$

Combining all these results, the self-inducing coupling coefficient for a body of revolution becomes

$$K(s_m, s_m) = -\frac{1}{2} + \frac{\Delta s_m}{4\pi R_m} - \frac{\Delta s_m}{4\pi R_m}\left\{\ln\frac{8\pi r_m}{\Delta s_m} - \frac{1}{4}\right\}\cos\beta_m \tag{4.22}$$

Although no rigorous proof of Ryan's model has yet been put forward the sample calculations which follow confirm its accuracy and suitability for a range of applications. Numerical calculations by Lewis & Sorvatziotis (1987), making use of sub-elements, also confirmed the accuracy of both Lamb's formula (4.20) and its equivalent for sheet vorticity (4.20a). These are among the studies to be covered in Chapter 5.

4.3 Flow past a body of revolution

In the case of two-dimensional aerofoils it was shown in Chapter 2 that back diagonal correction must be applied to the matrix to eliminate implied numerical leakage flux caused by inaccuracies in opposite point coupling coefficients. We shall return to this problem in Section 4.4 when applying the present analysis to annular aerofoils. On the other hand V. P. Hill (1975) has shown that such considerations are quite unnecessary for bodies of revolution, for which excellent predictions may be obtained with the foregoing equations as they stand. Furthermore, by analogy with the straight-line elements recommended in Chapter 1 for plane two-dimensional flows, straight conical elements (i.e. frustrums) are perfectly adequate for representations of axisymmetric bodies. Two sample calculations will now be presented for a sphere and a body of revolution.

Table 4.1. *Flow past a sphere*

Element no.	Case 1 $K(s_m, s_m) = -\frac{1}{2}$ v_s/W	Case 2 (Look-up) v_s/W	Case 3 (Look-up & interpolate) v_s/W	Exact solutions v_s/W
1, 35	0.067 654	0.067 707	0.067 298	0.067 297
2, 34	0.205 457	0.202 902	0.201 195	0.201 350
3, 33	0.343 026	0.333 706	0.333 521	0.333 781
4, 32	0.478 534	0.463 360	0.463 153	0.463 526
5, 31	0.610 640	0.589 065	0.589 063	0.589 538
6, 30	0.738 120	0.707 888	0.710 231	0.710 803
7, 29	0.859 839	0.826 679	0.825 680	0.826 345
8, 28	0.974 745	0.933 568	0.934 477	0.935 235
9, 27	1.081 840	1.037 443	1.035 731	1.036 594
10, 26	1.180 219	1.130 289	1.128 642	1.129 607
11, 25	1.269 110	1.214 329	1.212 515	1.213 526
12, 24	1.347 586	1.287 274	1.286 588	1.287 673
13, 23	1.415 290	1.350 830	1.350 284	1.351 453
14, 22	1.471 355	1.403 955	1.403 077	1.404 352
15, 21	1.515 508	1.442 190	1.444 636	1.445 944
16, 20	1.547 335	1.472 865	1.474 587	1.475 894
17, 19	1.566 554	1.489 500	1.492 665	1.493 962
18	1.572 981	1.497 755	1.498 714	1.500 000

4.3.1   Flow past a sphere

The exact solution for the potential flow past a sphere situated in a uniform stream W is well known and has been given by Lamb (1945), the surface velocity v_s being

$$\frac{v_s}{W} = \frac{3}{2} \sin \phi \qquad (4.23)$$

A comparison of this solution with various versions of the surface vorticity analysis is provided by the results shown in Table 4.1.

Three solutions are presented of increasing accuracy with 35 equal length elements. In view of symmetry about the mid(r, θ) plane of the sphere, results in all cases were identical for equivalent elements such as 1 and 35, 2 and 34 etc.

Case 1 is intended to check the effect of ignoring the influence of surface curvature upon the self-inducing coupling coefficient as just discussed in Section 4.2.3. For this case $K(s_m, s_m)$, equation (4.22), was reduced to a value of -0.5 for all elements. In consequence of this errors of about 5% were present. Look-up tables were used for obtaining elliptical integrals.

For Case 2 the curvature terms were introduced according to
(4.22) resulting in considerable improvement. The look-up table
procedure of Section 4.2.1 was also used in this case. As a final
refinement for Case 3, the look-up tables were interpolated linearly,
producing yet further marginal improvements, the accuracy now
lying within 0.2% for all elements. This example brings out the
great importance of taking into account the self-induced velocities
of the ring vorticity elements due to their double curvature and
confirms that the assumptions underlying the simple formulations of
Section 4.2.3 were well justified.

4.3.2 Flow past a body of revolution

The results of a second example are shown in Fig. 4.6 where the
surface velocity distribution for a body of revolution predicted by
the axisymmetric surface vorticity theory, using a 50-surface
element representation, is compared with experimental tests. The
potential flow prediction was excellent apart from the conical
after-body rear apex from which the real flow was separated.
Potential flow of course predicts a rear stagnation point with fully
attached flow. For this body, comprising a hemispherical nose and
conical afterbody joined by a cylindrical centre section, flow

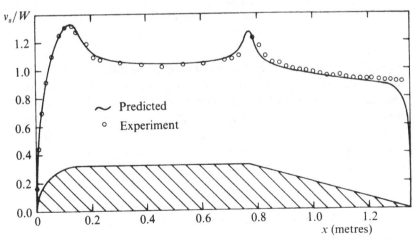

Fig. 4.6. Comparison of predicted and measured surface velocity distribution of a body of revolution.

Table 4.2. *Coordinates of test case body of revolution,*
Fig. 4.6

x	r	x	r
0.000 000	0.000 000	0.682 857	0.160 000
0.003 074	0.031 214	0.711 905	0.160 000
0.012 179	0.061 229	0.740 952	0.160 000
0.026 965	0.088 891	0.770 000	0.160 000
0.046 863	0.113 137	0.798 435	0.152 381
0.071 109	0.133 035	0.826 869	0.144 762
0.098 771	0.147 821	0.855 304	0.137 143
0.128 786	0.156 926	0.883 739	0.129 524
0.160 000	0.160 000	0.912 173	0.121 905
0.189 048	0.160 000	0.940 608	0.114 286
0.218 095	0.160 000	0.969 043	0.106 667
0.247 143	0.160 000	0.997 477	0.099 048
0.276 190	0.160 000	1.025 912	0.091 429
0.305 238	0.160 000	1.054 347	0.083 810
0.334 286	0.160 000	1.082 781	0.076 190
0.363 333	0.160 000	1.111 216	0.068 571
0.392 381	0.160 000	1.139 651	0.060 952
0.421 429	0.160 000	1.168 085	0.053 333
0.450 476	0.160 000	1.196 520	0.045 714
0.479 524	0.160 000	1.224 954	0.038 095
0.508 571	0.160 000	1.253 389	0.030 476
0.537 619	0.160 000	1.281 824	0.022 857
0.566 667	0.160 000	1.310 258	0.015 238
0.595 714	0.160 000	1.338 693	0.007 619
0.624 762	0.160 000	1.367 128	0.000 000
0.653 809	0.160 000		

accelerations and diffusions were experienced at the junctions, providing a severe test for the prediction technique. The diffusions were sufficiently mild to avoid boundary layer separation however and surface vorticity theory coped extremely well. A Pascal program axisym.pas to accomplish this calculation has been included in the Appendix, Program No. 4.2. For the benefit of readers who require a test case, the body coordinates were as given in Table 4.2.

4.4 Annular aerofoils or engine cowls

As the first step towards the modelling of a complete ducted fan or propeller unit, we shall consider next the flow past an axisymmetric duct, or engine cowl, which may alternatively be thought of as an annular aerofoil, Fig. 4.7. There is a fairly extensive literature for such flows based upon linearised aerofoil theory, which was ably reviewed by Weissinger & Maass (1968). Although the alternative source panel method was also well established in the 1960s, e.g., Smith & Hess (1966), surface vorticity modelling was still undeveloped for duct flows until the publication of Ryan's method in 1970. Since then a good deal has been built upon this early work. The equations already developed for bodies of revolution are directly applicable with two modifications only. Firstly it is usually important to apply the back diagonal correction to the coupling coefficient matrix, especially in the case of very thin ducts. Secondly the trailing edge Kutta condition must be applied in a manner analogous to that explained in Chapter 2, Section 2.4.

To apply the back diagonal correction to equations (4.17), we must replace the coupling coefficients on the backward sloping diagonal by the values

$$K(s_{M+1-m}, s_m) = -\frac{1}{\Delta s_{M+1-m}} \sum_{\substack{n=1 \\ n \neq M+1-m}}^{M} K(s_n, s_m)\, \Delta s_n \Bigg\} \qquad [2.19]$$

As for plane two-dimensional aerofoils the Kutta condition requires the imposition of zero pressure loading approaching the trailing edge of the duct. This can be accomplished quite simply by applying the restriction to the two elements adjacent to the trailing edge

$$\gamma(s_{te+1}) = -\gamma(s_{te}) \qquad [2.22]$$

As already illustrated in Section 2.4.3 for plane aerofoils, since the (te + 1)th equation is now redundant we may simply subtract column te + 1 from column te and likewise row te + 1 from row te, reducing the matrix dimensions to M-1. This also implies of course that the same treatment is given to the right hand side values rhs(te) and rhs(te + 1). The reader may check how this is achieved by comparing the Pascal codes for the body of revolution, axisym.pas and the duct, duct.pas given as Programs 4.2 and 4.3 in the Appendix.

V. P. Hill (1975), (1978) constructed the duct illustrated in Fig. 4.7 with considerable precision to obtain a really reliable experimental datum for annular aerofoils, including both axial and incident flows with angles of attack in the range 0°–15°. An aerofoil

161

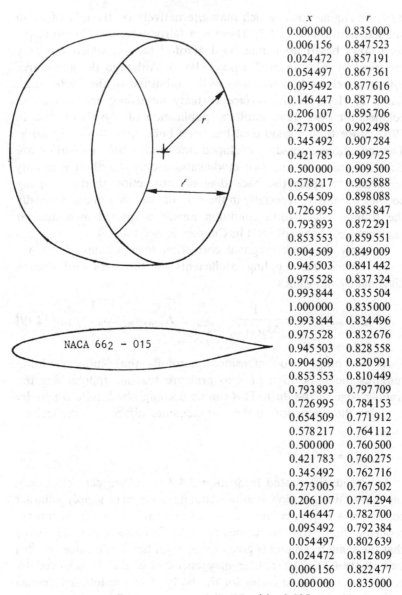

x	r
0.000 000	0.835 000
0.006 156	0.847 523
0.024 472	0.857 191
0.054 497	0.867 361
0.095 492	0.877 616
0.146 447	0.887 300
0.206 107	0.895 706
0.273 005	0.902 498
0.345 492	0.907 284
0.421 783	0.909 725
0.500 000	0.909 500
0.578 217	0.905 888
0.654 509	0.898 088
0.726 995	0.885 847
0.793 893	0.872 291
0.853 553	0.859 551
0.904 509	0.849 009
0.945 503	0.841 442
0.975 528	0.837 324
0.993 844	0.835 504
1.000 000	0.835 000
0.993 844	0.834 496
0.975 528	0.832 676
0.945 503	0.828 558
0.904 509	0.820 991
0.853 553	0.810 449
0.793 893	0.797 709
0.726 995	0.784 153
0.654 509	0.771 912
0.578 217	0.764 112
0.500 000	0.760 500
0.421 783	0.760 275
0.345 492	0.762 716
0.273 005	0.767 502
0.206 107	0.774 294
0.146 447	0.782 700
0.095 492	0.792 384
0.054 497	0.802 639
0.024 472	0.812 809
0.006 156	0.822 477
0.000 000	0.835 000

Fig. 4.7. NACA 662-015 annular aerofoil with $r/\ell = 0.835$.

162

section known to yield high performance in plane two-dimensional flows was selected, namely NACA 662-015, to minimise viscous effects, and the duct aspect ratio (trailing edge radius/chord) of 0.835 was chosen to be typical of pump jet or Kort nozzle applications. The profile was symmetrically distributed about a cylinder with zero camber. One difficult feature of this profile is its cusped trailing edge which places severe pressure upon the surface vorticity model. To ease manufacture of the experimental pressure tapped duct the trailing edge was in fact thickened to 2 mm.

Hill's solution by the surface vorticity method using 64 elements is compared in Fig. 4.8 with output from Program No. 4.3 using 40 elements based upon the (x, r) coordinates shown in Fig. 4.7. For case 1 the program was modified to leave out the back diagonal correction, revealing quite significant errors in predicted surface pressure coefficient. On the other hand with back diagonal correction, case 2, the two solutions were in excellent agreement both with one another and with the experimentally measured pressures, Fig. 4.8(b).

In fact Hill's computation was undertaken without back diagonal correction but using an alternative technique for optimum selection of pivotal points. Fig. 4.9 shows the results of a study undertaken by Hill of the accuracy of the opposite profile point coupling coefficient $K(s_{M+1-n}, s_n) = v_{sn}$ as a function of element length to profile thickness ratio $\Delta s_n / \Delta T_{mn}$. The outcome was that pivotal point vortex modelling of the opposite element is adequate provided $\Delta s_n / \Delta T_{mn} < 1.0$, eliminating the need for back diagonal correction in such circumstances. This result is in general agreement with the study undertaken in Section 2.3.1 with regard to thin non-lifting aerofoils indicated by Table 2.2. To combat this Hill employed a curve-fitting procedure to select optimum pivotal point locations in order to ensure that this criterion was met, resulting in this case in the need for 64 elements. However, using only 40 elements combined with back diagonal correction it is clear that accuracy may be retained with reduced computational requirements.

Prior to Hill's work Ryall & Collins (1967) produced a series of thinner profiled ducts for experimentation of linearised aerofoil theory for ducted propellers. Young (1969), (1971) also investigated a series of engine cowls designed for higher Mach number applications as part of a study programme on by-pass engine intakes, for which some comparisons with surface vorticity analysis have also been made by Ryan (1970). For the present purpose it is sufficient

163

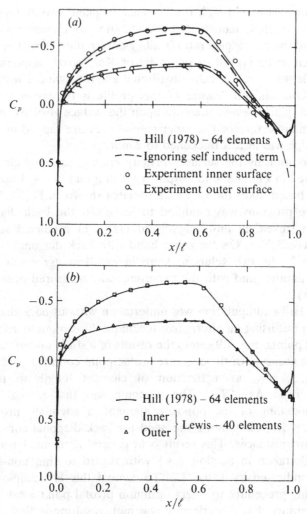

Fig. 4.8. Surface pressure distribution on annular aerofoil NACA 662-015, with $r/\ell = 1.2$.

to show sample results for the diffusing duct $B1$ and the accelerating duct $B3$ investigated by Ryall and Collins, Fig. 4.10 for which repeat experimental tests were completed for confirmation in the author's wind tunnel laboratory.

These aerofoils typify the levels of duct bound circulation and consequent internally induced velocity which one would seek to

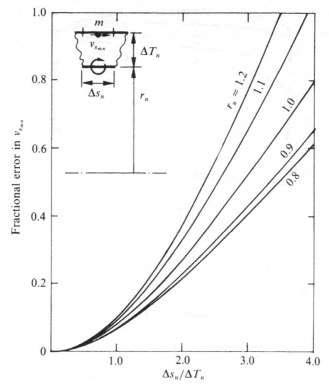

Fig. 4.9. Fractional error of predicted velocity at opposite element *m* of a body due to assumption of concentrated ring vortex at s_n.

introduce into a pump jet or Kort nozzle at the propeller plane. In the case of the diffusing duct the pressure on the inner surface is greater than that of the free stream, the reverse being true for the accelerating duct. Alternatively one can think of the duct net circulation as positive for *B*1 resulting in a bursting lift force and the opposite for duct *B*3. In both cases quite reasonable agreement was obtained between theory and experiment for these two extreme cases. The profile geometries are given by Ryall & Collins (1967) and Ryan (1970).

Once the surface vorticity is known it is quite a simple task to evaluate velocity components at any point (x_p, r_p) within the flow

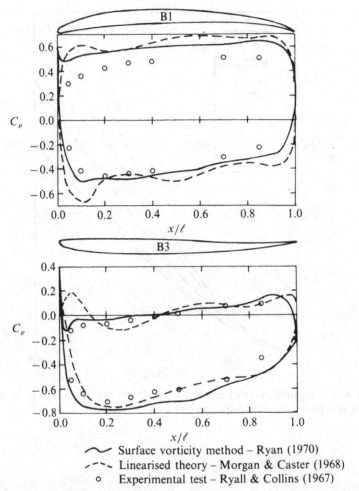

Fig. 4.10. Accelerating and diffusing ducts. (Reproduced from the Proceedings of the Institution of Mechanical Engineers by permission of the Council of the Institution.)

field, since we may write, following the strategy outlined in Section 1.8

$$u = U + \sum_{n=1}^{M} u_{pn} \gamma(s_n) \, \Delta s_n \Bigg\}$$

$$v = \sum_{n=1}^{M} v_{pn} \gamma(s_n) \, \Delta s_n$$

(4.24)

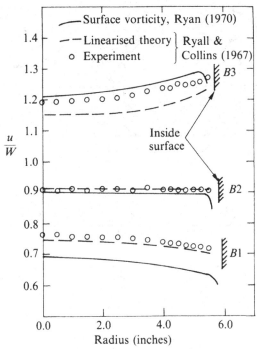

Fig. 4.11. Axial velocity traverses at duct mid-axial plane. (Reproduced from the Proceedings of the Institution of Mechanical Engineers by permission of the Council of the Institution.)

where the unit velocities (4.8) are applicable with

$$x = \frac{x_p - x_n}{r_n}, \qquad r = \frac{r_p}{r_n} \qquad (4.25)$$

The results of the application of this procedure to ducts $B1$, $B2$ and $B3$ are shown in Fig. 4.11 where experimental axial velocity traverses are compared with theoretical results both by the present theory and by linearised theory. The duct profile camber clearly has a strong effect upon axial velocity which is predicted competently by these methods, the surface vorticity model coping especially well with ducts $B2$ and $B3$.

4.5 The semi-infinite vortex cylinder

So far we have considered closed bodies of revolution and ducts of finite length located in open flow. We shall now extend this analysis

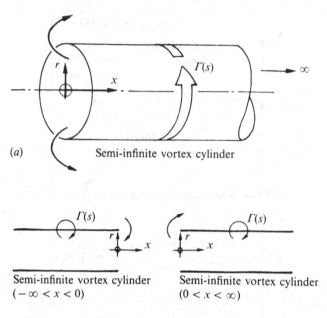

(a)   Semi-infinite vortex cylinder

Semi-infinite vortex cylinder
$(-\infty < x < 0)$

Semi-infinite vortex cylinder
$(0 < x < \infty)$

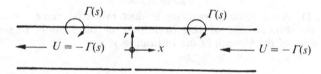

(b) Combined doubly-infinite vortex cylinder

Fig. 4.12. The semi-infinite and doubly-infinite vortex cylinders.

to flows which may be completely internal, such as the flow through contractions, diffusers or turbomachine annuli. In such cases it is necessary to arrange cylindrical inlet and exit sections which extend to infinity. To facilitate this let us first consider the flow field induced by a semi-infinite ring vortex cylinder of strength $\Gamma(s)$ per unit length and of radius r_m, extending axially over the distance $0 < x < \infty$, Fig. 4.12. The streamline pattern is directly analogous to the magnetic flux induced by a semi-infinite solenoid. Although there is some leakage through the cylinder walls close to the origin, the flow becomes progressively more parallel to the duct as $x \rightarrow \infty$, making this a suitable device to adopt as an inlet or exit tube for modelling a duct flow problem as we shall see. The velocity

168

The semi-infinite vortex cylinder

components for the vortex tube were given by Gibson (1972) as

$$
\left.
\begin{aligned}
U_c &= \frac{\Gamma}{2\pi}\left\{ A + \frac{x}{\sqrt{[x^2 + (r+1)^2]}}\left[K(k) - \left(\frac{r-1}{r+1}\right)\mathrm{II}(n, k)\right]\right\} \\
V_c &= \frac{2\Gamma}{\pi k^2 \sqrt{[x^2 + (r+1)^2]}}\left[E(k) - \left(1 - \frac{k^2}{2}\right)K(k)\right]
\end{aligned}
\right\} \tag{4.26}
$$

where x and r are dimensionless coordinates previously defined by equations (4.10) and where

$$
\left.
\begin{aligned}
A &= \pi && \text{if } r < 1 \\
&= \pi/2 && \text{if } r = 1 \\
&= 0 && \text{if } r > 1
\end{aligned}
\right\} \tag{4.27}
$$

The complete elliptic integrals of the first and second kind $K(k)$ and $E(k)$ have already been given by equations (4.11). $\mathrm{II}(n, k)$ is the complete elliptic integral of the third kind which is given by Dwight (1963) as

$$
\mathrm{II}(n, k) = \int_0^{\pi/2} \frac{\mathrm{d}\alpha}{(1 - n\sin^2\alpha)\sqrt{[(1 - k^2\sin^2\alpha)]}} \tag{4.28}
$$

where the additional parameter n is defined

$$
n = \frac{4r}{(1+r)^2} \tag{4.29}
$$

Two observations should be made at this point. Firstly $\mathrm{II}(n, k)$ is a function of two variables n and k which renders the use of look-up tables impractical. $\mathrm{II}(n, k)$ must be evaluated from (4.28) as needed unless available in software. Secondly $\mathrm{II}(n, k)$ is singular at $r = 1.0$. Fortunately, however, it can be shown that the product $(r - 1)\mathrm{II}(n, k) \to 0$ as $r \to 1$, so that we then have an alternative expression for U_c,

$$
U_c = \Gamma\left(\frac{1}{4} + \frac{xK(k)}{2\pi\sqrt{(x^2 + 4)}}\right) \quad \text{for} \quad r = 1 \tag{4.30}
$$

The velocity components induced by a semi-infinite ring vortex cylinder of strength $\Gamma = -1.0$ are shown in Fig. 4.13 for a wide range of (x, r) values, revealing some symmetrical characteristics of this flow. In particular we observe that $V_c(x) = V_c(-x)$. The behaviour of U_c is rather more complex and it is helpful to consider the situation illustrated in Fig. 4.12 where a second semi-infinite vortex cylinder has been introduced, extending over the axial range

169

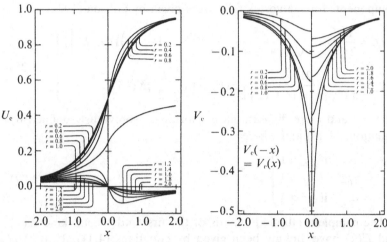

Fig. 4.13. Velocity components induced by a semi-infinite vortex cylinder.

$-\infty < x < 0$. By inspection we see that its induced radial velocities would be equal to $-V_c$ and would cancel those of the first semi-infinite vortex cylinder throughout the flow field. On the other hand the axial velocity components would cancel only for $r > 1$ but would have the value $U_c = -1.0$ for $r < 1$. Thus for the doubly infinite vortex tube of strength $\Gamma(s)$ shown in Fig. 4.12 the above solution, as we would expect, reduces to

$$\left.\begin{aligned} U_c &= 0 && \text{for} \quad r > 0 \\ U_c &= -\Gamma && \text{for} \quad r < 0 \\ U_c &= -\Gamma/2 && \text{for} \quad r = 1 \end{aligned}\right\} \tag{4.31}$$

which is of course the flow of a uniform stream through an infinite cylindrical duct.

4.6 Flow through a contraction

This last result suggests a suitable flow model for contractions or diffusers, Fig. 4.14. Consider the axisymmetric contracting duct lying between $0 < x < x_1$. Inlet and exit sections may now be represented by semi-infinite vortex cylinders of radius r_1 and r_2 respectively, extending over the ranges $-\infty < x < 0$ and $x_1 < x < \infty$. If the inlet velocity W_1 is specified, we may obtain Γ_1 directly and

170

Flow through a contraction

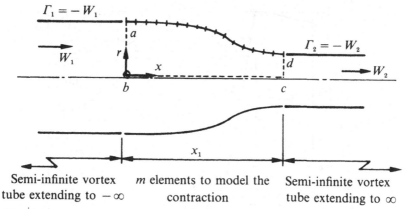

Semi-infinite vortex tube extending to $-\infty$ m elements to model the contraction   Semi-infinite vortex tube extending to ∞

Fig. 4.14. Surface vorticity element model for flow through a contraction.

refer to mass flow continuity to obtain Γ_2. Thus

$$\left.\begin{array}{l} \Gamma_1 = -W_1 \\ \\ \Gamma_2 = -W_2 = \left(\dfrac{r_1}{r_2}\right)^2 \Gamma_1 \end{array}\right\} \qquad (4.32)$$

The sign convection here is that W is positive in the x direction and Γ is positive when clockwise.

If we now represent the contraction by M surface elements, equations (4.17) may be applied directly if the coupling coefficients are chosen to represent the Dirichlet boundary condition of zero velocity parallel to the outer surface of the duct. The form of the general coupling coefficients $K(s_m, s_n)$ is thus unchanged, whereas the self-inducing coupling coefficients become

$$K(s_m, s_m) = +\frac{1}{2} - \frac{\Delta s_m}{4\pi R_m} - \frac{\Delta s_m}{4\pi r_m}\left\{\ln\frac{8\pi r_m}{\Delta s_m} - \frac{1}{4}\right\} \qquad (4.33)$$

The right hand side terms for this situation are

$$\text{rhs}_m = -(U_{c1} + U_{c2})\cos\beta_m - (V_{c1} + V_{c2})\sin\beta_m \qquad (4.34)$$

where suffixes 1 and 2 refer to the influence of the inlet and exit semi-infinite vortex cylinders respectively.

Pascal Program No. 4.4, contract.pas, which is listed in the Appendix, undertakes this computation for arbitrary contraction contour coordinates (x, r) which are read from file. Elliptic integrals $K(k)$ and $E(k)$ are evaluated by procedure look_up_and_interpolate from tabulated data also read from file. Elliptic integrals

171

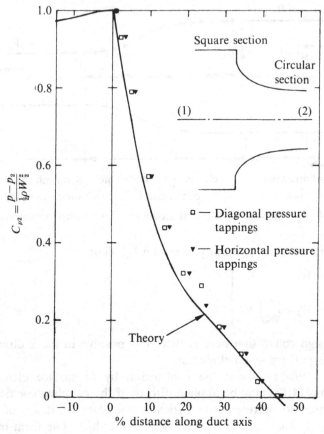

Fig. 4.15. Pressure distribution along wall of a wind tunnel nozzle contraction.

$\Pi(n, k)$ are evaluated by procedure third_kind which is called by procedure uv_tube which calculates the induced velocities (U_c, V_c) induced by the semi-infinite vortex cylinders.

Some results from this program are compared in Fig. 4.15 with experimental tests of the flow through a wind tunnel contraction reported by Gibson (1972). In this application the settling chamber upstream of the nozzle contraction was in reality not circular but an 11 ft square section, although the nozzle itself was fully axisymmetric. In view of this static pressure tappings were located on both horizontal and diagonal sections of the nozzle revealing only slight differences in the static pressure distribution. The predicted surface pressure distribution using Program 4.4 was in reasonable agreement with experimental values.

Fairbairn (1976) argued that an ideal test of the accuracy of this model is its application to the flow through a cylindrical duct, Fig. 4.16, for which case the solution should be that of constant velocity $\gamma(s) = W$ as we have already seen. Assuming $W = 1$, $r_m = 1$ let us consider the rather crude representation of the duct length $x = 2.0$ modelled by five elements only of equal length $\Delta s = 0.4$. The solution given by Program 4.4 is then as shown in the table.

Element No.	x	r	$v_s = \gamma(s)$
1	0.2	1.0	1.018 413
2	0.6	1.0	1.017 256
3	1.0	1.0	1.017 335
4	1.4	1.0	1.017 256
5	1.8	1.0	1.018 412

On average the solution was 1.77% in error in line with Fairbairn's findings, Fig. 4.16. If the duct length is doubled by taking 10 elements of the same length $\Delta s = 0.4$, it is found that the error level increases to 2.5%. Fairbairn derived error curves for various duct lengths expressed non-dimensionally by X/r_m. As shown by Fig. 4.16 errors increase with duct length and also with element size $\Delta s/r_m$. Since X/r_m is in any case fixed by the problem specification, it may be necessary to select large numbers of elements to contain acceptable errors. For example, for a duct length of $X/r_m = 5.0$ errors of 0.5% are obtained if we use 50 elements of length $\Delta s/r_m = 0.1$, a very reasonable requirement.

More to the point is the underlying reason behind this, namely the problem of numerical leakage flux. Back-diagonal correction as undertaken for annular aerofoils (4.24) is not a possibility here, although it would be feasible to apply the same principle to the circulation around some suitable interior contour such as *abcd*, Figs. 4.16 and 4.14. In this case we could replace $K(s_n, s_n)$ with

$$K(s_n, s_n) \rightarrow -\frac{1}{\Delta s_n} \left\{ \sum_{\substack{n=1 \\ n \neq M}}^{N} K(s_m, s_n) \Delta s_m + \int_{abcd} \mathbf{q}_s \cdot \mathrm{d}s \right\} \qquad (4.35)$$

In principle this is a possibility although, as we have already

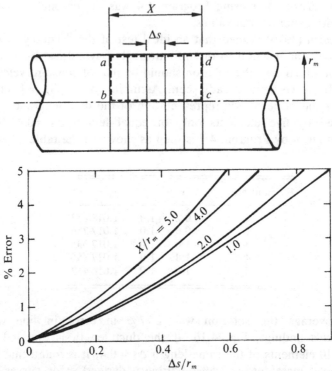

Fig. 4.16. Errors in numerical modelling of a cylindrical duct by ring surface vorticity elements.

established by error analysis, errors may be kept within acceptable bounds by judicious selection of dimensionless element lengths $\Delta s_m/r_m$ and element numbers M.

4.7 Flow through an annulus

The previous analysis was extended by Fairbairn (1976) to deal with turbomachine annuli by the introduction of an additional semi-infinite vortex tube to represent the hub section, Fig. 4.17. The related vortex sheet strengths then follow from a modified form of equations (4.32), namely

$$\left.\begin{aligned}
\Gamma_1 &= -W_1 \\
\Gamma_{2c} &= -W_2 = \left(\frac{r_1^2}{r_{2c}^2 - r_{2h}^2}\right)\Gamma_1 \\
\Gamma_{2h} &= -\Gamma_{2c}
\end{aligned}\right\} \tag{4.36}$$

174

If the contoured casing and hub sections are represented by M_c and M_h surface elements respectively, the equations take the form

M_c by M_c Mutual effect of casing elements on one another Block *A*	M_h by M_c Influence of hub elements on casing Block *B*
M_c by M_h Influence of casing elements on hub Block *C*	M_h by M_h Mutual effect of hub elements on one another Block *D*

$$\begin{pmatrix} \text{Casing } \gamma(s) \text{ values} \\ \text{Hub } \gamma(s) \text{ values} \end{pmatrix} = \begin{pmatrix} \text{Casing rhs values} \\ \text{Hub rhs values} \end{pmatrix}$$

$$(4.37)$$

The right hand side values at element m are now given by

$$\text{rhs} = -(U_{c1} + U_{c2} + U_{h2}) \cos \beta_m - (V_{c1} + V_{c2} + V_{h2}) \sin \beta_m \qquad (4.38)$$

where (U_{h2}, V_{h2}) are the additional velocities induced by the hub semi-infinite vortex tube and β_m is the profile slope at m. All coupling coefficients for $n \neq m$ are as previously given by (4.7) and (4.8). The self-inducing coupling coefficients for the casing are given by (4.33) and for the hub, requiring the internal rather than external boundary condition, by (4.22), previously derived for the body of revolution case. The only difference lies in the term $\pm\frac{1}{2}\gamma(s_m)$ which requires (+) for the casing and (−) for the hub.

A quite creditable prediction of the surface velocities for hub and casing was obtained by Fairbairn, Fig. 4.17, in comparison with experimental test, which was further improved by means of a boundary layer correction adding displacement thickness to the wall profile. In addition to these experiments Fairbairn undertook velocity traverses across the annulus at four stations to determine streamline locations. By means of a second integral equation similar to (4.28) using the $M_c + M_h$ known values of $\gamma(s)$ following solution of the system of equations given by equations (4.37), he was able to predict streamline locations. The outcome, shown also in Fig. 4.17, illustrates the good agreement obtained with experiment and the scope of this surface element method for obtaining detailed flow information within the main flow fields of turbomachine annuli.

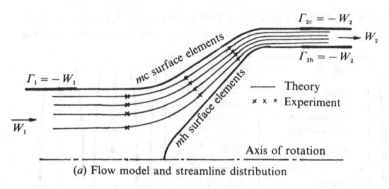

(a) Flow model and streamline distribution

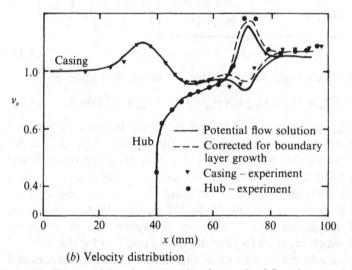

(b) Velocity distribution

Fig. 4.17. Flow model and computation for a mixed-flow fan annulus.

4.8 Source panel solutions for plane two-dimensional and axisymmetric flows

Source panels have been much more widely used than vortex panels for modelling three-dimensional flows because of the simpler formulation, as already described in Chapter 1. This is particularly the case for non-lifting bodies which can be modelled by source panels alone. Lifting bodies on the other hand require the introduction of bound vorticity also, either internal to the body profile or as an additional surface vortex panel distribution. In two-dimensional or axisymmetric flows the additional numerical effort demanded by the source panel method makes it unattractive, compared with the surface vorticity method. Thus, as explained in Section 1.7, not only are sub-elements required to gain acceptable accuracy, but a second integral must also be completed to derive the surface flow velocity once the panel source strengths are determined. The need to introduce additional bound vorticity for lifting body problems makes the source panel method even less attractive. Nevertheless it has been widely used and we shall conclude this chapter by applying source panel modelling to two-dimensional plane and axisymmetric flows. This will provide the reader with a basis for proceeding to three-dimensional flows for which source panel modelling is more easily adapted. We shall begin with the simpler situation of two-dimensional lifting aerofoils in order to experiment with various possible models of bound vorticity. Then we will extend the work to axisymmetric flows, with particular application to ducts.

4.8.1   Source panel modelling of lifting aerofoils

The source panel method was developed for plane two-dimensional non-lifting bodies in Section 1.7 resulting in the source boundary integral equation (1.36). For lifting bodies the numerical form of this (1.37) stating the Neumann boundary condition at element m, may be rewritten

$$\sum_{n=1}^{M} \bar{K}(s_m, s_n)\sigma(s_n) = U_\infty \sin \beta_m - V_\infty \cos \beta_m - \Gamma q_{n_m} \qquad (4.39)$$

where q_{n_m} is the normal outward velocity component at m induced by unit aerofoil bound vorticity. Giesing (1964) proposed such an extension of the source panel method to cascades of lifting aerofoils by the introduction of bound vorticity within the profile interior since there can be no net circulation about an aerofoil modelled by sources only. Three possible techniques of accomplishing this may be considered as illustrated in Fig. 4.18, namely,

(i) Introduction of a point vortex Γ somewhere within the body profile.
(ii) Introduction of a line vortex bound to the camber line.
(iii) A surface vorticity sheet distributed around the body surface but just inside the source panel sheet.

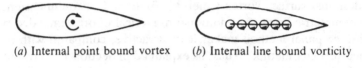

(a) Internal point bound vortex   (b) Internal line bound vorticity

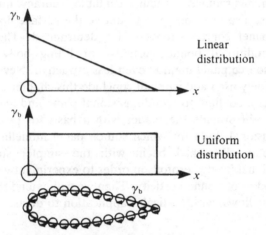

(c) Peripheral surface bound vorticity

Fig. 4.18. Alternative models for introducing bound vorticity into the source panel method.

To apply any of these models, the most economic numerical approach is to separate (4.39) into three unit equations by introducing the linear combination of sources

$$\sigma(s_n) = U_\infty \sigma_U(s_n) + V_\infty \sigma_V(s_m) + q_{n_m} \sigma_\Gamma(s_n) \qquad (4.40)$$

resulting in separate equations for unit velocities in the x and y directions and unit bound circulation

$$
\left.
\begin{aligned}
\sum_{n=1}^{M} \bar{K}(s_m, s_n)\sigma_U(s_n) &= \sin\beta_m \quad \text{unit } U_\infty \\
\sum_{n=1}^{M} \bar{K}(s_m, s_n)\sigma_V(s_n) &= -\cos\beta_m \quad \text{unit } V_\infty \\
\sum_{n=1}^{M} \bar{K}(s_m, s_n)\sigma_\Gamma(s_n) &= -q_{n_m} \quad \text{unit } \Gamma
\end{aligned}
\right\} \qquad (4.41)
$$

These equations have the same coupling coefficient matrix $|\bar{K}(s_m, s_n)|$, previously given by (1.38)–(1.41). Since the coupling coefficient prescribing the Neumann boundary condition for a unit source, $k(s_m, s_n)$, (1.36), is identical to that of a unit vortex prescribing the Dirichlet boundary condition (1.22), the leading diagonal terms are dominant. Solution by matrix inversion is straightforward and economic resulting in solutions for the unit source strengths.

For a unit vortex at (x_0, y_0) the outflow velocity normal to the profile at (x_m, y_m) is given by

$$q_{n_m} = \frac{1}{2\pi} \left\{ \frac{(x_m - x_0) \cos \beta_m + (y_m - y_0) \sin \beta_m}{(x_m - x_0)^2 + (y_m - y_0)^2} \right\} \tag{4.42}$$

This result can be adapted to models (ii) and (iii) with little difficulty making use of sub-elements to increase accuracy.

Following the strategy outlined in Section 1.9 the velocities parallel to the surface may then be obtained from the second set of integrals given by (1.42)–(1.44). For the three unit flows we then have for element m

$$
\left.
\begin{aligned}
v_{sUm} &= \sum_{\substack{n=1 \\ n \neq m}}^{M} \bar{L}(s_m, s_n)\sigma_U(s_n)\Delta s_n + \cos \beta_m \\[2mm]
v_{sVm} &= \sum_{\substack{n=1 \\ n \neq m}}^{M} \bar{L}(s_m, s_n)\sigma_V(s_n)\Delta s_n + \sin \beta_m \\[2mm]
v_{s\Gamma m} &= \sum_{\substack{n=1 \\ n \neq m}}^{M} \bar{L}(s_m, s_n)\sigma_\Gamma(s_n)\Delta s_n + q_{s\Gamma m}
\end{aligned}
\right\} \tag{4.43}
$$

where the coupling coefficient $\bar{L}(s_m, s_n)$ is given by (1.44). The last terms on the right hand sides represent the velocity parallel to the surface at element m due to the three disturbance flows $U_\infty = 1$, $V_\infty = 1$ and $\Gamma = 1$. The first two follow directly and the third one, q_{s_m}, needs special consideration for each type of bound vorticity model. We will deal with this shortly but first consider the trailing edge Kutta condition. Combining the solutions we have finally the surface velocity at element m

$$v_{sm} = U_\infty v_{sUm} + V_\infty v_{sVm} + \Gamma v_{s\Gamma m} \tag{4.44}$$

The most appropriate statement of the Kutta condition in this situation is method 2 described in Section 2.4.2. To unload the

179

trailing edge we may write for the adjacent element numbers te and
(te + 1),

$$v_{ste} = -v_{s(te+1)}$$

from which the required bound vortex strength follows directly,
namely

$$\Gamma = -\frac{U_\infty(v_{sUte} + v_{sU(te+1)}) + V_\infty(v_{sVte} + v_{sV(te+1)})}{v_{s\Gamma te} + v_{s\Gamma(te+1)}} \qquad (4.45)$$

The panel solution is now completely formulated apart from the
expressions for q_{s_m} which will now be derived for the three bound
vortex models which we have proposed.

Model (i) Internal point vortex
If a unit clockwise bound vortex is located at some point (x_0, y_0)
within the aerofoil profile, then the induced velocities at (x_m, y_m)
are given by

$$u_m = \frac{y_m - y_0}{2\pi r^2}, \qquad v_m = -\frac{x_m - x_0}{2\pi r^2} \qquad (4.46)$$

where $r = \sqrt{[(x_m - x_0)^2 + (y_m - y_0)^2]}$.
The velocity parallel to the surface is then

$$q_{s\Gamma m} = u_m \cos \beta_m + v_m \sin \beta_m \qquad (4.47)$$

As a test case the above method has been applied with 48
elements, to Joukowski profile number 2 of Fig. 2.10 for which the
exact solution has been derived in Section 2.5.3. With 10° angle of
attack the exact solution is first compared in Fig. 4.19(a) with the
surface vorticity method, which provided an extremely close prediction. The source panel method also produced an excellent prediction of surface pressure distribution in the leading and trailing edge
regions but was subject to considerable errors in the neighbourhood
of the bound vortex. To some extent these errors may be alleviated
by concentrating pivotal points in the neighbourhood of the vortex
but they do seem to represent an inherent error in the model. For
example, we can observe an inconsistency if we consider the
question of the centre of lift. The surface source elements make no
individual contributions to lift consequently the whole of the lift
force must apply at the location of the point vortex. From the exact
surface pressure distribution this is likely to be close to the quarter
chord position, Fig. 4.19(a) rather than the half chord location
chosen for Γ, Fig. 4.19(b). There is clearly an indeterminacy here

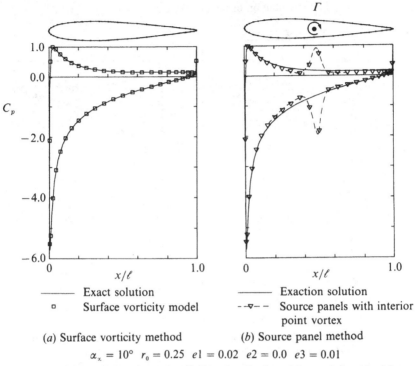

Fig. 4.19. Comparison of exact solution for Joukowski aerofoil with (a) Surface vorticity analysis. (b) Source panel model with interior point bound vortex.

imposed by the chosen model. However, a shift of Γ to the quarter chord position did not eliminate this error in predicted surface pressure which would seem to be related to the need to distribute the bound vorticity more correctly within the profile envelop.

Model (ii) Internal line distribution

The results shown in Fig. 4.20(a) were calculated with a uniform distribution of vorticity along the camber line between $0.25 < x/\ell < 0.75$. Since in practice this can be modelled by means of an array of point vortex sub-elements, (4.46) and (4.47) will suffice. As may be observed, considerable improvement was obtained by this method although there were still significant errors to be observed in the regions close to the ends of the bound vortex line. Over the region of the constant strength line vortex there was a tendency, as might be expected, for the lift per unit length to approach a constant value

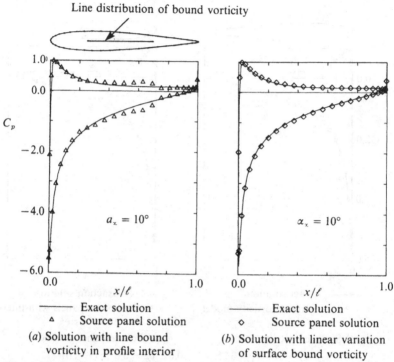

Line distribution of bound vorticity

——— Exact solution
△ Source panel solution

(a) Solution with line bound
vorticity in profile interior

——— Exact solution
◇ Source panel solution

(b) Solution with linear variation
of surface bound vorticity

Fig. 4.20. Comparison of source panel solutions for a Joukowski aerofoil with interior and surface bound vorticity to generate lift.

consistent with the local lift force per unit length on the enclosed vortex.

Model (iii) Surface distribution of bound vorticity

Fig. 4.20(b) illustrates the excellent predictions obtained using a surface distribution of bound vorticity $\gamma_b(s)$ of prescribed linear variation to produce zero loading approaching the trailing edge, namely

$$\gamma_b(s) = \gamma_0(1 - x/\ell) \tag{4.48}$$

For unity total bound vorticity, required to suit the above equations, $\gamma_b(s)$ may be integrated around the profile perimeter to determine γ_0. We then have finally

$$\gamma_b(s_m) = \frac{1 - x_m/\ell}{\sum\limits_{n=1}^{M} (1 - x_n/\ell)\,\Delta s_n} \tag{4.48a}$$

182

Table 4.3. *Prediction of bound vortex Γ and lift coefficient C_L*

Method	Description	Γ	C_L
(i)	Source panels with point vortex	0.604 320	1.208 641
(ii)	Source panels with line vortex	0.608 455	1.216 911
(iii)	Source panels constant peripheral vorticity	0.634 770	1.269 539
(iii)	Source panels linear peripheral vorticity	0.615 945	1.231 889
	Surface vorticity method	0.603 323	1.206 646
	Exact solution	0.610 996	1.211 430

Now, as already pointed out, the source coupling coefficient $\bar{K}(s_m, s_n)$ also represents the parallel velocity just outside element m due to a unit vortex at n. For this model the surface velocity q_{s_m} due to the unit peripheral bound vortex sheet may thus be obtained directly from the coupling coefficient matrix through

$$q_{s_m} = \sum_{n=1}^{M} \gamma_b(s_m)\bar{K}(s_m, s_n) \qquad (4.49)$$

Since sub-elements would normally be used in deriving $\bar{K}(s_m, s_n)$, back-diagonal correction is not required to reduce leakage flux, according to the studies undertaken in Section 2.3.

Although problems seem to arise in the prediction of surface velocity and pressure depending upon one's choice of bound vorticity distribution with the panel method, the estimate of total bound vortex strength Γ delivered by (4.45) is always good as indicated by the table of results.

Viewed overall the surface vorticity method produced the best prediction of both surface pressure distribution and lift coefficient, offering also enormously simpler computational requirements. Further refinements to the source panel model can lead to equally satisfactory results as illustrated by its extension to lifting cascades by Giesing (1964). Fig. 4.21 illustrates excellent results obtained for a cascade for which exact solutions were also available.

4.9 Source panel method for axisymmetric flows

The analysis for source panels in plane flow may be extended with relative ease to axisymmetric flows by taking advantage of the

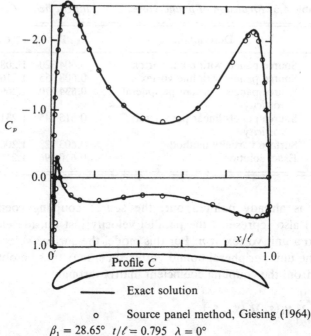

C_p

x/ℓ

Profile *C*

———— Exact solution

o   Source panel method, Giesing (1964)

$\beta_1 = 28.65°$ $t/\ell = 0.795$ $\lambda = 0°$

Fig. 4.21. Comparison of source panel method with an exact solution for a cascade, Giesing (1964).

expressions for ring source induced velocities given by Kuchemann & Weber (1953) and by Ryall & Collins (1967). Adopting cylindrical polar coordinates, the velocity components in the (x, r) directions at *m* due to a ring source of unit strength at *n* are given by

$$\left.\begin{array}{l} u\sigma_{mn} = \dfrac{1}{2\pi r_n \sqrt{[x^2 + (1+r)^2]}} \left\{ \dfrac{2xE(k)}{x^2 + (r-1)^2} \right\} \\[3mm] v\sigma_{mn} = \dfrac{1}{2\pi r_n \sqrt{[x^2 + (1+r)^2]}} \left\{ K(k) - \left[1 - \dfrac{2r(r-1)}{x^2 + (r-1)^2} \right] E(k) \right\} \end{array}\right\}$$

(4.50)

where the non-dimensional coordinates (x, r) are defined

$$x = \frac{x_m - x_n}{r_m}, \qquad r = \frac{r_m}{r_n} \tag{4.51}$$

and the elliptic integrals $K(k)$ and $E(k)$ are defined by (4.11). Let us first apply these to the simpler problem of non-lifting flow past a body of revolution, then consider the flow past engine cowls.

184

4.9.1   Source panel method for a body of revolution

Following the procedure laid down in Section 1.9 the governing integral equation expressing the Neumann boundary condition at point m on an axisymmetric body is similar to (1.39) for plane flow but with axial velocity U_∞ only, namely

$$\tfrac{1}{2}\sigma(s_m) + \oint \bar{k}(s_m, s_n)\sigma(s_n)\, \mathrm{d}s_n = U_\infty \sin \beta_m \qquad (4.52)$$

In numerical form this may be written, for representation by M source panels,

$$\sum_{n=1}^{M} \bar{K}(s_m, s_n)\sigma(s_n) = U_\infty \sin \beta_m \qquad (4.53)$$

where $\bar{K}(s_m, s_n)$ is the average coupling coefficient, expressing the induced normal velocity at n due to a unit strength source element at m, namely

$$\bar{K}(s_m, s_n) = \frac{1}{N}\sum_{i=i}^{N} K(s_m, s_i)$$

$$= \frac{1}{N}\sum_{i=1}^{N} \{-u\sigma_{mi}\sin \beta_m + v\sigma_{mi}\cos \beta_m\}\, \Delta s_n \qquad (4.54)$$

The average for the disturbing source panel at n has been evaluated here by introducing N sub-elements with centre locations

$$\left. \begin{aligned} x_i &= x_n + \{i - \tfrac{1}{2}(1 + N)\}\Delta s_n \cos \beta_n / N \\ y_i &= y_n + \{i - \tfrac{1}{2}(1 + N)\}\Delta s_n \sin \beta_n / N \end{aligned} \right\} \qquad (4.55)$$

For the special case of the self-inducing coupling coefficient, $n = m$, we must absorb the $\tfrac{1}{2}\sigma(s_m)$ term of (4.52), resulting in

$$\bar{K}(s_m, s_m) = \frac{1}{N}\sum_{i=1}^{N} \{-u\sigma_{mi}\sin \beta_m + v\sigma_{mi}\cos \beta_m\}\, \Delta s_m + \tfrac{1}{2} \qquad (4.54a)$$

For plane flows the first term, because of the symmetrical influence of the sub-elements, vanishes. For axisymmetric source panel modelling on the other hand, this is not the case and (4.52a) should be used.

Having obtained the source strengths $\sigma(s_n)$ required to satisfy the Neumann boundary condition by solution of equations (4.51), the surface velocity parallel to the surface follows from the second

185

Bodies of revolution, ducts and annuli

Table 4.4. *Flow past a sphere by the source panel method with 20 elements*

Element number	Exact solution v_s/U_∞	Surface vorticity	Source panels
1, 20	0.117 689	0.116 773	0.107 092
2, 19	0.350 168	0.348 684	0.344 796
3, 18	0.574 025	0.574 036	0.567 515
4, 17	0.781 748	0.781 598	0.776 752
5, 16	0.974 172	0.973 147	0.965 419
6, 15	0.140 609	0.137 980	1.130 830
7, 14	1.278 960	1.275 029	1.267 699
8, 13	1.385 819	1.382 178	1.373 222
9, 12	1.485 555	1.453 314	1.445 582
10, 11	1.495 376	1.492 219	1.482 193

direct integral equation analogous to (1.42), namely

$$v_{sm} = \oint \bar{\ell}(s_m, s_n)\sigma(s_n)\, ds_n + U_\infty \cos \beta_m \qquad (4.56)$$

or in numerical form

$$v_{sm} = \sum_{n=1}^{M} \bar{L}(s_m, s_n)\sigma(s_n) + U_\infty \cos \beta_m \qquad (4.56a)$$

where, once again, $\bar{L}(s_m, s_n)$ is the average velocity at m parallel to the body surface, induced by a unit strength source sheet covering element n. With N sub-elements this becomes

$$\bar{L}(s_m, s_n) = \frac{1}{N}\sum_{i=1}^{N} \{u\sigma_{mi} \cos \beta_m + v\sigma_{mi} \sin \beta_m\}\Delta s_n \qquad (4.57)$$

Program No. 4.5, named dnaxisym.pas completes this analysis using the elliptic integral table look-up procedure. Results are compared in Table 4.4 with the exact solution and with the surface vorticity method, Program No. 4.2, for a twenty-element representation. Twenty sub-elements were used for the source panel calculations. Although not as accurate as the surface vorticity analysis, the source panel method was accurate to about 1% compared with the exact solution (4.23).

4.9.2 Source panel method for an annular aerofoil or engine cowl

The annular aerofoil differs from the body of revolution in two important respects so far as flow modelling is concerned. Firstly practical cowls are usually thin, giving rise to strong interference effects between opposite surface elements and consequently the possibility of a dominant back diagonal of the coupling coefficient matrix. Secondly, such foils generate a radial lift force, requiring specification of a trailing edge Kutta–Joukowski condition. Let us now deal with these two problems.

When developing the surface vorticity method for lifting aerofoils in plane two-dimensional flow, the back-diagonal correction technique was introduced in Section 2.3.3 to handle the problem of the dominating influence of the opposite surface element. This was based upon the principle that the net circulation around the profile perimeter due to any surface vorticity element $\gamma(s_n)\Delta s_n$ must be zero. In the surface vorticity method it should be observed that the vorticity sheet lies just outside the surface boundary. If we now apply the analogous principle to the source panel model, for which the source sheet lies just inside the profile, as illustrated by Fig. 4.22, we can state the divergence theorem, namely

$$\left\{\begin{matrix}\text{The flux crossing the surface} \\ \text{due to the source ring element at } m\end{matrix}\right\} = \left\{\begin{matrix}\text{strength of source} \\ \text{element } m\end{matrix}\right\}$$

If we apply this to the case of a unit strength source sheet at s_m, $\gamma(s_m) = 1.0$

$$2\pi \sum_{n=1}^{M} r_n \bar{K}(s_n, s_m)\,\Delta s_n = \sigma(s_m)2\pi r_m\,\Delta s_m = 2\pi r_m\,\Delta s_m$$

To enforce the condition of zero 'numerical' leakage flux, we may therefore replace the opposite point coupling coefficient by the revised estimate dictated by this equation.

$$\bar{K}(s_{\text{opp}}, s_m) = -\frac{1}{r_{\text{opp}}\,\Delta s_{\text{opp}}}\left\{\sum_{\substack{n=1 \\ n\neq\text{opp}}}^{M} \bar{K}(s_n, s_m)r_n\,\Delta s_n - r_m\,\Delta s_m\right\} \quad (4.58)$$

where element opposite to m is $\text{opp} = M - m + 1$. The application of this strategy has been found to produce improvements although the use of sub-elements already tends to reduce leakage flux errors and was indeed proposed in Section 2.3 as an alternative to back-diagonal correction.

187

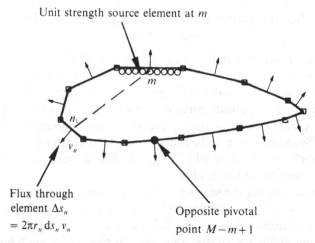

Fig. 4.22. Efflux from profile due to unit strength ring source element at m, for back-diagonal correction of an annular aerofoil modelled by ring source panels.

The second matter mentioned above which remains to be dealt with is that of the introduction of bound vorticity to account for the radial lift force. Taking advantage of the previous study of plane aerofoils in Section 4.8, we will select Method (iii) and distribute a surface vorticity sheet around the body contour just inside the surface source sheet. Two approaches to this were proposed in Fig. 4.18(c) namely (a) prescription of a linear shape function for the bound surface vorticity and alternatively (b) prescription of a constant value. The strategy for defining unit solutions and application of the Kutta condition is then identical to that explained for two-dimensional aerofoils in Section 4.8.1, (4.43)–(4.45). The only substantial difference lies in the evaluation of $q_{s\Gamma m}$ in (4.44), the velocity parallel to the annular aerofoil surface due to the bound vortex sheet. This is given by

$$q_{s\Gamma m} = \Gamma \sum_{n=1}^{M} K(s_m, s_n)\gamma_{\mathrm{b}}(s_n) \tag{4.59}$$

where $K(s_m, s_n)$ is the ring vortex coupling coefficient used in axisymmetric ring vortex analysis (4.7) with the exception that the self-inducing coupling coefficient $K(s_m, s_m) = 0.5$. $\gamma_{\mathrm{b}}(s_n)$ is the unit bound vorticity sheet local strength and Γ the scaling factor derived from application of the Kutta Joukowski condition (4.45). It is worth noting that some economy is possible if $q_{s\Gamma m}$ is evaluated

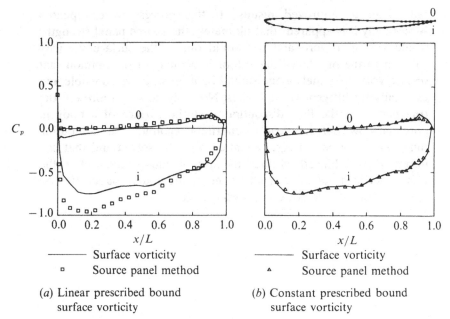

Fig. 4.23. Comparison of predicted surface pressure coefficient for A.R.L. Duct B3 by the surface vorticity and source panel methods.

during the procedure to calculate the source coupling coefficients since $K(k)$ and $E(k)$ are common to both the source and vorticity coupling coefficients. In this case (4.59) benefits from the use of sub-elements.

The author's program dnduct.pas has options for either linear or constant unit bound vortex loading. A comparison of these methods with surface vorticity theory is shown in Fig. 4.23 for A.R.L. duct *B*3 previously referred to in Section 4.4. This is a particularly interesting case since the *B*3 aerofoil was designed to achieve almost constant loading. It is perhaps not surprising therefore that option (*b*), namely $\gamma_b(s) = $ constant, gave a better prediction. Results were indeed in close agreement with surface vorticity theory. On the other hand the reader will observe that the linear bound vorticity shape function led to excessive load predictions in the leading edge region. In fact it is surprising that the erroneous prescribed loading distribution of option (*a*) should lead to results of any degree of credibility whatsoever. It is also clear that a second iteration employing the predicted surface velocity v_s to give a new estimate of the prescribed bound vorticity $\gamma_b(s) = v_s$ would undoubtedly lead to improved results. If this process were repeated successively it is apparent that ultimately the source panel strengths would vanish to zero and we would obtain the surface vorticity solution to the problem. The author is strongly of the opinion that surface vorticity panels alone should be used wherever possible and especially for lift-generating bodies. Not only do they represent the true nature of the flow discontinuity at the surface of a body in potential flow, but they lead to enormous economies in computing time since only one integral equation must be solved and that can normally be achieved without the use of sub-elements. For the axisymmetric problems considered here, the source panel method offers no identifiable advantages whatsoever.

CHAPTER 5

Ducted propellers and fans

5.1 Introduction

A range of flow computational techniques has been developed over many years to meet the design and analysis requirements of a wide range of rotodynamic machines, some of which were illustrated by Fig. 4.1. For dealing with turbomachine meridional or through flows, which are usually completely confined within a continuous duct annulus such as that of the mixed-flow fan depicted in Fig. 4.1(*a*), surface vorticity or panel methods have been proved less attractive in competition with grid based analyses such as the matrix through-flow method of Wu (1952) and Marsh (1966) and the more recent time marching analyses such as those of Denton (1974), (1982). Although the annulus boundary shape exercises important control over the flow through the blade regions, in all turbo-machines complex fluid dynamic processes occur throughout the whole flow field due to interactions between the S-1 and S-2 flows which were referred to in Chapter 3. Boundary integral methods based solely upon potential flow equations such as we have considered so far obviously cannot handle these interactions between the blade-to-blade and meridional flows, which involve detailed field calculations and spatial variations of properties best dealt with by the introduction of a grid strategically distributed throughout the annulus. Some attempts to achieve this with extended vortex boundary integral analysis will be outlined in Chapter 6, but generally speaking channel grid methods such as those referred to above have proved more fruitful to date for turbomachinery meridional analysis. In the case of ducted propellers on the other hand, not only are the duct and hub of finite length, but the duct in particular is a most important part of the propulsive system. For example the duct of a super-tanker Kort nozzle propulsor may contribute as much as 30% of the total thrust due to its interaction with both the free stream and the propeller. The duct is in effect an annular aerofoil whose interaction with the free stream generates important propulsive forces, while simultaneously providing the desired internal flow conditions in which the propeller system is to operate. The aforementioned channel matrix type methods suit this situation rather badly, whereas the surface vorticity technique has made considerable contributions towards design and analysis procedures. Source panel methods likewise are well suited to deal with this situation which combines the features of both internal and external aerodynamics.

The aim of this chapter is to introduce some of these design and analysis procedures for ducted propellers or fans, the groundwork for which was laid down in Chapter 4. We will begin by considering the 'pipe-flow rig' or 'sucked duct', a device sometimes used for developing engine intake cowls or for wind tunnel testing their behaviour in a given open stream with varying engine swallowing capacity. Next a model will be proposed to examine the interaction between two axisymmetric bodies, namely the propeller hub and duct. These items will then be assembled to form an analysis program for free-vortex ducted propellers for which the meridional flow is in fact irrotational permitting the use of potential flow modelling. For this case the propeller blade circulation is assumed to be shed in its entirety at the blade tips and is represented by a semi-infinite vortex tube. Following on from this we shall consider the non-free vortex ducted propulsor, involving first the use of lifting surface theory to model blade vortex shedding, then the meridional flow equations for its axisymmetric equivalent approximation. Fortunately this work has been backed up by careful experimental validation work to which some reference will be made.

5.2 The sucked duct or pipe flow engine intake facility

Ducted propellers or fan engines are required to operate for a range of ratios of swallowing velocity V_j to forward flight velocity W. One technique which has been adopted to check engine intake designs at an early stage, Gibson (1972), Young (1969), is the 'sucked duct' rig illustrated in Fig. 5.1. The intake cowl is mounted onto a downstream duct containing an axial fan which augments the swallowing capacity of the duct. The entire rig is located in a wind tunnel section providing an adjustable mainstream velocity W. This device

The sucked duct or pipe flow engine intake facility

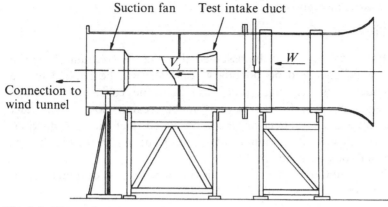

Fig. 5.1. The sucked duct or intake test facility.

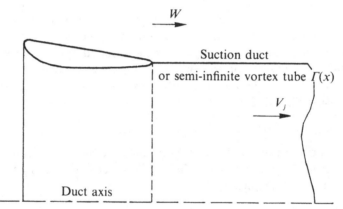

Fig. 5.2. Modelling of the Van Manen 19A Kort nozzle duct with downstream suction.

provides information about the influence of V_j/W upon the pressure distribution which is of special importance in the region of the inlet lip where the flow divides between the interior and exterior regimes.

The analyses for the duct and semi-infinite vortex cylinder outlined in Sections 4.4 and 4.5 of the previous chapter may be adapted to deal with this situation with minimal modifications as illustrated in Fig. 5.2. If we model the suction duct by means of a semi-infinite vortex cylinder located downstream of the duct trailing edge, then the appropriate duct vorticity strength a long way downstream is given by

$$\Gamma(x) = W - V_j \qquad (5.1)$$

For simplicity it will be assumed that $\Gamma(x)$ is constant along the entire cylinder and no boundary condition will be applied to the cylinder wall. Although there will be some consequent leakage through the cylinder, particularly in the region close to the duct trailing edge, the effect upon the predicted duct pressure distribution is found to be minimal in the important regions of the duct as we shall demonstrate.

The system of equations for this problem may be developed from (4.15) and (4.38) as follows:

$$\sum_{n=1}^{M} \bar{K}(s_m, s_n)\gamma(s_n) \, \Delta s_n = -W \cos \beta_m - U_c \cos \beta_m - V_c \sin \beta_m \qquad (5.2)$$

where U_c, V_c are the velocity components at m induced by the cylindrical vortex tube, and are given by equations (4.26). One other essential modification *vis-à-vis* the isolated duct is the Kutta condition statement for the duct trailing edge loading. For continuous loading at the junction with the semi-infinite vortex cylinder we must modify (2.22) to read

$$\gamma(s_{te+1}) = \Gamma - \gamma(s_{te}) \qquad (5.3)$$

The introduction of this restriction into the governing equations (5.2) results in an additional right hand side term accounting for the semi-infinite vortex cylinder.

$$\sum_{\substack{n=1 \\ n \neq te+1}}^{M} \bar{K}(s_m, s_n)\gamma(s_n) \, \Delta s_n = -W \cos \beta_m$$

$$- \Gamma(\bar{U}_c \cos \beta_m + \bar{V}_c \sin \beta_m + \bar{K}(s_m, s_{te+1}) \, \Delta s_{te+1}) \qquad (5.4)$$

where $\bar{U}_c$ and $\bar{V}_c$ are the velocity components induced by a semi-infinite vortex cylinder of unit strength.

Implied in the left hand side of (5.4) is the combination of columns te and te + 1 with the modified coupling coefficient.

$$\bar{K}(s_m, s_{te}) \, \Delta s_{te} := \bar{K}(s_m, s_{te}) \, \Delta s_{te} - \bar{K}(s_m, s_{te+1}) \qquad (5.5)$$

To restore the matrix to square, row te + 1 is subtracted from row te by analogy with the duct analysis of Section 4.4, so that

$$\bar{K}(s_{te}, s_n) \, \Delta s_n := \bar{K}(s_{te}, s_n) \, \Delta s_n - \bar{K}(s_{te+1}, s_n) \, \Delta s_n \qquad (5.6)$$

The sucked duct or pipe flow engine intake facility

One further simplification which speeds repeating computations for reselected W and V_j values, results from expressing $\gamma(s)$ in terms of units solutions

$$\gamma(s) = W\gamma'(s) + \Gamma\gamma''(s) \tag{5.7}$$

satisfying the separate equations independent of W and Γ, namely

$$
\left.
\begin{aligned}
&\sum_{n=1}^{M} \bar{K}(s_m, s_n)\gamma'(s_n)\,\Delta s_n = -\cos\beta_m \\
&\sum_{n=1}^{M} \bar{K}(s_m, s_n)\gamma''(s_n)\,\Delta s_n = -\bar{U}_c \cos\beta_m - \bar{V}_c \sin\beta_m \\
&\qquad\qquad\qquad\qquad - \bar{K}(s_m, s_{\text{te}+1})\,\Delta s_{\text{te}+1}
\end{aligned}
\right\} \tag{5.8}
$$

Computer program suckduct, included in the Appendix as Program 5.1, accomplishes this task for specified values of V_j/W. Output is shown in Fig. 5.3 for the Van Manen 19A duct, Van Manen & Oosterveld (1966), with $V_j/W = 1.5$, 2.0 and 2.5. This

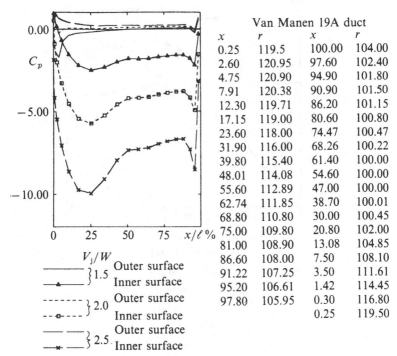

Van Manen 19A duct			
x	r	x	r
0.25	119.5	100.00	104.00
2.60	120.95	97.60	102.40
4.75	120.90	94.90	101.80
7.91	120.38	90.90	101.50
12.30	119.71	86.20	101.15
17.15	119.00	80.60	100.80
23.60	118.00	74.47	100.47
31.90	116.00	68.26	100.22
39.80	115.40	61.40	100.00
48.01	114.08	54.60	100.00
55.60	112.89	47.00	100.00
62.74	111.85	38.70	100.01
68.80	110.80	30.00	100.45
75.00	109.80	20.80	102.00
81.00	108.90	13.08	104.85
86.60	108.00	7.50	108.10
91.22	107.25	3.50	111.61
95.20	106.61	1.42	114.45
97.80	105.95	0.30	116.80
		0.25	119.50

V_j/W

———— $\Big\}$ 1.5 Outer surface
—▲—— Inner surface

- - - - - $\Big\}$ 2.0 Outer surface
- -□- - - Inner surface

— — — $\Big\}$ 2.5 Outer surface
—✶— — Inner surface

Fig. 5.3. Predicted surface pressure distributions for Van Manen 19A duct with downstream suction.

195

duct is typical of ducted propeller Kort nozzles and produces accelerating internal entry flow up to 50% of the duct chord L followed by slight diffusion leading into the suction duct. The surface pressure coefficient is defined here by

$$C_p = \frac{p - p_\infty}{\frac{1}{2}\rho W^2} \tag{5.9}$$

and the tabulated coordinates for the duct are given in Fig. 5.3.

Optimum inflow may be defined as the condition for which the stagnation point is attached to the leading edge of the duct entry lip. From this study a value of $V_j/W = 2.0$ was found to produce

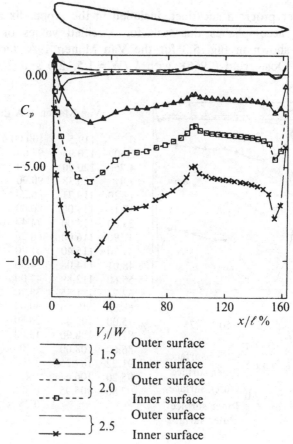

Fig. 5.4. Predicted surface pressure distribution for extended 19A duct with downstream suction.

optimum inflow, at which flow rate the pressure distribution on the whole of the duct outer surface was close to the local ambient pressure p_∞. This is a fair criterion for a well designed intake or ducted propeller cowl.

Bearing in mind that we have applied no boundary condition on the surface of the semi-infinite vortex suction tube, one might expect some errors due to flow leakage, particularly in the trailing edge region of the cowl, especially for an inlet duct of such low aspect ratio, $L/D = 0.5$. As a check upon this possibility results are shown in Fig. 5.4 for an extended 19A duct. The predicted pressure distributions are almost identical over the important region $x/L <$ 80%, indicating the suitability of this flow model for the design or analysis of inlet ducts, even of short length.

By means of the test rig illustrated in Fig. 5.1, Gibson (1972) investigated the B3 duct already referred to in Chapter 4, but in this

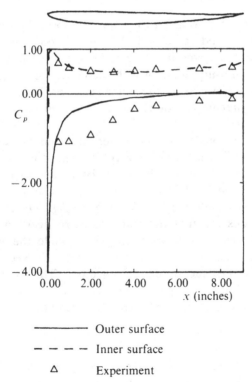

Outer surface

— — — — Inner surface

△   Experiment

Fig. 5.5. Pressure distribution on B3 duct with downstream suction for V_j/W of 0.553.

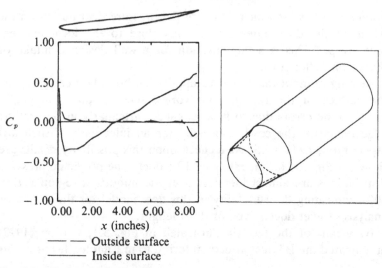

x (inches)

    – – – Outside surface

    ——— Inside surface

Fig. 5.6. B3 duct with 7° diffuser half angle and downstream suction.

case with the fan free running to generate a loss within the suction duct. For a measured value of $V_j/W = 0.553$ comparison with this theoretical model is given in Fig. 5.5. The duct is badly adapted to deal with this diffusing situation with the consequence that a major flow separation occurred on the outer surface. Despite this disturbance to the outer flow, a quite reasonable prediction of the inner flow was obtained.

A correctly adapted duct for either diffusion $(V_j < W)$ or acceleration $(V_j > W)$ should lie on a conical chord line surface as in the previous example of the 19A duct. By numerical experimentation a cone half angle of 7° was found to produce optimum inflow for the B3 duct profile with $V_j/W = 0.553$, Fig. 5.6. For this case also we may observe that the static pressure distribution on the outer surface was close to p_∞ throughout, while the inner static pressure decreased to -0.4 at the throat followed by steady diffusion to the final suction duct C_p of 0.694. This example serves to illustrate both the principles involved in intake selection and the usefulness of the sucked duct analytical techniques as an aid to design.

5.3 Free vortex ducted propeller

The foregoing analysis may now be extended quite easily to deal with the case of the free vortex Kort nozzle ducted propeller. In the

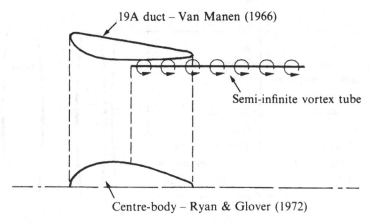

Free vortex ducted propeller

19A duct – Van Manen (1966)

Semi-infinite vortex tube

Centre-body – Ryan & Glover (1972)

Fig. 5.7. Free vortex ducted propeller model applied to N.S.M.B. 19A duct with centre-body.

first place we must introduce an additional centre body of revolution to represent the propeller boss. Secondly, a semi-infinite vortex tube will be introduced to represent the effect of the propeller tip vortex shedding. Later we shall refer to the non-free vortex model developed by Ryan & Glover (1972) which permits traditional propeller lifting surface models to be combined with axisymmetric duct/hub theory. For the moment the analysis will be restricted to that developed by Gibson (1972), Gibson & Lewis (1973), who applied axisymmetric actuator disc theory to a free vortex propeller for which there is no vorticity shedding except from the blade tips. Since the typical Kort nozzle has roughly cylindrical flow in the jet downstream of the propeller, the use of a single semi-infinite vortex tube emanating from the blade tips, has proved to be an excellent model for this type of ducted propeller.

The vorticity model for this situation is illustrated in Fig. 5.7 and the governing equation is the same as (5.2) for the sucked duct problem with the essential difference that the normal aerofoil trailing edge Kutta condition now applies. Equation (5.3) is thus replaced by

$$\gamma(s_{te+1}) = -\gamma(s_{te}) \tag{5.10}$$

Since two bodies are to be represented, the coupling coefficient matrix must be partitioned as follows.

199

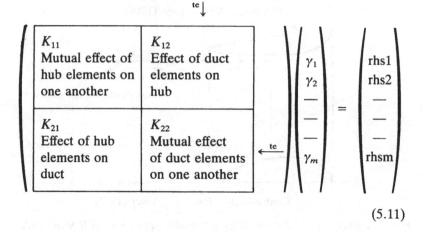

$$(5.11)$$

As recommended for annular aerofoil ducts in Section 4.4, (2.19), back-diagonal correction of the elements in sub-matrix K_{22} is required. Following this procedure, application of the trailing edge Kutta condition expressed by (5.10) results in elimination of one column due to the combination of columns te and te + 1. To return to a square matrix, row te + 1 must be subtracted from row te. The effect of this procedure upon the partitioned matrices K_{12} and K_{22} is fully acceptable, since these represent the influence of the duct vortex elements upon all body points and the influences of $\gamma(s_{te})$ and $\gamma(s_{te+1})$ are now correctly merged. The effect of the procedure upon partitioned matrix K_{21} is perhaps less easy to interpret physically since we appear to have involved unnecessarily the influence of the hub elements upon the duct trailing edge. In the event this produces no harmful side effects upon the solution.

Gibson (1972), Gibson & Lewis (1973) completed experimental investigations of the Ka 4-55 Kaplan blade type propeller combined with the Van Manen 19A duct yielding surface pressure distributions and duct thrusts. More recently Balabaskaran (1982), Lewis & Balabaskaran (1983) have undertaken a wider range of wind tunnel investigation of the same ducted propeller for which sample results are shown in Fig. 5.8 of predicted and measured surface pressure coefficients as defined by (5.9). Agreement with the surface vorticity model is excellent over the wide range of thrust coefficients considered.

In order to relate the vorticity strength Γ of the vortex wake to the thrust of the device it is necessary to undertake one-dimensional momentum and energy balances. Let us first adopt the dimensionless thrust coefficient usually defined by

$$C_T = \frac{T}{\frac{1}{2}\rho V_a^2 \frac{\pi}{4} D^2} \qquad (5.12)$$

where D is the propeller diameter, T is the total thrust of both duct and propeller and $V_a (= -W)$ is the advance velocity of the propulsor in stationary water. The duct and propeller thrust are related through the thrust ratio τ defined

$$\tau = \frac{\text{Propeller thrust}}{\text{Total thrust}} \qquad (5.13)$$

In addition, the advance coefficient J links forward velocity V_a to propeller revolutions per s, n, through

$$J = \frac{V_a}{nD} \qquad (5.14)$$

If $p_2 - p_1$ is the propeller static pressure rise at the r.m.s. radius, the propeller thrust τT may be approximated by

$$\left. \begin{aligned} \tau T &= (p_2 - p_1)\frac{\pi D^2}{4}(1 - h^2) \\ &= \frac{\rho \pi D^2}{8}(1 - h^2)(V_j^2 - V_a^2) \end{aligned} \right\} \qquad (5.15)$$

where the swirl velocity V_θ has been neglected. Referring this result to (5.12) we have finally an expression which relates Γ to τ, C_T and the propeller hub/tip ratio h.

$$\frac{\Gamma}{V_a} = 1 - \frac{V_j}{V_a} = 1 - \sqrt{\left(1 + \frac{\tau C_T}{1 - h^2}\right)} \qquad (5.16)$$

It is thus possible to prescribe the dimensionless jet wake vorticity Γ/V_a for any given operating condition, provided the propeller thrust coefficient τC_T is known. The measured characteristics for propeller Ka 4-55 with duct 19A in both open water and wind tunnel tests are shown in Fig. 5.9 for a very wide range of operation, from which it can be observed that both τ and C_T are functions of the advance ratio J. By introduction of the cascade analysis of Chapter 2 it is possible, as shown by Balabaskaran

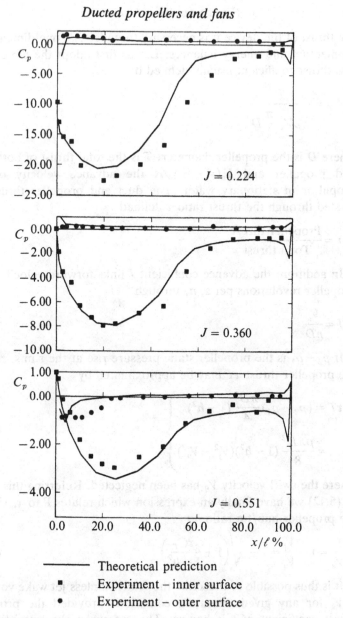

Fig. 5.8. Surface pressure distribution for 19A with Ka 4-55 N.S.M.B. propeller.

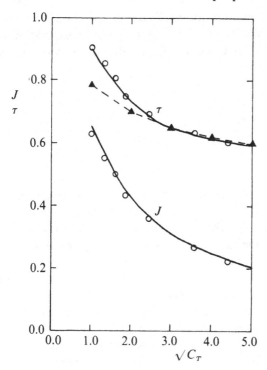

N.S.M.B. open water tests – Van Manen (1966)

○ Wind tunnel tests – Balabaskaran (1982), (1983)

– ▲ – – Predicted thrust ratio

Fig. 5.9. Predicted performance characteristics for 19A duct with Ka 4-55 propeller compared with open water and wind tunnel tests.

(1982), to obtain the $\tau C_T(J)$ characteristic. The present analysis also enables us to calculate the duct surface pressure distribution, and therefore the duct forward thrust coefficient $(1 - \tau)C_T$ for a prescribed propeller thrust τC_T. From this we may obtain also the $\tau(C_T)$ characteristic. Thus the duct thrust is given by

$$T_D = (1 - \tau)T$$

$$= -\oint 2\pi r(p - p_\infty) \sin \beta \, \mathrm{d}s$$

Ducted propellers and fans

Introducing the definitions of duct surface pressure coefficient (5.9) and thrust coefficient (5.12) we have finally an expression for duct thrust coefficient in terms of our predicted surface pressure coefficients C_p, namely

$$\left.\begin{aligned} C_{TD} &= (1 - \tau)C_T \\[2mm] &= -\frac{8}{D^2} \oint C_p r \sin\beta \, ds \end{aligned}\right\} \tag{5.17}$$

Making use of computer program ductprop.pas given in the Appendix as Program 5.2, the predicted $\tau(C_T)$ characteristic is also shown on Fig. 5.9 for the whole experimental range of C_T between 1.0 and 25.0, confirming the power of this relatively crude model of duct/propeller interaction using only a single vortex cylinder to model the propeller vortex wake. Precise predictions of thrust ratio were obtained for C_T values in excess of 6.0. For lower system thrusts the duct thrust is slightly under predicted as can be seen also from the pressure distribution for $J = 0.551$, Fig. [5.8].

For these calculations the vortex tube 'tip clearance' was set arbitrarily at 5% of the propeller radius. The vortex tube itself is of course a circumferentially averaged representation of the trailing vortices shed from the propeller tips and its location is unknown. However, the influence of the propeller upon the duct is totally accounted for in this model by the velocities induced at the duct surface by the semi-infinite vortex tube. Its radial location would be close to the duct inner surface for a free vortex propeller, but only at the design point. At off-design the blade bound circulation would vary radially, invalidating the present model. However, it is possible to correlate the tip clearance with C_T in order to obtain accurate prediction of thrust ratio from this simple single vortex tube model. The outcome is shown in Table 5.1 for five values of C_T.

5.4 Non-free vortex ducted propeller – lifting surface theory

As just pointed out, in most turbomachines or ducted propulsors the blade bound circulation will vary radially, resulting in the shedding of helical trailing vortex sheets extending to infinity

Table 5.1. *Selected tip clearances for semi-infinite vortex tube to correlate prediction for thrust ratio* τ

C_T	$\left(\dfrac{\text{Tip clearance}}{\text{Propeller radius}}\right)\%$	τ_{expt}	$\tau_{\text{predicted}}$
1.0	26.5	0.900	0.90020
4.0	9.5	0.735	0.73421
9.0	5.0	0.650	0.64978
16.0	3.7	0.612	0.61255
25.0	4.4	0.595	0.59456

downstream in the wake. The interactions between this trailing vortex system and the duct are crucial to the performance characteristics of a ducted propulsor, controlling the sharing of thrust between duct and propeller. Velocity components, U_c, V_c, accounted for in (5.2), are induced at the duct boundary by the propeller wake. Conversely velocity components U_d, V_d are induced in the propeller plane due to the duct and hub. Of these the axial velocity component is of particular importance since it influences the design pitch selection for the propeller. U_d can be calculated directly from the annulus surface vorticity through

$$U_d = \sum_{n=1}^{m\text{hub}} u_{mn}\gamma(s_n)\,\Delta s_n + \sum_{n=1}^{m\text{duct}} u_{mn}\gamma(s_n)\,\Delta s_n \tag{5.18}$$

where u_{mn} is the axial velocity induced by a unit ring vortex (4.8a). Including also the influence of the uniform stream W, the total velocity at the propeller plane is thus given by

$$V_{ap} = W + U_c + U_d \tag{5.19}$$

Analysis of the flow through a given propulsor requires an iterative procedure along the following lines to produce progressively improved estimates of U_c and U_d.

Ryan & Glover (1972) presented the first surface vorticity scheme to accomplish this iterative procedure by combining axisymmetric surface vorticity analysis for the hub and duct with lifting line theory to represent the propeller. The objective of the latter was to select the radial distribution of propeller loading or bound vorticity in such a manner that would minimise the shed vorticity energy dissipation in the wake, following the well established propeller

205

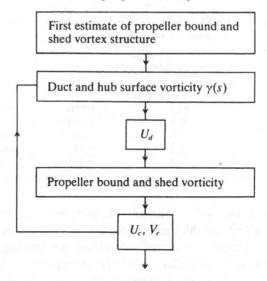

design method of Burrill (1955). Techniques for propeller profile selection or subsequent profile fluid dynamic analysis differ widely between authors. Some such as Glover (1970) and Burrill (1955) adopted lifting line theories and others, such as Pien (1961), Kerwin & Lee (1978), used, instead, more advanced lifting surface models. On the other hand the treatment of induced velocities due to the helical vortex sheets is common to all. We will therefore give consideration here to the analysis presented by Ryan & Glover.

The complete helical vortex sheet may be broken down into a discrete number of elementary helical vortices, distributed radially from hub to tip, and emanating from the propeller blade trailing edges. In reality the mutual convection influence of these elements will cause the sheets to distort in a manner similar to the rolling up of an aircraft wing tip vortex. Furthermore the vorticity is subject to viscous diffusion. In propeller analysis both effects are frequently ignored, the assumption being made that each vortex element proceeds along a cylindrical spiral path with fixed helix angle β_i, independently from all other elements. Adopting these assumptions let us consider the velocity field induced by one such helical trailing vortex element of strength $\Delta\Gamma$, extending from point P at (x_p, r_p, θ_p) to infinity downstream. Its vectorial sign is determined by the positive corkscrew rule, Fig. 5.10. Consider a unit vector $\mathbf{t}$ parallel to the vortex element at some location (x, r, θ). Then from

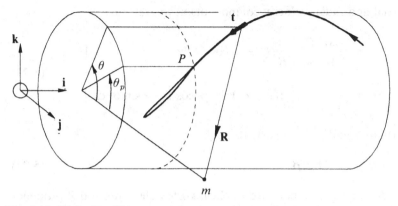

Fig. 5.10. Helical trailing vortex downstream propeller blade.

the Biot–Savart law (4.2) the velocity induced at m due to the elementary length ds of the helical vortex is given by

$$\mathrm{d}v = \frac{\Gamma \, \mathrm{d}s \, tX\mathbf{R}}{4\pi \, |R|^3} \qquad\qquad [4.2]$$

Introducing unit vectors, $\mathbf{i}$, $\mathbf{j}$ and $\mathbf{k}$ along the x, y, z axes, Fig. 5.10 we have

$$\mathbf{t} = -\mathbf{i} \sin \beta_i + \mathbf{j} \cos \beta_i \sin \theta - \mathbf{k} \cos \beta_i \cos \theta \qquad (5.20)$$

$$\mathbf{R} = -\mathbf{i}(x - x_m) + \mathbf{j}(r_m - r \cos \theta) - \mathbf{k}r \sin \theta \qquad (5.21)$$

A restriction has been imposed here upon the location of m which lies in the $(\mathbf{i}, \mathbf{j})$ plane at $(x_m, r_m, 0)$. Equation (4.2) may now be resolved into the three velocity components:

$$\left.\begin{aligned}
\mathrm{d}v_x &= -\frac{\Delta\Gamma}{4\pi R^3}[r - r_m \cos \theta]\, \mathrm{d}x \\[2mm]
\mathrm{d}v_r &= \frac{\Delta\Gamma}{4\pi R^3}[(x - x_m) \cos \beta_i \cos \theta - r \sin \theta]\, \mathrm{d}x \\[2mm]
\mathrm{d}v_\theta &= -\frac{\Delta\Gamma}{4\pi R^3}[r_m - r \cos \theta - (x - x_m) \cos \beta_i \sin \theta]\, \mathrm{d}x
\end{aligned}\right\} \qquad (5.22)$$

Now let m be a point on the duct or hub surface. The induced

207

Ducted propellers and fans

axial and radial velocity components at m are thus

$$
\left.
\begin{aligned}
\Delta v_x &= -\frac{\Delta\Gamma}{4\pi} \int_{\theta=\theta_p}^{\infty} \frac{1}{R^3} (r - r_m \cos\theta)\, dx \\
\Delta v_r &= \frac{\Delta\Gamma}{4\pi} \int_{\theta=\theta_p}^{\infty} \frac{1}{R^3} [(x - x_m) \cos\beta_i \cos\theta - r \sin\theta]\, dx
\end{aligned}
\right\}
\tag{5.23}
$$

where θ and x are related through

$$
x = x_p + r\theta \tan\beta_i
\tag{5.24}
$$

Assuming that there are I helical vortex elements and Z propeller blades, the induced velocities for the vortex system become

$$
\left.
\begin{aligned}
v_x &= -\frac{1}{4\pi} \sum_{p=1}^{Z} \sum_{i=1}^{I} \Delta\Gamma_i I_{xi} \\
v_r &= \frac{1}{4\pi} \sum_{p=1}^{Z} \sum_{i=1}^{I} \Delta\Gamma_i I_{ri}
\end{aligned}
\right\}
\tag{5.25}
$$

where

$$
\theta_{pi} = \theta_{p1} + (i-1)2\pi/Z
\tag{5.26}
$$

and where I_{xi} and I_{ri} are the definite integrals in equations (5.23).

As expressed here, v_x and v_r vary circumferentially due to the periodicity of the helical vortex sheet wakes emanating from Z blades. Because the duct and hub surface vorticity analysis is axisymmetric, Ryan & Glover (1972) took the pitchwise average of v_x and v_r integrating over the range $\theta_p = 0$ to $2\pi/Z$. The definite integrals in equations (5.25) then become

$$
\left.
\begin{aligned}
I_{xi} &= \frac{Zr_i \tan\beta_i}{2\pi} \int_0^{2\pi/Z} \int_0^{\infty} \frac{1}{R_{im}^3} (r_i - r_m \cos\theta)\, d\theta\, d\theta_p \\
I_{ri} &= \frac{Zr_i \tan\beta_i}{2\pi} \int_0^{2\pi/Z} \int_0^{\infty} \frac{1}{R_{im}^3} \\
&\quad \times [(x_i - x_m) \cos\beta_i \cos\theta - r_i \sin\theta]\, d\theta\, d\theta_p
\end{aligned}
\right\}
\tag{5.27}
$$

where

$$
R_{im} = \sqrt{[(x_i - x_m)^2 + (r_m - r_i \cos\theta)^2 + r_i^2 \sin^2\theta]}
\tag{5.28}
$$

5.4.1   Matching the helix angle

The foregoing equations provide the means for calculating the downwash velocities induced at the duct and propeller surfaces by the helical trailing vortex wake system. Their evaluation as stated presupposes that the helix angle β_i is known. In fact β_i is not known *a priori* but we will show that it can be expressed in terms of J, τ and C_T. Of these three performance parameters it would be usual in design practice to specify the advance coefficient J and propeller thrust τC_T as input and to calculate the consequent duct thrust $(1 - \tau)C_T$ from the preceding analysis for various duct shapes to achieve duct matching. When following such a procedure Ryan & Glover were faced with the dilemma that in reality β_i varies along the wake between the values

$$\left.\begin{array}{ll} \beta_i = \beta_2 = \arctan\left(\dfrac{U - v_{\theta2}}{V_\text{p}}\right) & \text{close to the propeller} \quad (a) \\[2em] \beta_i = \arctan\left(\dfrac{U - v_{\theta2}}{V_\text{j}}\right) & \text{same distance downstream} \quad (b) \end{array}\right\}$$

$$(5.29)$$

where U is a blade speed and $v_{\theta2}$ the swirl velocity leaving the propeller.

Here the assumption has been made that there is negligible variation of $v_{\theta2}$ due to the wake contraction or expansion, although further correction for this could be made if necessary. The first of these expressions must be used when undertaking velocity triangle analysis in the propeller plane. However, Ryan & Glover argue that since the jet flow settles down quickly in the rear field behind the duct, it is more reasonable to adopt (5.29b) when calculating duct and hub wake induced downwash velocities. We may eliminate U from this expression since the advance coefficient (5.14) may be written

$$J = \frac{V_\text{a}}{(U/2\pi r)D} = \frac{\pi V_\text{a}}{U}\left(\frac{r}{r_t}\right)$$

$$(5.30)$$

Furthermore an expression for the exit swirl velocity $v_{\theta2}$ follows from the Euler pump equation

$$\Delta p_0 = \rho U v_{\theta2}$$

$$(5.31)$$

Since usually $v_{\theta2} \ll U$, then $\Delta p_0 \approx p_2 - p_1$, and making use of

(5.15) we then have

$$v_{\theta 2} = \frac{V_a}{2\pi}\left(\frac{\tau C_T}{1-h^2}\right)\frac{J}{r/r_t}$$ (5.32)

We have already shown that

$$\frac{V_j}{V_a} = \sqrt{\left(1 + \frac{\tau C_T}{1-h^2}\right)}$$ [5.15]

so that combining all these results we obtain finally

$$\tan \beta_i = \frac{\dfrac{\pi}{J}\dfrac{r}{r_t} - \dfrac{J}{2\pi}\left(\dfrac{\tau C_T}{1-h^2}\right)\left(\dfrac{r_t}{r}\right)}{\sqrt{\left(1 + \dfrac{\tau C_T}{1-h^2}\right)}}$$ (5.33)

In addition to this we must also relate the blade bound vortex strength Γ to the system performance parameters. The circulation at radius r about one blade of a propeller with Z blades is given by

$$\Gamma = \frac{2\pi r v_{\theta 2}}{Z}$$ (5.34a)

which can be reduced through (5.31) to the dimensionless form

$$\frac{\Gamma}{DV_a} = \frac{J}{2Z}\left(\frac{\tau C_T}{1-h^2}\right)$$ (5.34a)

Equations (5.33) for β_i and (5.34) for Γ complete the specification of the problem in terms of J and τC_T prior to fluid dynamic analysis. Before leaving this section it will also be useful to derive an expression for the mean velocity in the propeller plane V_p for use in laying out velocity triangles. This follows from a one-dimensional thrust-momentum balance yielding for the system

$$T = \rho\left(V_p\frac{\pi D^2}{4}\right)(1-h^2)(V_j - V_a)$$ (5.35)

If we eliminate T from (5.15) then we obtain

$$\frac{V_p}{V_a} = \frac{1}{2\tau}\left(1 + \frac{V_j}{V_a}\right)$$

$$= \frac{1}{2\tau}\left(1 + \sqrt{\left[1 + \frac{\tau C_T}{1-h^2}\right]}\right)$$ (5.36)

210

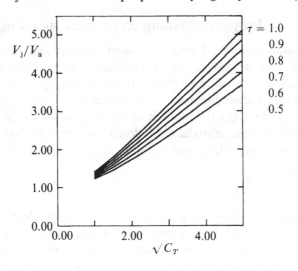

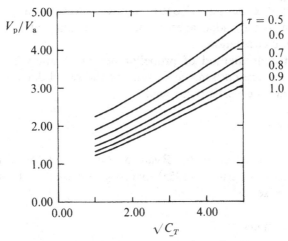

Fig. 5.11. Velocity in wake and propeller plane for Kort nozzle ducted propellers.

A comparison of V_p/V_a and V_j/V_a as given by (5.15) and (5.36) for a wide range of τ and C_T values, can be obtained from Fig. 5.11, from which one can check the operating conditions under which β_i will vary by great or small amounts along the wake according to the conditions discussed above.

211

5.4.2 Propeller loading and vortex shedding

The equations developed above for duct analysis and wake–duct interaction can be solved for a specified distribution of blade circulation Γ as a function of radius. Unfortunately (5.34b) appears to restrict the analysis to constant blade circulation, but this is only the result of the approximations in the derivation of $v_{\theta 2}$ in (5.32). Alternatively we may introduce the design propeller loading Δp_0 at r directly from the Euler pump equation (5.31) to obtain finally

$$\frac{\Gamma}{DV_a} = \frac{J}{2Z} \left\{ \frac{\Delta p_0}{\frac{1}{2}\rho V_a{}^2} \right\} \tag{5.37}$$

The blade circulation is thus proportional to the blade loading or energy input Δp_0. For free vortex loading Δp_0 and Γ are constant radially resulting in no vortex shedding apart from the blade tips where the entire circulation is shed as Z helical concentrated vortices of strength Γ. In this case (5.37) approximates to (5.34b) which is slightly more convenient to use for free vortex design. In all other vortex designs (5.34b) merely represents a good approximation to the radial average propeller loading and therefore average blade circulation.

Any particular method of propeller design or analysis may be linked into the present scheme by relating the radial derivative of Γ to the helical shed vorticity through

$$\Delta \Gamma_i = \frac{\mathrm{d}\Gamma}{\mathrm{d}r} \Delta r_i \tag{5.38}$$

An approach adopted by Ryan & Glover (1972) using the propeller theory of Burrill (1955) involves expansion of Γ as a half range Fourier series

$$\Gamma = \sum_1^\infty A_n \sin n\phi \tag{5.39}$$

where the variable ϕ is related to radius through

$$\frac{r - r_h}{r_t - r_h} = \tfrac{1}{2}(1 - \cos \phi) \quad \text{with} \quad 0 < \phi < \pi \tag{5.40}$$

$\Delta \Gamma_i$ may then be expressed as a series which is truncated to say ten terms involving the unknown coefficients A_1–A_{10}. Following Burrill, a principle of minimising wake trailing vortex energy dissipation is applied leading to ten simultaneous equations for A_b.

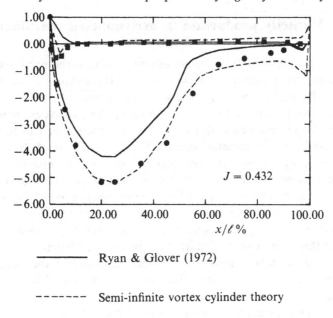

Fig. 5.12. Comparison of experimental surface pressure distribution on 19A duct with Ka 4-55 propeller and theoretical predictions for $J = 0.432$.

It is outside the present objectives and scope to provide further detail and indeed a variety of possible propeller blade profile selection methods are available ranging from lifting line and surface theories, Glover (1970), Pien (1961), Kerwin & Lee (1978), Weissinger & Maass (1968), to the cascade strip method covered in Chapter 2, which proves ideal for ducted propellers as demonstrated by Balabaskaran (1982), and Lewis & Balabaskaran (1983). To conclude, pressure distributions for the Ka 4-55 propeller in NSMB duct 19A calculated by Ryan & Glover are compared in Fig. 5.12 with output from the axisymmetric free vortex model of Section 5.3 for an advance coefficient J of 0.432. In fact the predictions using the latter simpler cruder model agree rather better with experimental test than the published results of Ryan & Glover using full helical vortex modelling, although both are indeed excellent.

5.5 Vorticity production in axisymmetric meridional flows

As a practical approach to design and analysis of turbomachines we have seen, in Chapter 3, how the true three-dimensional flow through the blade rows may be conveniently treated as a series of coupled two-dimensional flows. These comprise two types, namely blade-to-blade or cascade flows, the subject matter of Chapters 2 and 3, and an axisymmetric meridional through-flow. A frequent design strategy involves the prescription of blade loading for each cascade section, associated with the section bound circulation Γ. Variation of blade loading and therefore energy input between adjacent blade sections results in variation of bound circulation and consequent shedding of vorticity into the meridional flow. The effect of this is to produce disturbances to the meridional velocity q_s which must be taken into account when considering the velocity triangles and the consequent blade profile selection. That is the purpose of meridional analysis.

Let us begin by stating the equations of motion for axisymmetric flow in cylindrical coordinates. These consist of the continuity equation,

$$\frac{\partial v_x}{\partial x} + \frac{\partial v_r}{\partial r} + \frac{v_r}{r} = 0 \tag{5.41}$$

and the Eulerian equations. For the present purposes these are best expressed in terms of stagnation pressure.

$$\left. \begin{aligned} -F_x + \frac{1}{\rho}\frac{\partial p_0}{\partial x} &= v_r\omega_\theta - v_\theta\omega_r \\ -F_r + \frac{1}{\rho}\frac{\partial p_0}{\partial r} &= v_\theta\omega_x - v_x\omega_\theta \\ -F_\theta &= v_x\omega_r - v_r\omega_x \end{aligned} \right\} \tag{5.42}$$

where F_x, F_r and F_θ are distributed body forces and the vorticity components, also in axisymmetric flow, are defined

$$\left. \begin{aligned} \omega_x &= \frac{1}{r}\frac{\partial(v_\theta r)}{\partial r} \\ \omega_r &= -\frac{\partial v_\theta}{\partial x} \\ \omega_\theta &= \frac{\partial v_r}{\partial x} - \frac{\partial v_x}{\partial r} \end{aligned} \right\} \tag{5.43}$$

The Euler pump equation (5.31), which accounts for rotor energy input, completes the summary of basic governing equations for meridional incompressible flow for single rotor blade row turbomachines without prewhirl such as the ducted propeller to which we have just given consideration. By adopting appropriate forms of the Euler pump equation it is quite straightforward to extend the present analysis to multi-blade row machines. In order to reduce these general equations to a simpler and more appropriate form, it is helpful to derive Stoke's stream function from the continuity equation (5.41), namely

$$
\left.\begin{aligned}
v_x &= \frac{1}{r}\frac{\partial \psi}{\partial r} \\
v_r &= -\frac{1}{r}\frac{\partial \psi}{\partial x}
\end{aligned}\right\}
\tag{5.44}
$$

If these expressions are introduced into the equation for tangential vorticity (5.43c), we have Stoke's equation

$$
\frac{\partial^2 \psi}{\partial r^2} - \frac{1}{r}\frac{\partial \psi}{\partial r} + \frac{\partial^2 \psi}{\partial x^2} = -\omega_\theta r
\tag{5.45}
$$

This is the principal governing equation for axisymmetric incompressible meridional flow with prescribed tangential vorticity distribution ω_θ. All methods of meridional analysis in principle aim at the solution of this equation for prescribed boundaries. The ring vortex flow derived from the Biot–Savart law in Section 4.2, is a particular solution of this equation. The surface vorticity method for axisymmetric flow which we have developed over the past two chapters, achieves solutions for more complex flow regimes by simple use of the principle of superposition. However, in the cases considered so far we have assumed an irrotational flow with zero tangential vorticity. In all real situations vorticity is produced by the blade-to-blade/meridional flow interactions and an auxiliary equation is needed to relate ω_θ to these processes. Although the Eulerian equations as stated above in cylindrical coordinates could be used as auxiliary equations for flows which are almost cylindrical, Bragg & Hawthorne (1950) derived a simpler form of vorticity production equation which is applicable to annuli of arbitrary shape. By combining the momentum equations (5.42) for the case of zero body forces, these authors were able to obtain a single

Ducted propellers and fans

equation involving only total derivatives $d/d\psi$ as follows.

$$\omega_\theta = v_\theta \frac{d(rv_\theta)}{d\psi} - \frac{r}{\rho}\frac{dp_0}{d\psi} \qquad (5.46)$$

We see that the tangential vorticity in the annulus space downstream of a blade row is connected with gradients of angular momentum rv_θ and stagnation pressure p_0 normal to the meridional streamlines. Of particular interest is the fact that, as already noted, p_0 and $v_\theta r$ represent the initial blade loading variables selected by the designer and are linked through the Euler pump equation (5.31). In differential form with constant p_0 and rv_θ at inlet to the rotor, the Euler pump equation may be written

$$\frac{1}{\rho}\frac{dp_{02}}{d\psi} = \frac{d}{d\psi}(Uv_\theta) \qquad (5.31b)$$

where $U = r\Omega$ is the local blade speed. Combining this with (5.45) we have the following alternative forms of the full governing meridional flow equation

$$\frac{\partial^2\psi}{\partial r^2} - \frac{1}{r}\frac{\partial\psi}{\partial r} + \frac{\partial^2\psi}{\partial x^2} = (\Omega r^2 - rv_\theta)\frac{d(rv_\theta)}{d\psi} \qquad (5.47)$$

for prescribed rotor angular velocity Ω and downstream 'vortex' rv_θ.

By introducing the stream function into the Eulerian equations through the use of equations (5.44), Bragg & Hawthorne were also able to show that in the absence of body forces

$$\left.\begin{array}{l} v_\theta r = f_1(\psi) \\ p_0 = f_2(\psi) \end{array}\right\} \qquad (5.48)$$

The physical meaning of this result is of course that in the free annulus space downstream of a blade row the fluid angular momentum and stagnation pressure are conserved along the meridional streamlines, a result we would anticipate in the absence of body forces. Consequently all of these results are applicable to actuator disc type models of meridional flow, in which the blade row vortex shedding is assumed to be concentrated onto a representative plane, such as the disc mapped out by the blade centre lift line if circumferentially rotated. Although this model was originally restricted to classical solutions such as those of Bragg & Hawthorne (1950), Horlock (1952), Marble (1948) and Railly (1951), it is of more general applicability. As shown by Lewis and Horlock (1969),

216

the model may be extended to deal with axial distributions of vortex actuator discs to represent the effects of body forces within blade rows or even axial distributions of source actuator discs to represent the meridional disturbances caused by blade thickness. Classical solutions have also been derived for turbomachines with tapered annuli. However the actuator disc model is often implied in more general numerical schemes, including grid or channel methods for meridional flows, and has been successfully implanted into ducted propeller analysis codes by Gibson & Lewis (1973), Balabaskaran (1982) and others.

5.5.1   Streamwise and smoke-ring vorticity

Although the foregoing equations provide all the fundamentals required to proceed to a computational scheme, it is of importance to draw attention to the two types of vorticity shed by blade rows. We observe first that both terms on the right hand side of the vorticity production equation (5.46) vanish for the free vortex case previously considered in Section (5.3), for which the conservation equations (5.48) become

$$\left.\begin{array}{l} v_\theta r = \text{constant} \\ p_0 = \text{constant} \\ \text{and} \\ \omega_\theta = 0 \end{array}\right\} \tag{5.49}$$

For all other situations the tangential vorticity ω_θ has two independent components, one produced by gradients of angular momentum $d(v_\theta r)/d\psi$ and the other by gradients of stagnation pressure $dp_0/d\psi$. The individual characteristics of these components are highlighted if equations (5.48) are introduced into (5.46), whereupon

$$\omega_\theta = \frac{1}{r} F_1(\psi) + r F_2(\psi) \tag{5.50}$$

where the functions of ψ are linked to $v_\theta r$ and p_0 through

$$\left.\begin{array}{l} F_1(\psi) = v_\theta r \dfrac{d(v_\theta r)}{d\psi} \\[2mm] F_2(\omega) = -\dfrac{1}{\rho}\dfrac{dp_0}{d\psi} \end{array}\right\} \tag{5.51}$$

In cylindrical, or near-cylindrical meridional flow, the streamline radius r is constant so that the two components of ω_θ also remain constant in strength along the meridional streamlines. In all other cases such as flow through a conical shaped annulus, vortex stretching occurs which affects the two components in different ways. Thus if r increases moving downstream along a given ψ line, then the first component will decrease in strength while the second component will increase. Lewis & Horlock (1961) and Lewis (1964b) have explored these effects for non-cylindrical turbomachines and have also shown that the two vorticity components represent trailing or streamwise vorticity and smoke-ring vorticity respectively.

Thus for a constant stagnation pressure flow, the Eulerian equations (5.42), omitting body forces, reduce to

$$\frac{\omega_x}{v_x} = \frac{\omega_r}{v_r} = \frac{\omega_\theta}{v_\theta} = \frac{\omega}{\mathbf{v}} \tag{5.52}$$

In this case the local vorticity vector $\boldsymbol{\omega}$ lies parallel to the velocity vector $\mathbf{v}$, the definition of streamwise vorticity.

On the other hand, for flows with constant or zero angular momentum $(v_\theta r)$, equations (5.42) reduce to

$$\omega_\theta = -\frac{r}{\rho}\frac{\mathrm{d}p_0}{\mathrm{d}\psi} \tag{5.53}$$

In axisymmetric flow, ω_θ for this special case forms closed rings concentric with the axis which obey the stretching rules of smoke-ring vorticity, namely that ω_θ/r is conserved along streamlines. It should be pointed out that this component of vorticity appears to be missing from analyses which use helical vortex modelling in the form presented in Section 5.4. However this model should be correct if the helical vortex sheets are rotated with the angular velocity of the rotor.

5.6 Non-free vortex actuator disc model for axial turbomachines and ducted propellers

The analysis model for free vortex machines may easily be extended to axial turbomachines of arbitrary vortex by the introduction of a radial distribution of semi-infinite vortex cylinders to represent the meridional distribution of tangential vorticity ω_θ, Fig. 5.13. The

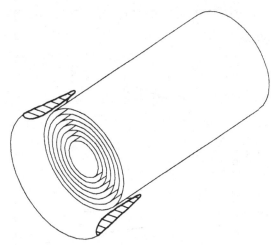

Fig. 5.13. Representation of propeller wake by a radial distribution of vortex cylinders.

strength of such an elementary cylinder at any radius r is then given by

$$\eta = \omega_\theta \, \Delta r$$

$$= (\Omega r - v_\theta) \frac{\mathrm{d}(rv_\theta)}{\mathrm{d}\psi} \Delta r$$

To a first approximation for near cylindrical flow (5.44) may be written $\mathrm{d}\psi \approx V_x r \, \mathrm{d}r$ where V_x is the radial mean meridional velocity ($V_x = V_\mathrm{j}$ for the Kort nozzle), whereupon

$$\eta \approx \left(\frac{\Omega r}{V_x} - \frac{v_\theta}{V_x} \right) \frac{1}{r} \Delta(rv_\theta) \tag{5.54a}$$

Introducing the blade circulation Γ, which may be calculated from cascade theory, this may be written in the alternative form, using (5.34a),

$$\eta = \left(\frac{\Omega r}{V_x} - \frac{v_\theta}{V_x} \right) \frac{Z}{2\pi r} \Delta\Gamma \tag{5.54b}$$

Balabaskaran (1982) applied surface vorticity cascade analysis to the Ka 4-55 propeller in the NSMB 19A duct, but did not include such corrections for the meridional disturbances due to distributed vortex shedding. His meridional velocities at the propeller plane were derived from (5.36). Despite this omission extremely good

219

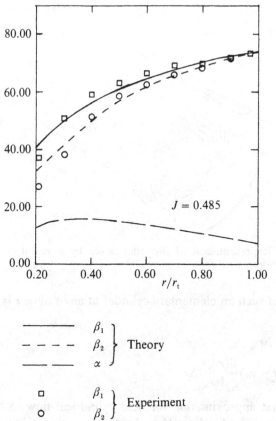

Fig. 5.14. Relative and absolute swirl angles downstream of Ka 4-55 propeller in 19A duct.

predictions of relative flow angle leaving the rotor blades were obtained. Experimental results showed little variation of relative inlet and exit angles β_1 and β_2 with advance coefficient over the range $0.266 < J < 0.652$ and the average of these results is compared with Balabaskaran's predictions for $J = 0.485$ in Fig. 5.14. The main reason for such good agreement in this case is reasonably constant but also light radial loading of this propeller. Thus the exit swirl angle α_2 was quite small, Fig. 5.13, as consequently was $v_{\theta 2}$. In other turbomachines with more heavily loaded blades meridional flow calculations are an essential part of the quasi-three-dimensional design or analysis procedures.

5.7 Models to deal with the induced effects of distributed ring vorticity in axisymmetric meridional flows

As we have seen the meridional flow inside a turbomachine is in general not irrotational but is influenced by a spatial distribution of tangential vorticity ω_θ generated by the blades. In the introduction to this chapter reference was made to the range of numerical methods such as finite element, finite difference, matrix throughflow or time marching schemes which are normally applied to the solution of such flow problems. These all involve the introduction of a grid throughout the flow domain and the solution of the governing equations at all grid points, possibly by successive iterations. Until recently the vorticity boundary integral method has not been explored as a potential tool to tackle this task. In fact it seems to present few disadvantages for dealing with turbomachine meridional flows and several advantages for open flow situations such as ducted propellers. In this section we will derive some of the foundation material and apply it to the flow of a thick shear layer over a body of revolution. At the conclusion of Chapter 6 we will return to this subject again but within the fuller context of turbomachinery meridional flows.

5.7.1 Numerical representation of rectangular and circular ring vortex elements

The first step for representation of a rotational meridional or shear flow region, Fig. 5.15, is the introduction of an appropriate grid system to reduce the distributed vorticity to a discrete number of ring vortex blocks, each assumed to be of constant strength. Three models will be considered here as follows.

Model 1. Rectangular ring vortex (RRV)
The most useful model is the rectangular ring vortex (RRV) representation of a single ring vortex block, Fig. 5.16(a) with sides ΔX and ΔY and therefore with total vortex strength

$$\Delta\Gamma = \omega_\theta \, \Delta X \, \Delta Y \qquad (5.55)$$

The velocity components Δu_m, Δv_m induced at any other location (x_m, r_m) may then be estimated by breaking the ring vortex down into PQ sub-elements represented by P cells in the X direction and

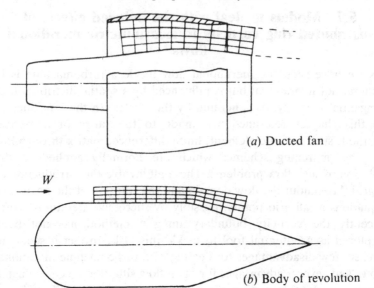

(a) Ducted fan

W

(b) Body of revolution

Fig. 5.15. Orthogonal grids for representation of (a) meridional flow through ducted fan and (b) shear layer on a body of revolution.

Q cells in the Y direction. As illustrated in Fig. 5.16 the ring vortex blocks may be orientated at angle γ_s to the x axis. For convenience the rotated coordinate system (X, Y) is therefore defined relative to the sides of the vortex block. Making use of the expressions for a unit concentrated ring vortex derived in Section 4.2, (4.8), we then have for the RRV,

$$
\left.
\begin{aligned}
\Delta u_m &= \frac{1}{2\pi} \sum_{p=1}^{P} \sum_{q=1}^{Q} \frac{\gamma(p, q)}{r(p, q)\sqrt{\bar{x}^2 + (\bar{r} + 1)^2}} \\
&\quad \times \left\{ K(k) - \left[1 + \frac{2(\bar{r} - 1)}{\bar{x}^2 + (\bar{r} - 1)^2} \right] E(k) \right\} \\
\Delta v_m &= \frac{1}{2\pi} \sum_{p=1}^{P} \sum_{q=1}^{Q} \frac{\gamma(p, q)\bar{x}/\bar{r}}{r(p, q)\sqrt{\bar{x}^2 + (\bar{r} + 1)^2}} \\
&\quad \times \left\{ K(k) - \left[1 + \frac{2\bar{r}}{\bar{x}^2 + (\bar{r} - 1)^2} \right] E(k) \right\}
\end{aligned}
\right\}
\tag{5.56}
$$

where the cellular vortex strength is

$$
\gamma(p, q) = \Delta\Gamma/(PQ)
\tag{5.57}
$$

222

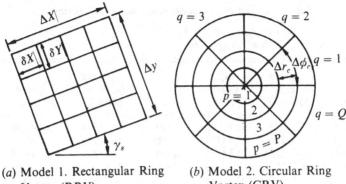

(a) Model 1. Rectangular Ring   (b) Model 2. Circular Ring
          Vortex (RRV)   Vortex (CRV)

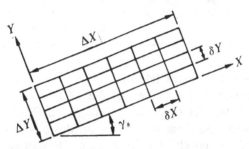

(c) Model 3. Flat Ring Vortex (FRV)

Fig. 5.16. Ring vortex grid models.

and the dimensionless coordinates $(\bar{x}, \bar{r})$ are defined

$$\bar{x} = \frac{x_m - x(p, q)}{r(p, q)}, \qquad \bar{r} = \frac{r_m}{r(p, q)} \tag{5.58}$$

Model 2. Circular ring vortex (CRV)

A second model, which will be used in Section 5.7.1 to check this numerical technique against Lamb's formula for the self propagation velocity of a ring vortex, is the circular ring vortex, Fig. 5.16(b). Assuming constant strength vorticity spread over the CRV core, (5.56) and (5.58) are still applicable with the modified cellular vortex strength, now dependent upon cell area and given by

$$\gamma(p, q) = \frac{\Delta \Gamma}{\pi a^2} r_{\rm c}(p, q) \, \Delta r_{\rm c} \, \Delta \phi_{\rm c} \tag{5.59}$$

where a is the CRV radius.

223

Table 5.2. *Self-propagation of a smoke-ring vortex*

Radius of ring vortex $r_m = 1.0$ $P = 16$
Total vortex strength $= 1.0$ $Q = 16$

a	v_a using (C.R.V.) Model 2	v_a Lamb (1945)	a	v_a using (C.R.V.) Model 2	v_a Lamb (1945)
0.095	0.35273	0.33290	0.050	0.40484	0.38398
0.090	0.35710	0.33720	0.045	0.41340	0.39236
0.085	0.36173	0.34175	0.040	0.42291	0.40173
0.080	0.36663	0.34657	0.035	0.43383	0.41236
0.075	0.37187	0.35171	0.030	0.44627	0.42463
0.070	0.37747	0.35720	0.025	0.46008	0.43913
0.065	0.38348	0.36310	0.020	0.47654	0.45689
0.060	0.38998	0.36947	0.015	0.49895	0.47978
0.055	0.39706	0.37639	0.010	0.53122	0.51205
			0.005	0.58636	0.56721

Model 3. Flat ring vortex (FRV)
Although basically the same as the RRV, obeying equations
(5.56)–(5.58), the flat ring vortex, designated FRV, is particularly
useful for modelling shear layers. Again, this model will be used in
Section 5.7.2 to check the Ryan/Lamb formulation derived in
Section 4.2.3 for the self-induced velocity of a surface vorticity
element in axisymmetric flow.

5.7.2 Check on self-propagation of smoke-ring vortex

Sorvatziotis (1985) made use of the classical solution of Lamb
(1945) for the self-propagation of a smoke-ring vortex (4.20), as a
datum check for the numerical cellular model using the CRV
model. The results with 256 cells are shown in Table 5.2.

The results of this study are as much a validation of Lamb's
approximate formulation as of the numerical model and show
encouraging agreement in the predicted trends of increasing self-
propagation velocity with decreasing core radius. As expected
Lamb's equation is in best agreement with the CRV model for the
smaller values of core radius a. This study confirmed both the
accuracy of Lamb's equation (4.20) and the likely success one might
anticipate in applying the RRV and FRV models to meridional and
shear layer flows.

Table 5.3. *Self-propagation of sheet element ring vortex – Comparison of Lamb/Ryan with F.R.V. model no. 3*

Radius of ring vortex $r_m = 1.0$
$DY = DX/12,\ P = 12,\ Q = 12$
Total vortex strength $= 1.0$

DX	F.R.V.	Lamb/Ryan	DX	F.R.V.	Lamb/Ryan
0.100	0.40205	0.41991	0.050	0.22832	0.23753
0.095	0.38572	0.40279	0.045	0.20905	0.21755
0.090	0.36928	0.38547	0.040	0.18943	0.19713
0.085	0.35262	0.36792	0.035	0.16940	0.17621
0.080	0.33572	0.35013	0.030	0.14888	0.15472
0.075	0.31860	0.33210	0.025	0.12777	0.13256
0.070	0.30136	0.31381	0.020	0.10571	0.10960
0.065	0.28368	0.29522	0.015	0.08271	0.08563
0.060	0.26555	0.27634	0.010	0.05837	0.06031
0.055	0.24716	0.25712	0.005	0.03194	0.03292

5.7.3 Self-propagation of a sheet ring vortex element

Sorvatziotis (1985) made use of the FRV, model No. 3, to check the accuracy of the Lamb/Ryan formula (4.20a), for the self-propagation velocity of a typical surface vorticity ring vortex element. Results are shown in Table 5.3 for an inclination γ_s of zero and for a wide range of values of DX/r_m, with DY set at $DX/12$ and with 144 cells. Once again the two methods of estimation are in good agreement. It is gratifying to find that the very simple Lamb/Ryan formulation is proved quite adequate to deal with the self-inducing coupling coefficients $K(s_m, s_m)$ in the axisymmetric Martensen method of Section 4.2.3, (4.22).

5.7.4 Induced velocities close to a rectangular ring vortex

In meridional analysis it is necessary to calculate the induced velocity at the grid centres and at the edges. The previous sub-section served to illustrate that the velocity components at the centre of a (12×12) RRV will be small and accurately represented. Velocity components at the centres of sides A and B of a RRV are shown in Fig. 5.17 for the likely practical range of element aspect ratio $\Delta X/\Delta Y$ for a small element of size $\Delta X/r_m = 0.03$ with $\Delta\Gamma = 1.0$. v_B is of course zero and the normal velocity u_A small in

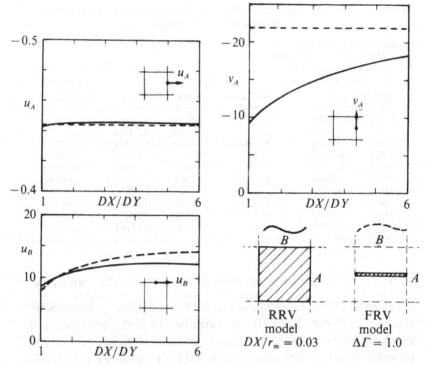

Fig. 5.17. Induced velocities at core edge of a RRV, compared with estimates using FRV approximation.

comparison with the components u_B and u_A parallel to the faces of the element. Also shown for comparison are estimates obtained if we use a FRV (12×1) model across the horizontal diameter, the only reason being the consequent twelve-fold reduction in computational requirements. Although u_A is adequately represented by this simple model it is clear that u_B and v_A can be in considerable error. The full RRV should therefore apparently be used for small grid sizes.

As a further check for fairly flat grids such as those needed to simulate a thick shear layer as in the next section, u_B and v_A values are given for an aspect ratio $DX/DY = 4.0$ for a range of DX/r_m values in Fig. 5.18(a). It is apparent that for DX/r_m values greater than 0.1, the FRV estimate is adequate for u_B but still unsuitable for v_A. On the other hand, as shown in Fig. 5.18(b), the induced velocities at the centres of neighbouring grid elements using the

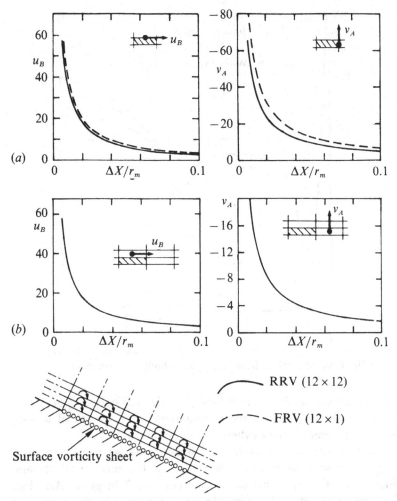

Fig. 5.18. Velocities induced by a flat ring vortex (*a*) at element edge, (*b*) at centre of adjacent element.

FRV (1×12) approximation are indistinguishable from those employing the full RRV model.

5.7.5 Flow of a shear layer past a body of revolution

Sorvatziotis (1985) developed two computational schemes for simulation of the flow of a thick shear layer past the body of revolution previously considered in Section 4.3.2, Table 4.2. Sorvatziotis's aim

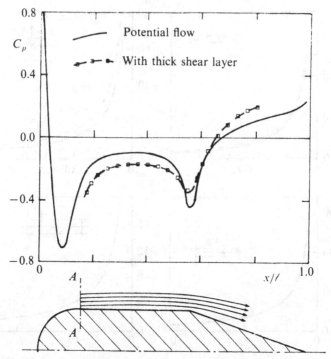

Fig. 5.19. Flow of a thick shear layer past a body of revolution.

was to model the shear layer flow past the tail cone region which would result from the upstream boundary layer growth of a body with a very much longer cylindrical mid-section.

His first scheme involved an adjustable (s, n) curvilinear grid derived by successive iterations to follow the meridional stream-lines. The grid was initiated at element No. 7 in plane AA, Fig. 5.19, where the shear layer profile was categorised by the power law profile

$$\frac{u}{U_s} = \left\{ \frac{r - r_m + \varepsilon}{\delta + \varepsilon} \right\}^{1/n} \tag{5.60}$$

where r_m is the body radius at AA and δ the boundary layer thickness. The term ε was introduced to produce finite velocity at the body surface and therefore sufficient kinetic energy for the shear layer to survive diffusion in the tail cone region. U_s, the velocity at the outer edge of the boundary layer, was taken as the value given by the potential flow solution and the initial boundary layer thickness δ was set at 4.5% of the body length. With $\varepsilon/\delta = 0.288$ and $n = 7$ the resulting profile crudely simulated a turbulent boundary layer, with velocity $0.6U_s$ close to the wall and therefore with residual surface vorticity akin to a viscous sub-layer. Inviscid flow was however assumed. At entry into the shear layer the tangential vorticity may be calculated directly from (5.50), (5.51) and (5.60) assuming constant static pressure normal to the surface. Thus

$$\omega_\theta(r) = \frac{r}{\rho}\frac{dp_0}{d\psi} = \frac{\partial u}{\partial r}$$

$$= \frac{U_s}{n(\delta + \varepsilon)}\left(\frac{r - r_m}{\delta + \varepsilon}\right)^{(1/n)-1} \tag{5.61}$$

Since we are dealing here with smoke-ring vorticity only, the vorticity strengths at any two locations on a given streamline ψ are related through (5.50), namely,

$$F_2(\psi) = \frac{\omega_\theta(r_1)}{r_1} = \frac{\omega_\theta(r_2)}{r_2} \tag{5.62}$$

This enables us to calculate the vorticity strength at the centre of all grid elements for a known streamline distribution.

Sorvatziotis's second model employed the same strategy for defining entry vorticity into the shear layer but adopting a fixed grid system for distribution of the shear layer vorticity downstream of *AA*. This approach results in considerable economy since coupling coefficient matrices may be set up once and for all to represent (a) the influence of all grid vorticity upon the boundary elements and vice versa and (b) the mutual convective velocities induced by all grid elements.

The grid system adopted by Sorvatziotis consisted of FRV elements constructed in the manner illustrated in Fig. 5.18 with 8 grid spaces normal to the body and 22 along the body. Normals were drawn as bisectors at the point of adjoining surface elements. Equal FRV heights ΔY were constructed along each normal to define the curvilinear but fixed shear layer vorticity grid. The mid-point widths of each element were then taken to calculate an appropriate value of ΔX and (5.56)–(5.58) were used to evaluate induced velocities due to each FRV element. Following the guide lines of the previous section, velocity components at the edge of a given element were calculated using the full RRV model. Induced velocities elsewhere employed either the FRV (12×1) model or concentrated ring vortex model for more distant points of application. It should be emphasised that in implementing any similar scheme the reader should experiment to decide which approximations can be made to reduce computational requirements while retaining accuracy.

One slight complication of the fixed grid method is the problem of estimating the distribution of vorticity strength over the grid region since the latter no longer follows the streamlines. This is achieved numerically by evaluating the function $F_2(\psi)$ in (5.62) and storing it as a table of F_2 versus ψ. During successive iterations as illustrated below, ψ may be evaluated from the velocity distribution u_X. Thus with reasonable approximation the change in ψ in crossing an element at height position j located in grid column i is given by

$$\Delta \psi_{ij} = u_{Xij} r_{ij} \, \Delta Y_{ij} \tag{5.63}$$

where u_{Xij} is the velocity in direction X at the centre of r_{ij} of grid element (i, j) and ΔY_{ij} is its 'normal' height. Assuming zero stream function on the body surface we then have at the outer edge of element (i, j)

$$\psi_{ij} = \sum_{p=1}^{j} u_{Xip} r_{ip} \, \Delta Y_{ip} \tag{5.63a}$$

So far we have referred only to the vorticity induced velocities due to the shear layer and its self-convection effects. The velocity u_{Xip} must of course include contributions due to all vorticity elements in the field, both surface and grid elements. The axisymmetric 'Martensen' equations for satisfaction of the surface boundary condition must likewise include all such influences and will thus take the general form for body point m,

$$\sum_{n=1}^{M} \bar{K}(s_m, s_n) \gamma(s_n) \, \Delta s_n = -W \cos \beta_m$$

$$- \sum_{i=1}^{I} \sum_{j=1}^{J} (\Delta u_{mij} \cos \beta_m + \Delta v_{mij} \sin \beta_m) \tag{5.64}$$

where Δu_{mij}, Δv_{mij} are the velocity at m induced by grid element (i, j) and are given by equations (5.56).

The computational sequence is shown in the flow diagram.

The predicted streamline pattern and surface pressure distribution are shown in Fig. 5.19. Of particular interest is the influence of blockage caused by the shear layer, resulting in lower surface

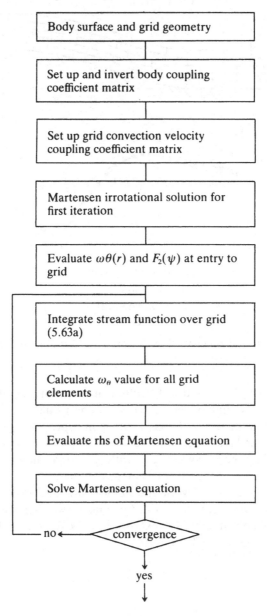

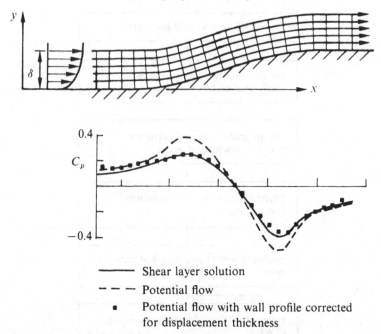

Fig. 5.20. Grid and predicted surface pressure distribution for flow of a shear layer past a wall.

pressure along the cylindrical section and a more gentle suction peak at the beginning of the tail cone as one would expect. Similar effects were predicted by Lewis (1983b) in a previous study of the flow at a shear layer past a curved wall in plane two-dimensional flow, including good agreement with potential flow for a wall corrected to allow for the displacement thickness of the shear layer, Fig. 5.20. In this latter work the author also published a cellular viscous boundary layer method for plane two-dimensional flow which is probably suitable for extension to axisymmetric flow. This will be summarised in Chapter 10.

CHAPTER 6

Three-dimensional and meridional flows in turbomachines

6.1 Introduction

Our principal aim so far has been to lay down the foundations of surface vorticity analysis for a series of progressively more advanced turbomachinery flow problems. Although a brief outline of three-dimensional flow analysis was presented in Chapter 1, specific applications have been limited to problems which are two-dimensional in the strict mathematical sense. Unlike the source panel method, which has been extensively applied to three-dimensional flows, serious application of the surface vorticity analysis has been limited to few such engine problems. The aim of the first part of this chapter will be to expand on the basic foundation theory for dealing with the flow past three-dimensional objects by surface vorticity modelling and to consider two such problems in turbine engines which have received some attention. These will include the prediction of engine cowl intake performance at angle of attack and the behaviour of turbine cascades exhibiting sweep.

As discussed in Chapter 3 the flow through turbomachinery blade passages is in general three-dimensional, although the design or analysis problem may be tackled in a practical way by reference to a series of superimposed equivalent interacting two-dimensional flows. The two models usually adopted, which are equivalent in some respects, are the S-1, S-2 surfaces of Wu (1952) and the superposition of blade-to-blade (S-2 type) flows upon an assumed axisymmetric meridional flow. We concluded Chapter 5 with a derivation of the meridional flow equations for ducted propellers, indicating that the blade-to-blade/meridional interactions result in vorticity production within the mainstream. Relatively simple treatments were given there to deal with this matter for the special case of the Kort nozzle propulsor. As originally propounded, the surface vorticity method was essentially a representation of purely incompressible potential flow. In turbomachine meridional flows on the other hand these spatial distributions of vorticity must also be accounted for as discussed in Sections 5.4 and 5.5. Further treatment of these meridional disturbances will be the subject of the second half of this chapter, with indications also about the introduction of compressibility effects. Once again the aim is not to produce a comprehensive treatment to meet all situations of three-dimensional flow, but to provide the essential foundations upon which readers may proceed to build practical schemes to suit their own particular applications.

6.2   Three-dimensional flow past lifting bodies

Generalised surface singularity equations for three-dimensional flow have been given by Ribaut (1968), Hunt & Hewitt (1986), Hess & A. M. O. Smith (1962), (1966) and Hess (1971), (1974) and Lewis & Ryan (1972) and others in relation to non-lifting and lifting bodies in either external or internal flows. Hess (1974) in particular delivered a well argued discussion of the adoption of surface vorticity sheets to model lifting aerofoils or wing/body combinations in three-dimensional flow, although within the previously established framework of the source panel method. Some of his conclusions confirm the author's findings of Sections 4.8 and 4.9 regarding the arbitrariness of this technique, calling for careful judgment in prescription of the chordwise shape factor for the bound vorticity. Ribaut expressed more general non-potential flow fields in terms of both boundary and spatial distributions of source and vortex singularities with a view to the analysis of turbomachine meridional flows, including the use of curvilinear coordinate representation of surface singularities. Although he did acknowledge the sufficiency of surface vorticity to model a potential flow, Ryan & Lewis were alone among these authors in expressing the full three-dimensional potential flow problem entirely in terms of a surface vorticity distribution expressed in curvilinear coordinates (u_1, u_2) covering the body surface. A summary of this analysis was given in Section 1.12 which we are now ready to develop further.

We have seen that the surface of a body in inviscid potential flow is in fact covered with a vortex sheet of infinitesimal thickness, which is in general two-dimensional, Fig. 6.1. The vorticity at element n may thus be resolved along local curvilinear coordinate directions into two components, which, for convenience, are

Three-dimensional flow past lifting bodies

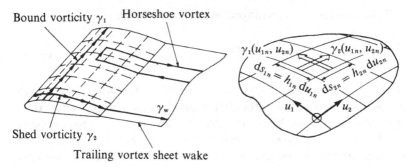

Bound vorticity γ_1 Horseshoe vortex

Shed vorticity γ_2

Trailing vortex sheet wake

(a) Aerofoil at angle of attack   (b) Surface vorticity elementary panel

Fig. 6.1. Surface and trailing vorticity sheets in three-dimensional flow.

sometimes referred to as the bound vorticity $\gamma_1(u_{1n}, u_{2n})$ and the shed or trailing vorticity $\gamma_2(u_{1n}, u_{2n})$. At the outset of tackling a specific problem an arbitrary choice of coordinate axes must be made, which in many situations can be chosen such that γ_1 and γ_2 are in fact closely related to what are physically considered to be the bound and shed vorticities. This is not essential for non-lifting flows but is particularly helpful when modelling lifting surfaces such as the finite aerofoil at angle of attack illustrated in Fig. 6.1, which sheds a trailing vortex sheet γ_W into the wake flow downstream of its trailing edge. The trailing vortex filaments are then continuous with the surface vorticity components γ_2 shed from the elements adjacent to the aerofoil trailing edge on both the upper and lower surfaces.

Various numerical schemes are possible for expressing the inter-relationship between the bound and shed vorticities within the frame-work of practical computations and we will refer to these later. For the moment let us focus upon the local flow at a particular element n. The change in bound vorticity $\gamma_1(u_{1n}, u_{2n})$ in crossing the element in the u_{1n} direction results in the shedding of a trailing vorticity component $\gamma_2(u_{1n}, u_{2n})$, Fig. 6.1(b). As shown in Chapter 1, these local bound and shed surface vorticities obey a continuity equation, which is in fact a statement of Helmholtz's vortex theorem applied to the vorticity sheet. In curvilinear coordinates we then have

$$\frac{\partial}{\partial u_{1n}}[h_{2h}\gamma_1(u_{1n}, u_{2n})] + \frac{\partial}{\partial u_{2n}}[h_{1n}\gamma_2(u_{1n}, u_{2n})] = 0 \qquad (6.1a)$$

235

If this equation is rearranged to read as follows,

$$d[h_{1n}\gamma_2(u_{1n}, u_{2n})] = -\frac{\partial}{\partial u_{1n}}[h_{2n}\gamma_1(u_{1n}, u_{2n})]\, du_{2n} \qquad (6.1b)$$

$\gamma_2(u_{1n}, u_{2n})$ can thus always be expressed in terms of $\gamma_1(u_{1n}, u_{2n})$ for any element of a three-dimensional body surface grid. It is apparent then that if M surface elements are prescribed to represent the body, the analysis problem reduces to the derivation of M initially unknown values of $\gamma_1(u_{1n}, u_{2n})$ since the M initially unknown values of $\gamma_2(u_{1n}, u_{2n})$ can in fact be expressed directly in terms of $\gamma_1(u_{1n}, u_{2n})$ through Helmholtz's theorem, (6.1b). After solution the surface velocity components then follow directly from the surface vorticity components strengths, namely

$$\left.\begin{aligned} v_{21n} &= \gamma_2(u_{1n}, u_{2n}) \\ v_{s2n} &= \gamma_1(u_{1n}, u_{2n}) \end{aligned}\right\} \qquad (6.2)$$

By analogy with (1.1) and (1.2) of Chapter 1, the implicit assumption here is that absolute fluid velocity on the interior side of the surface vorticity sheet is zero. As in two-dimensional flows this requirement is met completely by application of the Dirichlet boundary condition of zero parallel velocity on the vorticity sheet inner surface. If the Dirichlet boundary condition is now stated for element m, making use of the Biot–Savart law to account for the velocity contributions due to all vorticity elements $\gamma_1(u_{1n}, u_{2n})$ and $\gamma_2(u_{1n}, u_{2n})$ at surface locations n, the vector boundary integral equation (1.61) is obtained. A full derivation up to this point was given in Section 1.12. If the vector equation (1.61) is now resolved into its components in the curvilinear coordinate directions u_1 and u_2 we obtain the following separate integral equations applicable to element m.

$$\left.\begin{aligned} &\tfrac{1}{2}\gamma_1(u_{1m}, u_{2m}) + \oint\!\!\oint \{\gamma_1(u_{1n}, u_{2n})K_{nm}{}' \\ &\qquad + \gamma_2(u_{1n}, u_{2n})L_{nm}{}'\}\, ds_{n1}\, ds_{n2} + v_{t2m} = 0 \\[6pt] &-\tfrac{1}{2}\gamma_2(u_{1m}, u_{2m}) + \oint\!\!\oint \{\gamma_1(u_{1n}, u_{2n})K_{nm}{}'' \\ &\qquad + \gamma_2(u_{1n}, u_{2n})L_{nm}{}''\}\, ds_{n1}\, ds_{n2} + v_{t1m} = 0 \end{aligned}\right\} \qquad (6.3)$$

In any three-dimensional flow the motion adjacent to a boundary must clearly be two-dimensional. The Dirichlet boundary condition requires that the resulting two velocity components parallel to the surface vorticity sheet interior must be zero. This condition has thus been completely satisfied here by stating that the internal velocity components in the u_{1m} and u_{2m} directions are independently zero, demanding two equations, one for each coordinate direction at element m. Although these represent the application of independent boundary conditions for u_{1m} and u_{2m}, it is clear that the local vorticity components at element n, $\gamma_1(u_{1n}, u_{2n})$ and $\gamma_2(u_{1n}, u_{2n})$ both make contributions to the surface parallel velocity at element m and in both coordinate directions. Consequently the kernels of both boundary integral equations involve contributions to the Dirichlet condition due to bound and shed vorticities with appropriate coupling coefficients K_{nm}', K_{nm}'' to deal with bound vorticity and L_{nm}', L_{nm}'' to deal with shed vorticity.

This is apparently all just as we would expect when applying the Biot–Savart law in vectors in a three-dimensional flow with the constraint of two-dimensional motion at all body surfaces. However several qualifying remarks need to be made as follows:

(i) The two integral equations cannot be solved independently since they each contain the $2M$ unknown vortex strengths $\gamma_1(u_{1n}, u_{2n})$ and $\gamma_2(u_{1n}, u_{2n})$. In numerical form they are each represented by M linear but coupled equations.

(ii) In view of this coupling they could be combined to form a single set of $2M$ equations assuming that $\gamma_1(u_{1n}, u_{2n})$ and $\gamma_2(u_{1n}, u_{2n})$ may be treated as independent of one another.

(iii) However this assumption seems to be ill-founded since we have just shown from Helmholtz's theorem (6.1a) and (6.1b) that the bound and trailing surface sheet vorticities are not independent but obey an explicit relationship of the form $\gamma_2 = f(\gamma_1)$. In view of this, it seems as already stated that (6.1b) should be used to remove all of the shed vorticity unknowns $\gamma_2(u_{1n}, u_{2n})$ from the kernel of say (6.3a), leaving only the $\gamma_1(u_{1n}, u_{2n})$ unknowns. Equation (6.3a) alone is then apparently a completely sufficient statement of the Dirichlet boundary condition. Equation (6.3b) may be discarded.

(iv) Where downstream trailing vortex sheets are shed finally into the body wake, their induced velocities must also be included in either the second or third terms of (6.3a) and will normally be also expressible in terms of the cumulatively shed vorticity components on the body itself.

237

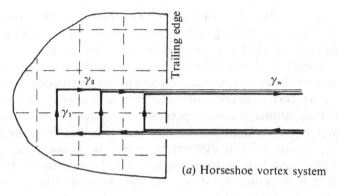

(a) Horseshoe vortex system

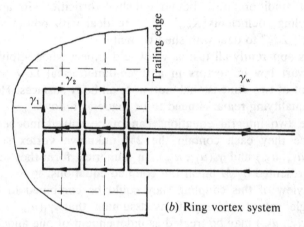

(b) Ring vortex system

Fig. 6.2. Horseshoe and ring vortex systems for modelling surface vorticity.

V. P. Hill (1978) drew attention to these qualifications when applying the above generalised equations to the specific problem of the flow past annular aerofoils and ducted propellers. His analysis, which will be dealt with in some detail in Section 6.3, was based upon the horseshoe vortex principle illustrated in Figs. 6.1 and 6.2 where various possible numerical models are portrayed. These may be summarised as follows:

(a) Individual vortex element model
When developing the surface vorticity method for plane flow past an ellipse, Section 1.11, it was pointed out that symmetry about the major and minor axes resulted in fourfold redundancies, that is massive over-specification. Though wasteful this did not create any

238

special difficulties in derivation of the solution. Likewise one could ignore observation (iii) above and use both of equations (6.3), treated as a single set to solve for $\gamma_1(u_{1n}, u_{2n})$ and $\gamma_2(u_{1n}, u_{2n})$ independently and directly under the full two-dimensional boundary conditions. Helmholtz's equation (6.1b) should then be satisfied automatically but could be used as a check upon accuracy. The main advantage of this approach is its numerical simplicity and its main disadvantage the double size of matrix required. There are many possible variations in numerical procedure to avoid this disadvantage such as iterative schemes which solve say equation (6.3a) only for the $\gamma_1(u_{1a}, u_{2n})$ values, inserting successively up-graded estimates of $\gamma_2(u_{1n}, u_{2n})$ derived from (6.2) between itera-tions. In other words equations (6.3a) and (6.2) may be treated as coupled or simultaneous equations. In fact analytical elimination of $\gamma_2(u_{1n}, u_{2n})$ from (6.3a) by direct substitution from (6.2) is a possibility but results in extremely complex recurrence relationships due to cumulative growth of the trailing vorticity as one proceeds downstream from one element to the next.

(b) Horseshoe vortex model
The problem just referred to can be eliminated if each bound vortex element $\gamma_1(u_{1n}, u_{2n}) \Delta s_{n1}$ is treated as part of a 'horseshoe' vortex by the introduction of trailing vortices of equal strength extending to and from downstream infinity. The trailing vortices must be made to lie along the $\gamma_2(u_{1n}, u_{2n})$ tracks element by element until the trailing edge is reached and then to follow some prescribed direction to infinity downstream, such as that of the main stream velocity vector or improvements on this derived iteratively. This is the most popular technique, advocated by Hill (1978) and Turbal (1973), since all shed vorticity is accounted for without the need for reference to $\gamma_2(u_{1n}, u_{2n})$. Equation (6.3a) can then be adapted by deleting the L_{nm}' coupling coefficients and replacing K_{nm}' by horseshoe vortex equivalents. These would involve Biot–Savart law contributions from unit strength vortex elements for all contributors to the horseshoe vortex on the body downstream of n plus the two wake trailing vorticity filaments, Fig. 6.2(a).

(c) Ring vortex model
An alternative method for imposing vorticity conservation element by element is to split each γ_1 and γ_2 filament into two pieces, recombining them into square shaped ring vortex elements as

depicted in Fig. 6.2(*b*). This is equivalent to terminating each horseshoe vortex of method (b) at the next element downstream and beginning a new one. Although this method seems to offer an extremely neat model involving an identical form of coupling coefficient for all elements, there are certain inherent difficulties caused by the attempt to avoid specific modelling of the cumulative vortex shedding process. The residual bound and shed vorticity values must be obtained finally by subtracting the ring vortex strengths of adjacent elements. The ring vortex strengths themselves are cumulative as one proceeds downstream from the leading edge which may endanger accuracy when dealing with large numbers of elements.

On the whole, of these three proposed models the horseshoe vortex method is to be recommended. It is of particular interest to note that the source panel method presents no such problems for non-lifting bodies since no attention to vortex shedding is required in such situations. On the other hand vortex sheets must be introduced into all lifting body models including the panel method and the prescribed methods for introducing bound and trailing vorticity all seem to generate inherent difficulties. Let us now consider two engine applications of surface vorticity modelling to three-dimensional flows.

6.3 Three-dimensional flow past annular aerofoils and engine cowls

V. P. Hill (1978) proposed the vortex system illustrated in Fig. 6.3 to model the flow past an engine cowl or annular aerofoil set at an angle of incidence to a uniform stream W_∞. The duct surface is covered by bound and shed vorticity sheets handled numerically by discretisation onto a curvilinear grid, together with a tubular trailing vortex sheet extending to infinity downstream. Turbal (1973) adopted a similar surface vorticity model to simulate the flow past a non-axisymmetric duct, extending his theory also to include the influence of a propeller located inside the duct modelled as an actuator disc. In both cases as illustrated by Fig. 6.3, an appropriate coordinate system (s, θ) was chosen to represent the annular aerofoil surface, where θ is the angular coordinate and s is the distance along the profile perimeter measured from the leading edge for any given (r, θ) meridional plane.

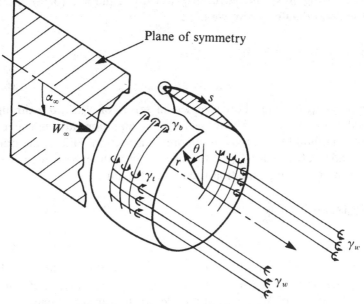

Fig. 6.3. Vortex system for an angular aerofoil in a uniform stream with angle of attack (Reproduced from the Proceedings of the Institution of Mechanical Engineers by permission of the Council of the Institution.)

Expressing the vorticity sheet continuity equation (6.2) in this coordinate system we have the relationship

$$\Delta\gamma_t(s, \theta) = -\frac{1}{r}\frac{\partial\gamma_b(s, \theta)}{\partial\theta}\,\Delta s \tag{6.4}$$

where, following Hill's notation, $\gamma_b(s, \theta)$ and $\gamma_t(s, \theta)$ are the local bound and trailing vorticity components respectively, equivalent to γ_1 and γ_2 above. Due to the cross-wind $W_\infty \sin\alpha_\infty$, the axisymmetric annular aerofoil sheds trailing vortex sheets from both inner and outer surfaces which eventually combine at the duct trailing edge to form the tubular vortex sheet wake γ_w as illustrated in Fig. 6.3. For each single surface element the additional contribution $\Delta\gamma_t(s, \theta)$ to the trailing vorticity is given by (6.4). At any given location m on the surface, the local trailing vorticity is thus the sum of all such contributions delivered by upstream elements,

$$\gamma_t(s_m, \theta_m) = \sum_{n=1}^{m} \Delta\gamma(s_n, \theta_m) \tag{6.5}$$

241

Expressing now the governing integral equation (6.3a) in the chosen coordinate system we have

$$\tfrac{1}{2}\gamma_b(s_m,\ \theta_m) + \oint_s \oint_\theta \{\gamma_b(s_n,\ \theta_n)K_{mn}$$

$$+ \gamma_t(s_n,\ \theta_n)L_{mn}\}r_n\ \mathrm{d}\theta_n\ \mathrm{d}s_n + v_{tm} = 0 \qquad (6.6)$$

where $\gamma_t(s_n,\ \theta_n)$ may be expressed in terms of $\gamma_b(s_n,\ \theta_n)$ through equations (6.4) and (6.5). In this form the equations are sufficient to follow method (a) outlined in the previous section. Hill on the other hand advocates method (b), the use of horseshoe vortices, whereupon the governing equation reduces to the form

$$\tfrac{1}{2}\gamma_b(s_m,\ \theta_m) + \oint_s \oint_\theta \gamma_b(s_n,\ \theta_n)K_{mn}(s_m,\ s_n,\ \theta_m,\ \theta_n)r_n\ \mathrm{d}\theta_n\ \mathrm{d}s_n$$

$$+ v_{tm}(s_m,\ \theta_m) = 0 \qquad (6.7)$$

which states the Dirichlet boundary condition in the s direction for any point on the aerofoil surface $(s_m,\ \theta_m)$. The term v_{tm} is of course the component of W_∞ resolved parallel to the surface, namely

$$v_{tm} = U_\infty \cos \beta_m + V_\infty \sin \beta_m \cos \theta_m \qquad (6.8a)$$

and the coupling coefficient K_{mn} now absorbs all vorticity induced velocities due to γ_b, γ_t and the wake trailing vortex filaments γ_w. As an alternative to this we could remove the influence of γ_w from the coupling coefficient and introduce its induced velocities (u_w, v_w) into v_{tm} instead, rewriting (6.8a)

$$v_{tm} = (U_\infty + u_w) \cos \beta_m + (V_\infty \cos \theta_m + v_w) \sin \beta_m \qquad (6.8b)$$

Turbal (1973) tended towards this type of approach which offers some computational attractions, especially when extending the analysis to ducted propellers to include other influences such as those imposed by ship's hull interference. (u_w, v_w) must then be derived iteratively by successive approximations.

To further simplify computation, Hill (1978) recommended that sufficient accuracy will be retained if the tubular vortex wake is assumed to be a cylindrical surface coaxial with the annular aerofoil. His comparisons with experimental test agreed extremely well on this basis below the stalling angle of attack of the duct (say $\alpha_\infty = \pm 15°$).

Broadly speaking there are two numerical approaches to the solution of the foregoing equations.

242

(i) Series expansion of the bound vorticity $\gamma_b(s_m, \theta_m)$ in the θ direction leads to some useful reductions in computational requirements

(ii) Direct solutions of the integral equation (6.7) for the given discrete surface elements, Fig. 6.3.

We shall now summarise some of the work completed along these two directions.

6.3.1 Numerical scheme using circumferential series expansions

Linearised methods for aerofoils, annular aerofoils and non-axisymmetric flow past ducts such as those by Morgan (1961) and George (1976), (1978) provide a fund of ideas for the implementation of useful functions or series expansions to reduce analysis. Such analytical devices are frequently ignored in surface vorticity analysis in favour of purely numerical models. However V. P. Hill (1975), (1978) applied such techniques to the present problem to great effect by proposing a series expansion to the bound vorticity in the θ direction. For mathematical simplicity Hill assumed that the $\gamma_b(s_n, \theta_n)$ could be expressed in the form

$$\gamma_b(s_n, \theta_n) = \bar{\gamma}_b(s_n)f_n(\theta_n) \tag{6.9}$$

where $f_n(\theta)$ is a function of general form whose coefficients vary only with s_n. In particular Hill chose the Fourier series

$$\gamma_b(s_n, \theta_n) = W_\infty(A_{0n} + A_{1n} \cos \theta_n + A_{2n} \cos 2\theta_n$$
$$+ \ldots + A_{Pn} \cos P\theta_n) \tag{6.10}$$

Because of symmetry about the plane $\theta = 0$ it is necessary here to specify only the half range cosine series. Indeed the choice of Fourier expansion rather than any other type of series was not without good reason. Recalling the potential flow solution for a circular cylinder in a cross-flow (1.27) the character of the trailing vorticity on, say, a body of revolution due to the transverse velocity component $W_\infty \sin \alpha_\infty$ would be of the form

$$\gamma_t \approx 2W_\infty \sin \alpha_\infty \sin \theta \tag{6.11}$$

Bearing in mind (6.4) the character of the bound vorticity in diagonal flow past such a body would be

$$\gamma_b \approx K \cos \theta \tag{6.12}$$

Thus for bodies of revolution with incidence one would expect the first term in the Fourier series to characterise the circumferential variations extremely well. For annular aerofoils there is less certainty although one might still expect the first term to dominate. Indeed Hill (1978) noted that these characteristics of cross-flow could be observed from the experimental tests carried out by Bagley & Purvis (1972) for RAE cowl number 3. Based upon his own experimental and theoretical investigations of a NACA 66_2–015 annular aerofoil, Hill observed that the surface circumferential pressure distribution exhibited, like that of Bagley & Purvis, a gradual transition from near $\cos \theta_n$ variation close to the leading edge, to $\cos 2\theta_n$ variation further downstream. Because the first two terms of the Fourier series characterise the bound vorticity circumferential variations so well, only a few terms in the series need be retained resulting in enormous reduction in the number of unknown coefficients.

Introducing (6.9) into the integral equation results in the modified general form

$$\tfrac{1}{2}\bar{\gamma}_b(s_m)f_m(\theta_m) + \oint_s \gamma_b(s_n)\bar{K}_{mn}(s_m, s_n, \theta_m)\,\mathrm{d}s_n + v_t(s_m, \theta_m) = 0$$

(6.13)

where the new coupling coefficient is given by

$$\bar{K}_{mn}(s_m, s_n, \theta_m) = \int_0^{2\pi} f_n(\theta_n)K_{mn}(s_m, s_n, \theta_m, \theta_n)r_n\,\mathrm{d}\theta_n$$

(6.14)

Since the integral in this expression may be evaluated for any selected fixed circumferential location θ_m, θ_n ceases to remain a variable. We may then write the governing integral equation in the simplified form

$$\oint_s \bar{\gamma}_b(s_n)\bar{K}_{mn}(s_m, s_n)\,\mathrm{d}s_n + v_{tm}(s_m) = 0$$

(6.15)

where the local velocity discontinuity term $\tfrac{1}{2}\bar{\gamma}_b(s_m)f(\theta_m)$ has been absorbed into the self-inducing coupling coefficient $\bar{K}_{mm}$. Here it should be noted as before, that the coefficients $\bar{K}_{mn}$ represent the sum total velocity at any element located at (s_m, θ_m) due to the complete set of periodic vortex rings at n implied by the Fourier expansion of $\bar{\gamma}_b(s_n)$, together with their complementary trailing vortex sheets extending from s_n to infinity downstream. Such a model then satisfies all of the qualifications (i) to (iv) listed in

244

Section 6.2. Furthermore for zero angle of attack (6.15) simply reduces to the previous derivation for axisymmetric flow in Chapter 4, namely (4.1).

Returning to further consideration of the Fourier expansion for $\gamma_b(s_n, \theta_n)$, (6.10) several observations may be made. Firstly we take note that velocity and therefore line vorticity scale has been introduced by the common multiplier W_∞, so that the series coefficients A_{pn} are dimensionless. Secondly we observe that the first term is independent of θ involving the constant A_{0n} only and thus caters entirely for the axisymmetric part of the solution due to the mainstream velocity component $W_\infty \cos \alpha_\infty$ parallel to the duct axis. Introducing the series, the coupling coefficient becomes

$$
\left.
\begin{aligned}
\bar{K}_{mn}(s_m, s_n) = A_{0n}\int_0^{2\pi} K_{mn}(s_m, s_n, \theta_m, \theta_n)r_n\,\mathrm{d}\theta_n \\[2ex]
+ \sum_{p=1}^{P} A_{pn}\int_0^{2\pi} \cos p\theta_n K_{mn}(s_m, s_n, \theta_m, \theta_n)r_n\,\mathrm{d}\theta_n
\end{aligned}
\right\}
\tag{6.14a}
$$

If for the moment we ignore the presence of the trailing vorticity, the circumferential integrals in this equation may be evaluated term by term for the known expressions for K_{mn} derived from the Biot–Savart law applied to unit strength vortex elements coincident with $\gamma_b(s_n, \theta_n)$. Thus the coupling coefficient $\bar{K}_{mn}(s_m, s_n)$ reduces to a series of the form

$$
\bar{K}_{mn}(s_m, s_n) = A_{0n}I_0 + A_{1n}I_1 + A_{2n}I_2 \ldots A_{Pn}I_P
\tag{6.16a}
$$

where the integrals $I_0, I_1 \ldots I_P$ are explicitly stated in terms of the prescribed geometry of the duct. If the aerofoil profile in a given (x, θ) meridional plane is represented by M elements, then a total of $P \times M$ unknown coefficients are to be determined at as many locations on the duct surface, resulting in the required selection of P meridional sections with M pivotal points on each.

Thirdly, returning to the need to include the influence of shed vorticity γ_t, one advantage of Hill's series expansion method is the analytic simplicity which obtains when (6.10) is substituted into the vortex sheet continuity equation (6.4). The trailing vorticity is then given explicitly in terms of the coefficients A_{pn} through

$$
\begin{aligned}
\gamma_{tm} &= \frac{W_\infty}{r_m} \sum_{p=1}^{P} \left\{ \sum_{n=m+1}^{M/2} \Delta s_n p A_{pn} \sin p\theta_n \right. \\
&\qquad \left. + \tfrac{1}{2} \Delta s_m p A_{pn} \sin p\theta_m \right\} \quad \begin{array}{l} \text{outer} \\ \text{surface} \end{array} \\
&= \frac{W_\infty}{r_m} \sum_{p=1}^{P} \left\{ \sum_{n=M/2+1}^{M} \Delta s_n p A_{pn} \sin p\theta_n \right. \\
&\qquad \left. + \tfrac{1}{2} \Delta s_m p A_{pm} \sin p\theta_m \right\} \quad \begin{array}{l} \text{inner} \\ \text{surface} \end{array}
\end{aligned}
\right\} \tag{6.17}
$$

which takes the form of series expansion in $\sin p\theta_m$. The coupling coefficient $\bar{K}_{mn}(s_m, s_n)$ as expressed in (6.16a) could then be modified to include the effects of trailing vorticity as follows

$$
\bar{K}_{mn}(s_m, s_n) = \sum_{p=0}^{P} A_p(I_p + J_p) \tag{6.16b}
$$

where, by analogy with I_p,

$$
J_p = \int_0^{2\pi} \sin p\theta_m L_{mn}(s_m, s_n, \theta_m, \theta_n) r_n \, \mathrm{d}\theta_n
$$

and L_{mn} is the coupling coefficient due to a unit strength trailing vortex parallel to $\gamma_t(s_n, \theta_n)$.

Further analysis would be unproductive for present purposes and the reader is referred to V. P. Hill (1978) for full details of the derivation of complete coupling coefficients. Clearly, as stated the governing integral equation (6.15) is over-simplified since it is unable to express the individual Biot–Savart law influence of both bound and trailing vorticities. Even though they are related directly through the same set of coefficients A_{pn}, separate unit coupling coefficients K_{mn} and L_{mn} are required in the kernel of integral equation (6.15) to reflect their individual effects. In practice this requirement is easier to express directly in a numerical scheme specification than by analytical statements.

V. P. Hill (1975) undertook extensive experimental tests for an annular aerofoil of chord/diameter ratio 0.6, typical of pump jet dimensions and employing the symmetrical section NACA 66_2–05, chosen because of its 'roof top' stable surface pressure distribution. His cowl was precision manufactured in perspex and provided with 64 surface pressure tappings. This commendable piece of high

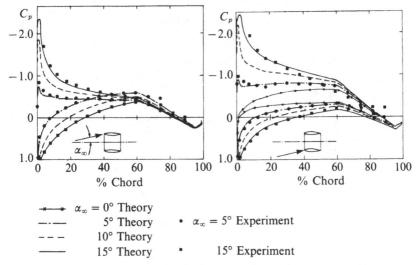

Fig. 6.4. Surface pressure distributions on plane of symmetry of annular aerofoil NACA 66_2-015 at angle of attack (Reproduced from the Proceedings of the Institution of Mechanical Engineers by permission of the Council of the Institution.)

quality experimental work now provides a valuable benchmark for validation of theoretical calculations. Details of his duct profile coordinates have already been given in Chapter 4, Fig. 4.7, and a comparison between experiment and theory for zero angle of attack in Fig. 6.4. Extremely good agreement was obtained. Further comparisons with Hill's predictions are shown in Fig. 4.19 for measurements taken on opposite sides of the duct in the plane of symmetry $\theta = 0$. Theoretical results are given for $\theta = 0°$, 5°, 10° and 15° together with experimental results for $\theta = 5°$ and 15°. Extremely good agreement was obtained in all cases. Equally valid results were predicted for other meridional planes confirming the capabilities of Hill's series expansion method using only a few terms of the series. In particular the more complex flow over the first two-thirds of the duct profile, which is responsible for important angle of attack effects such as lift generation and the sorting out of entry flow, is predicted with remarkable accuracy. At higher angles of attack separation occurred on the downstream surface of the duct when that surface was diffusing most of the way.

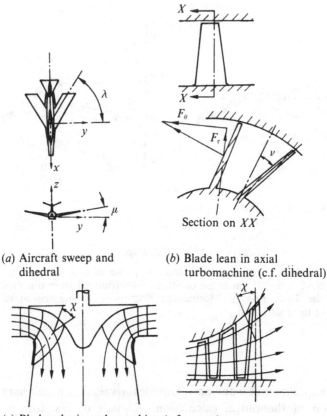

(a) Aircraft sweep and dihedral

(b) Blade lean in axial turbomachine (c.f. dihedral)

(c) Blade rake in turbomachine (c.f. sweep)

Fig. 6.5. Aircraft liting surface sweep and dihedral and blade rake and sweep equivalents in turbomachines. (Reproduced from the Proceedings of the Institution of Mechanical Engineers by permission of the Council of the Institution.)

6.4 Sweep and dihedral in turbomachine blade rows

Sweep and dihedral are well known functional design parameters of aircraft wings, Fig. 6.5(a). Dihedral improves lateral stability while sweep permits higher flight Mach numbers prior to the onset of shock induced drag. The sweep angle λ is defined in the plan view as the angle between the stacking line and the y axis. Dihedral μ is defined in the projection at right angles to this viewed along the flight path, again as the angle between the stacking line and the y axis. λ and μ are thus independent design variables for fixing the orientation of the stacking line.

Sweep and dihedral in turbomachine blade rows

Turbomachinery blade rows also exhibit analogous geometrical properties. To agree with the aircraft notation, true sweep and dihedral should be defined by viewing the blade stacking line normal to the surface mapped out by the vector mean velocity* W_∞ in the case of λ and along this surface for μ. We return to these definitions shortly when considering blade cascades. The more usual practice however when selecting blade row stacking lines is to introduce two alternative angles of orientation of more convenience to the designer, Fig. 6.5(b) and (c). The meridional sweep or rake angle χ is defined in the meridional (s, n) plane as the angle between the stacking line and the local normal to the meridional streamline and is closely related to sweep. The blade lean angle v is defined in the (r, θ) plane in the view taken along the x axis and is normally the angle between the stacking line and the (s, n) plane, which is strongly related to dihedral. In the literature χ and v are often referred to as sweep and dihedral but they clearly do differ from the aircraft definition and we shall use the terms rake and lean instead.

Sometimes rake and lean are the inescapable by-products of mechanical constraints such as blade twist or profile stacking. At the present time on the other hand there is much interest in the deliberate use of blade lean as a design variable. Thus the accidental introduction of local blade rake is inevitable in low specific speed Francis turbines, Fig. 6.5(c), due entirely to the geometrical constraints of the annulus and blade stacking requirements. Likewise in low pressure multistage axial turbines, casing flare, to accommodate decreasing density, can result in rake angles as great as 45°. In these applications blade lean, as a by-product of blade stacking, is normally small, but is now commonly introduced deliberately in stators to generate radial force components F_r, Fig. 6.5(b), which provide some design control over the radial distribution of mass flow.

These problems were firstly seriously addressed with regard to turbomachines by L. H. Smith & Yeh (1963) who demonstrated that blade rake is a primary cause of S-1 stream surface twist†. We will consider this effect in the following section (6.4.1) for cascades of infinite aspect ratio. Disturbances of the circumferentially averaged meridional flow are also generated by rake and blade lean and

---

* See Section 2.6.2 for definition of vector mean velocity in a cascade.
† See Section 3.1 and Fig. 3.1 for a discussion of the S-1, S-2 surface model of Wu (1952).

Three-dimensional flows in turbo-machines

these were investigated for incompressible flow by Lewis & J. M. Hill (1971) and J. M. Hill (1971), (1975) by extensions of actuator disc theory for incompressible flow. More recently Potts (1987) applied time marching models to this problem for high speed subsonic flows, investigating also major meridional disturbances which are caused by blade thickness blockage identified previously by Lewis & Horlock (1969). The study of these circumferentially

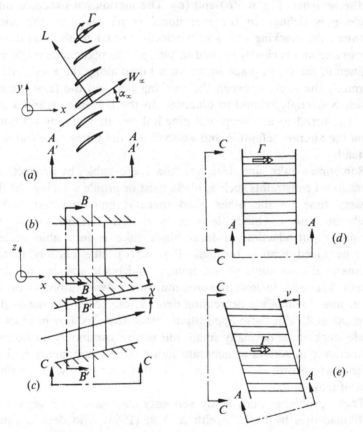

(a) Section on CC in (x, y) plane

(b) Section on AA in (x, z) plane – zero rake

(c) Section on $A'A'$ in (x, z) plane – rake angle χ

(d) Section on BB in (y, z) plane – zero blade lean

(e) Section on $B'B'$ in (y, z) plane – blade lean ν

Fig. 6.6. Definition of cascade rake and blade lean. (Reproduced from the Proceedings of the Institution of Mechanical Engineers by permission of the Council of the Institution.)

250

averaged meridional disturbances led to the development of a three-dimensional surface vorticity model for swept cascades by Graham (1972) and Graham & Lewis (1974) linked to experimental validation experiments by J. M. Hill and Lewis (1974). This analysis will be reviewed briefly in Section 6.4.3. More recently, Thompkins & Oliver (1976), J. D. Denton (1974) and Potts (1987) have investigated this problem analytically by time-marching methods, while Whitney *et al.* (1967) published experimental investigations of a 45° raked gas turbine blade with thick profiles and of low aspect ratio, which are valuable for validation purposes.

In view of the geometrical complexity presented by turbomachine annular blade rows, a clearer perception of blade rake and lean and their relationships to sweep and dihedral, may be obtained by considering the simpler situations of the rectilinear cascade, Fig. 6.6. Rake χ and lean v are then defined independently in the (x, z) and (y, z) planes respectively as illustrated by Figs. 6.6(c) and (e). The true sweep angle λ is defined as the angle between the stacking line and the z axis viewed parallel to the (x, y) plane and normal to W_∞. The true dihedral is defined as the angle between the stacking line and the z axis viewed parallel to the (x, y) plane but in the direction of W_∞. As a simple approximation the bound vortex may be substituted for the stacking line in this case.

6.4.1 Swept aerofoils and cascades of infinite aspect ratio

As already mentioned, the strongest effect induced by sweep or rake is the twisting of the $S-1$ surfaces, a well known feature of both single aerofoil and cascade flows. This is best illustrated by tracing out the path of the stagnation streamline as it divides between the upper and lower surfaces of the aerofoils. Fig. 6.7 illustrates the consequent pattern viewed in the direction along the line of a compressor cascade. For simplicity infinite aspect ratio is assumed here to avoid the additional problems of end effects. Although the swept flow is three-dimensional it consists essentially of a two-dimensional blade-to-blade flow on the (x, y) planes induced by the uniform stream component $W_\infty \cos \chi$ and a superimposed translational velocity $W_\infty \sin \chi$ in the z direction. Because of the greater surface velocity component on the upper surface in the (x, y) plane due to the blade-to-blade flow, it is clear that the

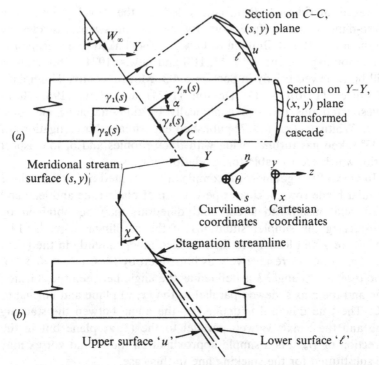

Fig. 6.7. Surface streamlines on a raked (swept) cascade with infinite aspect ratio. (Reproduced from the Proceedings of the Institution of Mechanical Engineers by permission of the Countil of the Institution).

stagnation streamline in a given meridional plane will not close again at the trailing edge but that the flow will remain twisted downstream. Thus even though this flow is essentially two-dimensional a residual three-dimensional distortion of the meridional stream surfaces originally lying in the (s, y) plane is introduced. As shown by Lewis & J. M. Hill (1971) and illustrated by surface vorticity calculation and experiment, Fig. 6.8, this streamline shift can be of similar magnitude to the throat dimension of a typical steam turbine tip section nozzle and thus represents a significant departure from the assumed quasi-two-dimensional modelling of surfaces of revolution proposed in Section 3.2.

The stream surface twist can be thought of also in terms of surface vorticity modelling with reference to Fig. 6.7(*a*). Following the designers strategy of Section 3.2, the normal design requirement is to define blade profile geometry on the meridional (s, y) surfaces

252

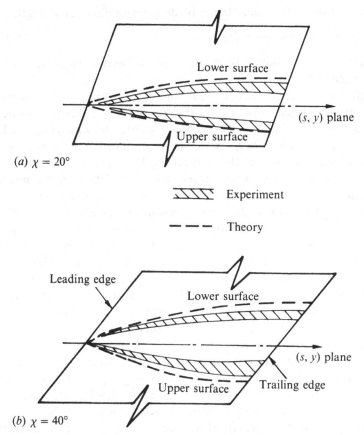

(a) $\chi = 20°$

XXXX Experiment

$- - - -$ Theory

(b) $\chi = 40°$

Fig. 6.8. Surface streamline shifts due to meridional stream surface twist in two raked (swept) turbine nozzle cascades. (Reproduced from the Proceedings of the Institution of Mechanical Engineers by permission of the Council of the Institution.)

to achieve prescribed fluid deflection in the y direction. This will be associated with bound vorticity strength $\gamma_n(s)$ normal to the (s, y) plane. It is evident from the stream surface twist that trailing vorticity $\gamma_s(s)$ will also be present on the blade surfaces but not in the wake downstream. Although the problem appears complex for surface vorticity analysis it is in fact extremely simple as can be seen if we resolve the surface vorticity into the (x, y) plane and the z direction. Thus in the (x, y) plane we would have a constant surface vorticity of known strength $\gamma_2(s) = W_\infty \sin \chi$ equal to the translational velocity. In the z direction we would have a local component

253

of bound vorticity to handle the blade-to-blade flow, of strength

$$\gamma_1(s) = \gamma_n(s) \cos \alpha + \gamma_s(s) \sin \alpha \qquad (6.18)$$

located along the line of intersection of the meridional surface with the cascade blades.

After some reflection it is clear that $\gamma_1(s)$ is exactly the same as the surface vorticity strength for the transformed cascade as viewed on section $Y-Y$, Fig. 6.7. This projection, originally proposed by L. H. Smith & Yeh (1963), permits the application of plane two-dimensional analysis in the projection $Y-Y$ plane to derive the exact solution of the twisted three-dimensional flow in the neighbourhood of the meridional planes. Although strictly limited to infinite aspect ratio blade rows, Graham (1972) and Lewis (1978) have shown experimentally by annular cascade tests that the design technique is remarkably successful even for aspect ratios of less than 5 with 40° of rake. Perhaps more important, this study shows the value of forethought in the early stages of surface vorticity (or other) flow modelling, which in this case led to the reduction of a seemingly three-dimensional flow into a straightforward two-dimensional equivalent for which there is available existing methodology.

To conclude this section a comparison is shown in Fig. 6.9 for the surface pressure distribution and outlet angle predictions for the turbine nozzle cascade given by Graham (1972) according to the proposed projection technique. Also shown is the surface pressure distribution predicted ignoring sweep, by surface vorticity analysis applied directly in the meridional (s, y) plane (conventional method). The influence of sweep is made plain by the two theoretical calculations and likewise the improvements in prediction achieved by the surface vorticity project method. Of particular importance to the turbine designer is the evidence that outlet angle decreases markedly with rake. For large rake angles of say 40° the decrease in outlet angle for the zero α_1 nozzle (Case 1) is about 7.2° and the conventional blade to blade theory is 2° in error compared with the projection method. Published experimental results for this cascade, J. M. Hill & Lewis (1974), show close agreement for α_2 by the projection method, and likewise the test results of Graham (1972) for an annular cascade. The predicted surface pressure distribution, Fig. 6.9, is likewise quite different from that predicted by the traditional method of blade to blade analysis or the meridional surface of revolution intersection ignoring rake. This

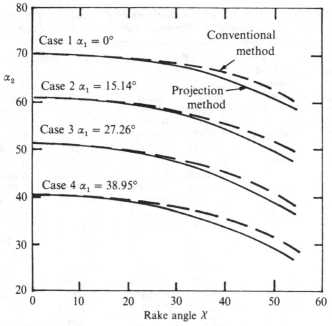

(a) Outlet angle versus rake for four turbine cascades

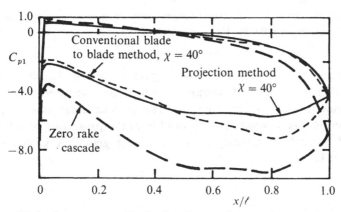

(b) Surface pressure distribution for nozzle cascade (case 1)

Fig. 6.9. Turbine cascade outlet angle and surface pressure distributions with rake. (Reproduced from the Proceedings of the Institution of Mechanical Engineers by permission of the Council of the Institution.)

Table 6.1. *Blade profiles used for blade to blade analysis of four swept turbine cascades with infinite aspect ratio, Figs. 6.8 and 6.9*

Chord position x/chord	Thickness of base profile y_t/chord	Parabolic camber lines y_c/chord			
		Case 1 nozzle	Case 2	Case 3	Case 4 impulse
1	0	0	0	0	0
0.975	0.00695	0.01058	0.01182	0.01230	0.01248
0.95	0.00775	0.02087	0.02333	0.02427	0.02463
0.925	0.0088	0.03087	0.03452	0.03590	0.03643
0.9	0.0097	0.04058	0.04536	0.04718	0.04788
0.85	0.01251	0.05904	0.06600	0.06864	0.06967
0.8	0.0154	0.07617	0.08516	0.08856	0.08988
0.7	0.023	0.10604	0.11855	0.12328	0.12512
0.6	0.0315	0.12921	0.14446	0.15023	0.15247
0.5	0.0398	0.14444	0.16148	0.16793	0.17044
0.4	0.0482	0.15000	0.16770	0.17440	0.17700
0.3	0.0495	0.14347	0.16040	0.16680	0.16929
0.2	0.0455	0.12117	0.13546	0.14087	0.14297
0.15	0.0418	0.10237	0.11445	0.11902	0.12080
0.1	0.036	0.07708	0.08618	0.08962	0.09096
0.075	0.0319	0.05817	0.06503	0.06763	0.06864
0.05	0.0268	0.04373	0.04889	0.05085	0.05161
0.025	0.01906	0.02335	0.02611	0.02715	0.02756
0	0	0	0	0	0

L.E. radius/chord = 0.008. T.E. radius/chord = 0.006.

leads to major errors in the throat region, the projection technique indicating more modest suction pressures as compared with the conventional method of analysis. The experimental studies just referred to again confirmed the validity of the projection technique linked to surface vorticity analysis. Profile geometry is given in Table 6.1.

6.4.2 Swept cascade of finite aspect ratio

So far we have considered only cascades of infinite aspect ratio for which the blade-to-blade solution and local stream surface twist are identical for all spanwise locations. In actual blade rows terminating at the annulus walls, end-wall interference produces additional disturbances of both meridional and blade to blade flows due to the

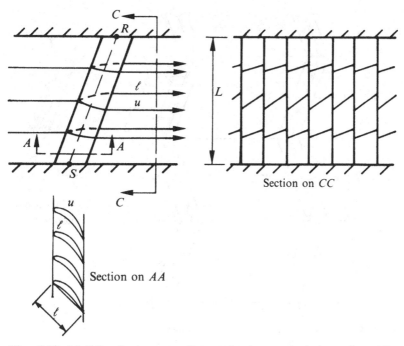

Fig. 6.10. Meridional stream surface twist in a raked (swept) turbine cascade. (Reproduced from the Proceedings of the Institution of Mechanical Engineers by permission of the Council of the Institution.)

constraint imposed upon the stream surface twist. The trailing vorticity $\gamma_s(s)$ is suppressed in this region, reducing to zero at the wall, Fig. 6.10. Meridional flow disturbances arising from this cause may be included in meridional analysis schemes for circumferentially averaged flow such as those developed later in this chapter. Actuator disc models backed by finite difference calculations were used by Lewis & Hill (1971) to explore these meridional effects for swept cascades, revealing significant disturbances as indicated earlier by L. H. Smith & Yeh (1963). The latter authors also investigated blade-to-blade end effects for swept cascades of thin flat plates with an assumed elliptical chordwise loading made constant for all spanwise positions, using a mirror image system similar to that shown in Fig. 6.11(a). These solutions gave some indication of end effects for lightly loaded cascades, although they were not truly representative of untwisted swept cascades for which the bound vorticity would vary along the span with associated trailing vorticity. For this reason Graham & Lewis (1974) examined a progression

257

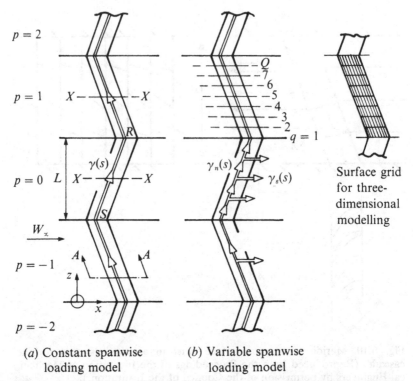

(a) Constant spanwise
loading model

(b) Variable spanwise
loading model

Fig. 6.11. Mirror image system for representing a raked (swept) cascade. (Reproduced from the Proceedings of the Institution of Mechanical Engineers by permission of the Council of the Institution).

from model (a) of Fig. 6.11 with constant spanwise loading to model (b) with full three-dimensional surface vorticity modelling. In the first and simpler of these models the blade profile in the plane XX is represented by M elements in the normal manner for two-dimensional modelling but in this case with zig-zag elements of constant strength $\gamma_n(s)$. Consequently only M equations are required, restricting application of the Dirichlet boundary condition to say the centre section XX and ruling out any representation of vortex shedding. In model (b) on the other hand a number of (x, y) planes are introduced to define a surface grid composed of $M.Q$ parallelogram shaped elements, Fig. 6.11(b), thus permitting full three-dimensional modelling with both bound and shed vorticity accounted for.

In either model the relevant coupling coefficients must be derived from the Biot–Savart law (1.8) which for a line vortex element of

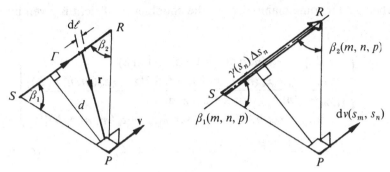

Fig. 6.12. Velocity induced by a line vortex or line vorticity element of a swept cascade.

strength Γ and of finite length, yields an induced velocity at P equal to

$$v = \frac{\Gamma}{4\pi d}(\cos \beta_1 + \cos \beta_2) \tag{6.19}$$

where d is the shortest distance between P and the vortex and β_1 and β_2 are as defined in Fig. 6.12 and v is normal to the plane PRS.

6.4.3 Analysis with constant spanwise loading

If we consider first the simpler model of constant spanwise loading, the induced velocity at any point s_m on section XX, Fig. 6.11(a), can be expressed as an infinite series involving all the mirror-image reflections of the finite length vortex SR between the cascade end walls. Thus for the pth reflection, Fig. 6.12

$$d\mathbf{v}(s_m, s_n) = \frac{\gamma(s_n)\, ds_n}{4\pi d(m, n, p)}(\cos \beta_1(m, n, p) + \cos \beta_2(m, n, p)) \tag{6.20}$$

Resolving this parallel to the blade surface in a direction parallel to the cascade side walls and combining contributions for a cascade of B blades, the elementary velocity at m due to a cascade of B zig-zag vortices of strength $\gamma(s_n)\, \Delta s_n$ becomes

$$\Delta v_m(s_m, s_n) = \sum_{b=1}^{B} \Delta v(s_m, s_n) = \sum_{b=1}^{B} \sum_{p=-\infty}^{\infty} K_1(s_m, s_n)\gamma(s_n)\, \Delta s_n \tag{6.21}$$

259

where, after some manipulation, the coupling coefficient is given by

$K_1(s_m, s_n)$

$$= \left\{ \frac{\begin{array}{l} \cos \beta_m [y_m - y(n, b)](\cos \lambda + \sin \lambda \tan \lambda) \\ -\sin_m[(x_m - x_w(m, n, p, b)] + \tan \lambda[z_m - z_w(m, n, p, b)] \end{array}}{\begin{array}{l} \{[y_m - y(n, b)]^2(\cos \lambda + \sin \lambda \tan \lambda)^2 \\ + [(x_m - x_w(m, n, p, b) - \tan \lambda[z_m - z_w(m, n, p, b)]^2\}^{\frac{1}{2}} \end{array}} \right\}$$

$$\times \left\{ \frac{\cos \beta_1(m, n, p, b) + \cos \beta_2(m, n, p, b)}{d(m, n, p, b)} \right\} \qquad (6.22)$$

and where β_m is profile slope in the meridional (x, y) plane.

The governing (Martensen) integral equation for this situation thus becomes

$$-\tfrac{1}{2}\gamma(s_m) + \frac{1}{2\pi} \sum_{p=1}^{B} \sum_{p=-\infty}^{\infty} K_1(s_m, s_n)\gamma(s_n) \, \Delta s_n$$

$$+ W_\infty(\cos \beta_m \cos \alpha_\infty \cos \lambda + \sin \beta_m \sin \alpha_\infty) = 0 \quad (6.23)$$

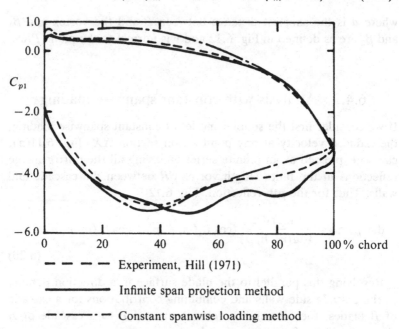

- - - - - Experiment, Hill (1971)

——————— Infinite span projection method

— · —— · — Constant spanwise loading method

Fig. 6.13. Comparison of experiment with theory for (a) infinite span projection method and (b) constant spanwise loading model. (Reproduced from the Proceedings of the Institution of Mechanical Engineers by permission of the Council of the Institution.)

A comparison of the solution of these equations with experimental tests conducted by J. M. Hill (1971) for the 40° sweep cascade, Case 2 of Table 6.1, is shown in Fig. 6.13, exhibiting a considerable improvement upon the infinite span theory for this particular cascade which had a fairly low aspect ratio L/l of 2.2.

6.4.4 Analysis with variable spanwise loading in three-dimensional flow

D. G. Graham (1972) undertook a complete three-dimensional flow analysis of a swept turbine cascade using the model illustrated in Fig. 6.14. Each zig-zig bound vortex $\gamma_n(s_n, q)$ at profile location s_n is broken down into Q elements across the blade span each of different strength. This system is then reflected a large number of times in each side wall. The total number of bound vortex elements whose strengths are to be determined is thus $M \cdot Q$, each one of which has sufficient wall reflections (say 8 in each wall) to render the side walls as plane stream surfaces with sufficient accuracy. In addition a total number $M(Q-1)$ of shed vortex elements $\gamma_s(s_n, q)$ must be accounted for, so that the boundary integral equation for

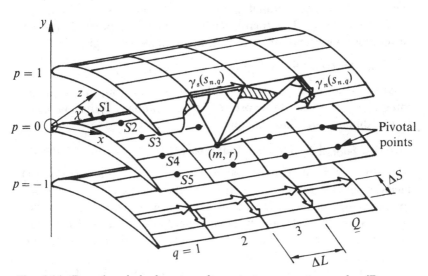

Fig. 6.14. Bound and shed vortex elements on a swept cascade. (Reproduced from the Proceedings of the Institution of Mechanical Engineers by permission of the Council of the Institution)

the Dirichlet boundary condition at element (m, r) takes the form

$$-\frac{\gamma}{2}(s_m, r) + \frac{1}{4\pi} \oint \sum_{q=1}^{Q+1} K_n(s_m, s_n, q)\gamma_n(s_n, q)\,\mathrm{d}s_{nq}$$

$$+ \frac{1}{4\pi} \oint \sum_{q=1}^{Q} K_s(s_m, s_n, q)\gamma_s(s_n, q)\,\mathrm{d}L_{nq}$$

$$+ W_\infty(\cos \beta_{mq} \cos \alpha_\infty \cos \lambda + \sin \beta_{mq} \sin \alpha_\infty) = 0 \quad (6.24)$$

The coupling coefficient for the bound vortex elements $K_n(s_m, s_n, q)$ is identical in form to $\sum_{b=1}^{B} \sum_{p=-\infty}^{\infty} K_1(s_m, s_n)$ as determined by (6.22) to include all blades and all reflections, but in this case for Q spanwise elements. A similar geometrical statement can be derived for the shed vortex element coupling coefficients $K_s(s_m, s_n, q)$. A detailed analysis such as that given by D. G. Graham (1972) is extremely complex and beyond our present purpose, which is to present the primary features of the model and the leading equations. In this respect reference must be made to a further simplification which enables us to eliminate the shed vorticity values $\gamma_s(s_n, q)$ from (6.24). As already discussed in Section 6.1 Helmholtz's theorem may be applied to achieve this end although in this case care is needed since the elements are not curvilinear. By reconsideration of the condition of irrotationality on the surface of the parallelogram shaped elements, Graham has shown that the 'shed' vorticity as defined may be expressed in terms of the bound vorticity of neighbouring elements through

$$\gamma_s(s_n, q) = (\cos \lambda + \sin \lambda)[\gamma_n(s_n, q) - \gamma(s_n, q - 1)] \quad (6.25)$$

After substitution into (6.24) and further reduction, the boundary integral equation can be expressed in the standard form analogous to two-dimensional flow, namely

$$[K(s_m, s_n, q)][\gamma(s_n, q)] = -W_\infty[G(s_m)] \quad (6.26)$$

In this case of course the coupling coefficient $K(s_m, s_n, q)$ includes all the multiple reflections of the vortex element $\gamma(s_n, q)$ for all blades in the cascades and is extremely complex. A fuller summary of the relevant working equations is given by D. G. Graham & Lewis (1974) but the reader should refer to D. G. Graham (1972) for a full treatment.

Predictions for two of the cascades tested by J. M. Hill (1971) with rake angles of 20° and 40° are shown in Fig. 6.15. For convergence to say 0.1% accuracy, studies by Graham suggested

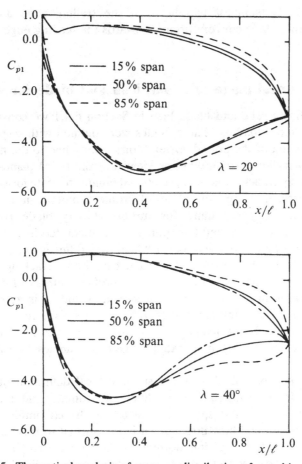

Fig. 6.15. Theoretical analysis of pressure distributions for turbine cascades with rake angles of 20° and 40° assuming variable spanwise loading, D. G. Graham (1972).

that six to eight wall reflections were required and about 11 blades, results being calculated for the centre blade. Computer constraints however limited his model to only five blades and two reflections in each wall. His surface grid likewise was rather coarse permitting only $Q = 5$ spanwise elements with $M = 40$ profile pivotal points. Despite this the results shown in Fig. 6.15 reveal the trends extremely well and show the increasing spanwise variation of surface pressure distribution as the sweep angle is raised from 20° to 40°. In addition we observe from the point of minimum pressure that an important effect of sweep in turbine cascades is a shift of the

263

throat section (smallest gap between the trailing edge and the neighbouring blade convex surface) towards the leading edge.

6.5 Local blade rake and lean and blade forces

When defining rake and blade lean in Section 6.4.1 we considered only the overall loading. These angles were defined with respect to the orientation of the bound vortex Γ or stacking line to the normal z to the meridional streamlines as viewed in the (x, z) planes, Fig. 6.6. This approach was completely adequate to categorise rectilinear cascades comprising identical profile geometry in all (x, y) planes. On the other hand for turbomachinery blade rows in general, such as the mixed-flow pump rotor illustrated in Fig. 6.16, local values of rake and lean occur which may differ for each blade location due to blade twist, a by-product of profile stacking. Thus even though we may be able to stack the profile centres of lift to lie on a normal n in the (s, n) meridional plane resulting in zero rake and lean of the stacking line and therefore of the blade row overall, elsewhere towards the leading and trailing edges the local bound vorticity vector $\Gamma(s, n)$ may exhibit significant angles of rake or lean.

For the purpose of obtaining a suitable local definition of rake and lean, let us construct a curvilinear coordinate grid (s, n, θ) throughout the annular space, in which s is the circumferentially averaged meridional flow direction, n is the normal to this and θ is the circumferential direction normal to both s and n. The blade local bound vorticity vector $\Gamma(s, n)$ may be assumed to lie along the camber surface $\theta(s, n)$ in a direction roughly parallel to the leading and trailing edges and to comprise the three components Γ_s, Γ_n and Γ_θ along the local curvilinear coordinate axes. The local rake angle χ is then defined in the (s, n) planes, Fig. 6.16(a) and (d), as the angle between the normal n and OA, the projection of $\Gamma(s, n)$, namely

$$\chi = \arctan \Gamma_s / \Gamma_n \qquad (6.27)$$

Local blade lean v is defined in the (n, θ) plane, Fig. 6.15(b) and (d), as the angle between the normal n and the camber surface CC, namely

$$v = \arctan(\Gamma_\theta / \Gamma_n) \qquad (6.28)$$

Local blade rake and lean and blade forces

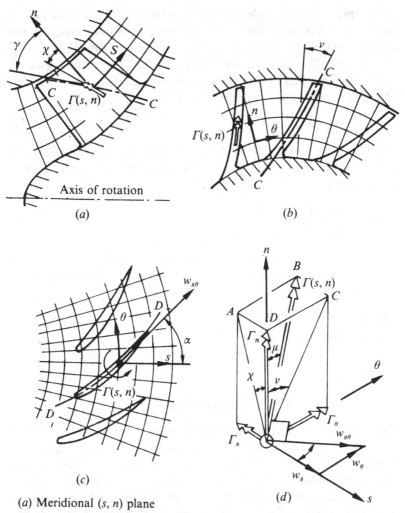

(a) Meridional (s, n) plane
(b) (n, θ) section through blade
(c) Blade-to-blade (s, θ) plane
(d) Velocity and vorticity components

Fig. 6.16. Definitions of local rake and blade lean with respect to camber surface CC. (Reproduced from the Proceedings of the Institution of Mechanical Engineers by permission of the Council of the Institution.)

The true dihedral angle μ may then be obtained by taking the projection of $\Gamma(s, n)$ viewed along the line DD of intersection of the camber surface with the (s, θ) meridional surface of revolution, Fig. 6.15(c) and (d); that is in the direction of the local fluid velocity, which may be taken as the mean of the velocities on the upper and lower surfaces relative to the moving blades

$$w_{s\theta} = \tfrac{1}{2}(w_{su} + w_{sl}) \tag{6.29}$$

Now $w_{s\theta}$ follows the local blade camber angle α in the (s, θ) surface and is normal to the bound vorticity $\Gamma(s, n)$ and its normal component Γ_n. Thus we may write

$$\tan \alpha = \frac{w_\theta}{w_s} = \frac{\Gamma_\theta}{\Gamma_s} \tag{6.30}$$

Thus finally from geometrical considerations expressed by Fig. 6.15(c), we have the following relationships between the angles α, v, χ and μ.

$$\left.\begin{array}{l} \tan \chi = \tan v \cot \alpha \\ \tan \mu = \tan v \operatorname{cosec} \alpha \end{array}\right\} \tag{6.31}$$

6.5.1 Local blade forces

The above discussion relates primarily to the orientation of the local bound vorticity vector in the blade space. Of special importance from the draughting viewpoint are the angles defining the intersections of the blade envelope with the (s, n, θ) coordinate system, required for manufacturing instructions. Closely related to these and of importance for expressing the local blade forces are the angles of intersection of the camber surface $\theta(s, n)$ with the (s, n, θ) coordinate system, namely α, v and γ where the additional angle γ is defined as the angle between the normal n and the intersection CC of the camber surface with the (s, n) meridional plane, Fig. 6.16(a). The camber surface has already been expressed in the form $\theta(s, n)$ i.e. in terms of two independent variables. The lean and rake angles likewise can therefore be expressed uniquely as $v(s, n)$ and $\gamma(s, n)$ respectively and similarly the blade deflection angle $\alpha(s, n)$. To specify the meridional flow equations for the region within the blades it is necessary next to relate the distributed body forces, F_s, F_n, F_θ required for an axisymmetric or circumferentially averaged flow to these camber surface angles.

To begin with we can state that the normal stress $\boldsymbol{\sigma}$ or surface pressure at the blade surface (assumed here to be the camber surface) must also be normal to the fluid velocity relative to the surface, namely

$$\mathbf{w} \cdot \boldsymbol{\sigma}_s = 0$$

Thus in (s, n, θ) coordinates we have

$$w_s \sigma_s + w_\theta \sigma_\theta = 0$$

If we now assume that the distributed body force is coincident with $\boldsymbol{\sigma}$ (i.e. normal to the camber line), we also have the relationship

$$\frac{F_s}{F_\theta} = -\frac{w_\theta}{w_s} = -\tan \alpha \qquad (6.32)$$

A relationship between F_n and F_θ may also be obtained in terms of intersection angle γ by reference to Fig. 6.17 which shows the body force components F_s, F_n in the (s, n) meridional plane. Since the body force $\mathbf{F} = \mathbf{i}F_s + \mathbf{j}F_n + \mathbf{k}F_\theta$ is normal to the blade camber surface, then the component of this $\mathbf{F}_{sn} = \mathbf{i}F_s + \mathbf{j}F_n$ must also be normal to the line of intersection CC of the camber surface and the

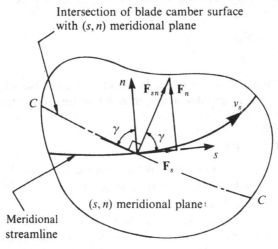

Fig. 6.17. Blade force components lying in the (s, n) meridional plane related to intersection with the blade camber surface.

267

(s, n) plane. Consequently we may write

$$\frac{F_n}{F_s} = \tan \gamma$$

or

$$\frac{F_n}{F_\theta} = -\tan \alpha \tan \gamma \qquad (6.33)$$

These equations have been derived in coordinates relative to a rotor spinning with angular velocity Ω. For later use we may express the absolute fluid velocities in terms of (w_s, w_θ) through

$$\left.\begin{array}{l} v_s = w_s \\ v_\theta = w_\theta + r\Omega \end{array}\right\} \qquad (6.34)$$

where Ω is positive for rotation in the θ direction. We note that for the selected curvilinear grid which follows the streamlines w_n and v_n are both zero.

6.6 Equations of meridional flow for bladed regions

In Section 5.5 it was shown that the principal governing equations for incompressible axisymmetric flow in the bladeless regions of turbomachines may be expressed in terms of Stoke's stream functions as follows

$$\frac{\partial^2 \psi}{\partial r^2} - \frac{1}{r}\frac{\partial \psi}{\partial r} + \frac{\partial^2 \psi}{\partial x^2} = -\omega_\theta r \qquad (5.45)$$

where the tangential vorticity ω_θ is related to gradients of stagnation pressure p_0 and angular momentum or 'vortex' (rv_θ) through

$$\omega_\theta = v_\theta \frac{\mathrm{d}(rv_\theta)}{\mathrm{d}\psi} - \frac{r}{\rho}\frac{\mathrm{d}p_0}{\mathrm{d}\psi} \qquad (5.46)$$

Meridional flow analysis involves the solution of (5.45) by one of many possible methods, subject to a stated spatial distribution of $\omega_\theta(x, r)$ which satisfies (5.46).

Now Stoke's equation (5.45) is true under all circumstances whereas the auxiliary equation (5.46) as stated is not applicable to regions occupied by the blade rows. The objective of this section is to derive an alternative auxiliary equation for such situations which

268

allows for the interaction between the blades and the meridional flow. This follows from a consideration of the equations of motion (5.42) rewritten in (s, n, θ) coordinates. Following Lewis & Hill (1971) we have

$$
\left.
\begin{aligned}
-\frac{1}{\rho}\frac{\partial p_0}{\partial s} + F_s &= v_\theta \omega_n \\[2mm]
-\frac{1}{\rho}\frac{\partial p_0}{\partial n} + F_n &= v_s \omega_\theta - v_\theta \omega_s \\[2mm]
F_\theta &= -v_s \omega_n
\end{aligned}
\right\}
\tag{6.35}
$$

where the vorticity components ω_s and ω_n for axisymmetric flow are given by

$$
\left.
\begin{aligned}
\omega_s &= \frac{1}{r}\frac{\partial (v_\theta r)}{\partial n} \\[2mm]
\omega_n &= -\frac{1}{r}\frac{\partial (v_\theta r)}{\partial s}
\end{aligned}
\right\}
\tag{6.36}
$$

The tangential vorticity may then be derived from (6.35b), namely

$$
\omega_\theta v_s = F_n - \frac{1}{\rho}\frac{\partial p_0}{\partial n} + \frac{v_\theta}{r}\frac{\partial (v_\theta r)}{\partial n}
\tag{6.37}
$$

Since the (s, n) coordinates lie along and normal to the meridional streamlines of our assumed axisymmetric flow, the equivalent to equations [5.44] become

$$
\left.
\begin{aligned}
v_s &= \frac{1}{r}\frac{d\psi}{dn} \\[2mm]
v_n &= 0
\end{aligned}
\right\}
\tag{6.38}
$$

For the special case of zero distributed blade force, for which we have proved also that $(v_\theta r)$ and p_0 are functions of ψ only (5.48) it is clear that (6.37) reduces to (5.46) as expected. Within a blade row on the other hand $(v_\theta r)$ and p_0 may vary both normal to and along the meridional streamlines. Indeed, the purpose of a turbomachine is to produce changes in p_0 in the s direction within the rotor by imposing changes in $(v_\theta r)$ through the blade-to-blade profile interaction with the meridional flow. This may be expressed analytically if we first introduce (6.32) and (6.34) into the third equation of motion (6.35), whereupon we have

$$
F_s = -F_\theta (v_\theta - r\Omega)/v_s = (v_\theta - r\Omega)\omega_n
\tag{6.39}
$$

269

Three-dimensional flows in turbo-machines

Introducing this into the first equation of motion (6.35a) we then have

$$\frac{1}{\rho}\frac{\partial p_0}{\partial s} = \Omega \frac{\partial(v_\theta r)}{\partial s} \tag{6.40}$$

This is the well known Euler pump (or turbine) equation, applicable independently along each meridional streamline. Upon integration the more familiar form for incompressible flow is obtained, relating the local stagnation pressure p_0 at any point inside the blade row to the value p_{01} at entry to the blade row and the change in vortex strength $(v_\theta r - v_{\theta 1} r_1)$, namely

$$\frac{1}{\rho}(p_0 - p_{01}) = \Omega(v_\theta r - v_{\theta 1} r_1) \tag{6.41}$$

If we now take the partial derivative of this with respect to n we have a companion equation to (6.40) for the quantity $\partial p_0/\partial n$.

$$\frac{1}{\rho}\frac{\partial p_0}{\partial n} = \frac{1}{\rho}\frac{\partial p_{01}}{\partial n} + \Omega\frac{\partial(v_\theta r)}{\partial n} - \Omega\frac{\partial(v_{\theta 1} r_1)}{\partial n}$$

After substitution into (6.3), the tangential vorticity may then be expressed in terms of the inlet conditions p_{01} and $v_{\theta 1} r_1$ and the local swirl velocity v_θ inside the blade through

$$\omega_\theta v_s = -\frac{1}{\rho}\frac{\partial p_{01}}{\partial n} + \Omega\frac{\partial(v_{\theta 1} r_1)}{\partial n} + F_n + \frac{(v_\theta - r\Omega)}{r}\frac{\partial(v_\theta r)}{\partial n}$$

From (6.32) and (6.34) we may introduce

$$v_\theta - r\Omega = v_s \tan \bar{\alpha}$$

where strictly speaking here $\bar{\alpha}$ is the local swirl angle of the circumferential average of the blade-to-blade flow, resulting in

$$\omega_\theta v_s = -\frac{1}{\rho}\frac{\partial p_{01}}{\partial n} + \Omega\frac{\partial(v_{\theta 1} r_1)}{\partial n} + F_n + \frac{v_s \tan \bar{\alpha}}{r}\frac{\partial}{\partial n}(r\Omega + v_s \tan \bar{\alpha})r$$

If F_n is now eliminated by introducing (6.33) and (6.35c), we have finally

$$\omega_\theta = -\frac{1}{\rho v_s}\frac{\partial p_{01}}{\partial n} + \frac{\Omega}{v_s}\frac{\partial(v_{\theta 1} r_1)}{\partial n} - \frac{\tan \alpha \tan \gamma}{r}\frac{\partial}{\partial s}(r^2\Omega + rv_s \tan \bar{\alpha})$$

$$+ \frac{\tan \bar{\alpha}}{r}\frac{\partial}{\partial n}(r^2\Omega + rv_s \tan \bar{\alpha}) \tag{6.42}$$

270

To summarise, we have now expressed the tangential vorticity explicitly in the following form

$$\omega_\theta = f(\underbrace{p_{01}, (r_1 v_{\theta 1})}_{\substack{\text{Inlet} \\ \text{conditions}}}, \underbrace{\Omega, \alpha, \gamma,}_{\substack{\text{Rotor} \\ \text{design} \\ \text{data}}} \underbrace{\alpha, v_s}_{\substack{\text{Blade-to-blade} \\ \text{design}}}) \qquad (6.42c)$$

where the last four variables are all known functions of (s, n). Since the meridional velocity v_s appears in both of the governing equations (5.45) and (6.42), they are clearly coupled and must be solved iteratively. The usual procedure is as indicated in the flow diagram.

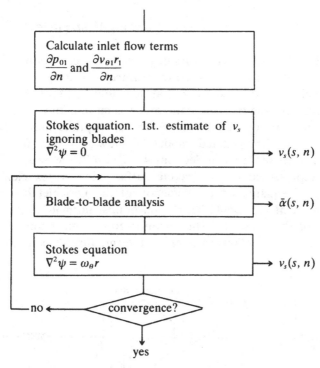

6.7 Axisymmetric meridional flows in mixed-flow turbomachines

To conclude this chapter we will consider a surface vorticity boundary integral meridional flow analysis for mixed-flow tur-

271

bomachines recently developed by Mughal (1989) and published in brief by Lewis & Mughal (1989). Although this work was applied to the flow in bladeless spaces, its extension to meridional flow in bladed regions should present no special problems regarding the boundary integral formulation, although the programming of equations (6.42) implies considerable geometrical analysis to determine the blade camber surface interactions with the proposed (s, n, θ) coordinate system.

Instead of Stoke's differential equation (5.45), the problem may be stated as a boundary integral equation of the following form

$$\oint K(m, n)\gamma(s_n)\,\Delta s_n - \tfrac{1}{2}\gamma(s_m) + \oiint L(m, ij)\omega_\theta(m, ij)\,\mathrm{d}A_{ij}$$

$$+ \oint M(m, s)\eta_s = 0 \quad (6.43)$$

This equation states the Dirichlet boundary condition on the annulus wall at point m due to (i) the surface vorticity $\gamma(s_n)$, (ii) the spatially distributed vorticity ω_θ at locations ij throughout the annulus and (iii) the contributions from S semi-infinite vortex tubes of strength η_s introduced to induce the appropriate mass flow at inlet and exit. The related model is shown in Fig. 6.18. As explained in more detail in Sections 4.6 and 4.7, the annulus wall may be represented by M discrete surface ring vortex elements providing M equations for satisfaction of the Dirichlet boundary conditions at M representative pivotal points, normally the centre locations of the elements. The equation for pivotal point m then transforms to the following numerical form, directly suitable for

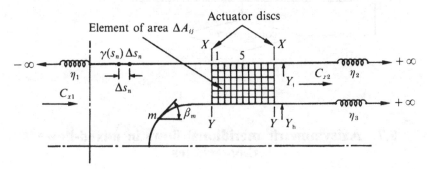

Fig. 6.18. Axisymmetric meridional flow model using grid structure in boundary integral analysis.

computation,

$$\sum_{n=1}^{M} K(m, n)\gamma(n)\,\Delta s_n + \sum_{i=1}^{I}\sum_{j=1}^{J} L(m, ij)\omega_{\theta ij}\,\Delta A_{ij} + \sum_{s=1}^{S} M(m, s)\eta_s = 0$$

$$(6.44)$$

The Kernel terms have the following physical significance and analytical form. $K(m, n)$ is the velocity parallel to the annulus surface at m due to a unit ring vortex at n, namely

$$K(m, n) = U_{mn}\cos\beta_m + V_{mn}\sin\beta_m \tag{6.45}$$

where (U_{mn}, V_{mn}) are the axial and radial velocity components due to a unit strength ring vortex at the pivotal point n, (4.8)–(4.11).

Kernel $M(m, s)$ is the parallel velocity at m induced by a semi-infinite vortex tube of unit strength, namely

$$M(m, s) = u_{cms}\cos\beta_m + v_{cms}\sin\beta_m \tag{6.46}$$

where (u_{cms}, v_{cms}) are defined by (4.26)–(4.30) with $\Gamma = 1$.

Kernel $L(m, ij)$ represents the velocity parallel to the wall at m due to a discrete ring vortex at location (i, j) of a grid introduced to contain and break down the distributed tangential vorticity ω_θ of the meridional flow. If the vorticity strength in cell (i, j) is $\omega_{\theta ij}$ and the cell area is ΔA_{ij}, we have

$$L(m, ij) = \Delta u_{mij}\cos\beta_m + \Delta v_{mij}\sin\beta_m \tag{6.47}$$

where, for increased accuracy, the velocities Δu_{mij} and Δv_{mij} may be evaluated by introducing PXQ sub-elements following the technique outlined in detail for ring vortex 'blocks' in Section 5.7, namely

$$\Delta u_{mij} = \frac{1}{PQ}\sum_{p=1}^{P}\sum_{q=1}^{Q} U_{mpq}, \quad \Delta v_{mij} = \frac{1}{PQ}\sum_{p=1}^{P}\sum_{q=1}^{Q} V_{mpq} \tag{6.48}$$

where U_{mpq}, V_{mpq} are again given by (4.8)–(4.11).

6.7.1 Flow through an actuator disc in a cylindrical annulus

Fig. 6.18 illustrates the annulus geometry of a test case considered by Mughal (1989) for which an axial blade row is assumed to shed a prescribed distribution of vorticity $\omega_\theta(\psi)$ from an equivalent actuator disc XX. Following Horlock (1952), (1958), classical actuator disc theory yields the following solution for the axial velocity profile

$$
\begin{aligned}
c_x &= C_x + c_{x\infty}'(\tfrac{1}{2}e^{kx/H}) && x < x_0 \\
&= C_x + c_{x\infty}'(1 - \tfrac{1}{2}e^{-kx/H}) && x > x_0
\end{aligned} \Bigg\} \tag{6.49}
$$

where $c_{x\infty}'$ is the perturbation from the mean C_x of the radial equilibrium velocity profile. Assuming zero radial velocities, this can be related to the precribed tangential vorticity through (5.43c), with zero radial velocity for radial equilibrium flow, whereupon

$$
c_{x\infty} = C_x + c_{x\infty}' = K - \int_{r_h}^{r_t} \omega_\theta \, dr \tag{6.50}
$$

The constant K follows from the mass flow continuity equation for the annulus, namely

$$
\dot{m} = 2\pi\rho \int_{r_h}^{r_t} c_{x\infty} r \, dr = \rho\pi C_x (r_t^2 - r_h^2) \tag{6.51}
$$

For the numerical calculations the meridional vorticity was represented by the rectangular ring vortex blocks shown in Fig. 6.18. The following guide lines of Mughal (1989) were used for sub-elements.

(i) To compute velocities on the face of an adjacent RRV use 10×10 sub-elements.
(ii) To compute velocities induced at the centres of neighbouring RRVs use 5×5 sub-elements.
(iii) To compute velocities further distant from a cell, no sub-elements are needed. Use a concentrated ring vortex at the cell centre position.

For this particular study, for a hub/tip ratio $h = 0.5$, $k = 3.1967$ (Horlock (1958)). A constant strength vorticity $\omega_\theta = 1.0$ was introduced throughout the grid region resulting in the following radial equilibrium velocity profile,

$$
c_{x\infty} = C_x + \left[\frac{2}{3} \frac{(r_t^3 - r_h^3)}{(r_t^2 - r_h^2)} - r \right] \omega_\theta \tag{6.52}
$$

Axial velocity profiles are compared in Fig. 6.19.

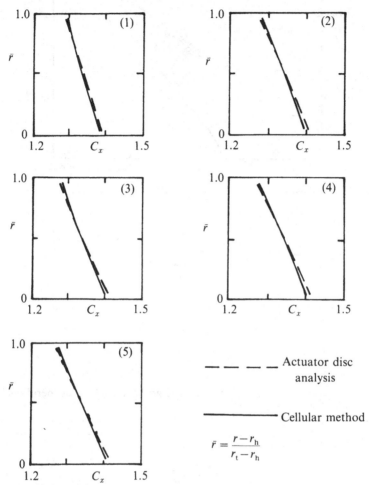

Fig. 6.19. Comparison of cellular method with actuator disc analysis of flow through axial blade row.

6.7.2 Meridional flow through a mixed-flow turbomachine

Fig. 6.20 illustrates the extension of this boundary integral method to a mixed-flow fan where a cellular grid is introduced into the region of blade row activity. The solution of this test case is divided into two stages. In stage 1 the solution is obtained for the bladeless annulus as described in full for this particular fan annulus in Section 4.7. This solution corresponds to the case of $\omega_\theta(m, ij) = 0$

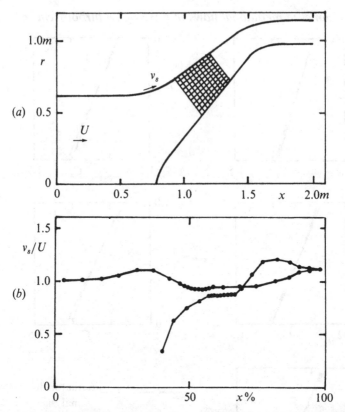

Fig. 6.20. (a) Mixed-flow fan annulus and grid for analysis of meridional flow. (b) Surface velocity distribution.

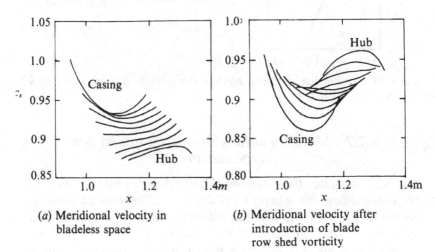

(a) Meridional velocity in bladeless space

(b) Meridional velocity after introduction of blade row shed vorticity

Fig. 6.21. Meridional analysis of rotational flow through mixed flow fan by boundary integral + cell method.

in the governing integral equation (6.43). In stage 2 the solution is obtained with distributed vorticity. In this test case a value $\omega_\theta(i, j) = 1.0$ was spread throughout the grid region uniformly for the purposes of illustration only. In a full turbomachine computational scheme $\omega_\theta(i, j)$ must be determined from equations (6.42) by successive approximations over a series of iterative calculations to account for the correct meridional/blade-to-blade interactions. A comparison of the predicted meridional velocities is given in Fig. 6.21, illustrating the major meridional disturbances caused by the presence of tangential vorticity in the grid region.

PART II

Free shear layers, vortex dynamics and vortex cloud analysis

CHAPTER 7

Free vorticity shear layers and inverse methods

7.1 Introduction

So far we have considered only the case of fully attached inviscid steady flows, for which the introduction of a surface vorticity sheet of appropriate strength and of infinitesimal thickness, together with related trailing vorticity in three-dimensional flows, is completely adequate for a true representation. As pointed out in Chapter 1, where the justification of this model was argued from physical considerations, the surface vorticity method is representative of the infinite Reynolds number flow of a real fluid in all but one important respect, namely the problem of boundary layer separation. Real boundary layers involve complex mechanisms characterised by the influence of viscous shear stresses and vorticity convections and eddy formation on the free stream side. Depending upon the balance between these mechanisms and the consequent transfer of energy across a boundary layer, flow separation may occur when entering a rising pressure gradient, even at very high Reynolds numbers. Flow separation at a sharp corner will most certainly occur as in the case of flow past a flat plate held normal to the mainstream direction.

For a decade or so the development of computational fluid dynamic techniques to try to model these natural phenomena has attracted much attention and proceeded with remarkable success. The context of a good deal of this work has fallen rather more into the realm of classical methods than that of surface vorticity modelling, and is often classified by the generic title Vortex Dynamics. Contrary to this the author's aim during this period has been specifically to extend the surface vorticity method, with its distinctively physical basis and numerical flexibility, into the realm of separated flows by taking further advantage of the properties of free vorticity shear layers. The aim of Part II of this book is to outline these developments of the vorticity method as a continuation and extension of the surface vorticity method towards the modelling of real fluid effects. In the present chapter we begin by considering the classical free streamline model of flow separation. This leads naturally into the notion of inverse design methods in which the surface vorticity sheet is treated as if it were a flexible or free shear layer of prescribed strength able to adopt its own natural shape to accommodate a prescribed surface velocity or pressure distribution. Finally we return to the free shear layers to consider some problems of self-convection and instability pertinent to later chapters dealing with full vortex cloud theory.

7.2 The free-streamline model

Free-streamline theory as presented in standard texts, e.g. Lamb (1945), Vallentine (1967), deals primarily with plane two-dimensional separated flows making use of conformal transformations and complex potential analysis. The consequent underlying assumption of constant stagnation pressure throughout the flow field renders this a rather primitive flow model for representing bluff body wake flows but nevertheless provides a useful starting point. A physical interpretation may be gained from Fig. 7.1 which illustrates free-streamline flow separation from the crest of a ridge. According to this model the wake is filled with motionless fluid with constant pressure equal to the stagnation pressure. Consequently the static pressure and therefore fluid velocity are constant along the entire outside edge of the separation streamline ψ_0 and equal to the ambient conditions p_∞ and U_∞. Furthermore the velocity discontinuity across the separation streamline implies the presence of a vorticity sheet extending from the separation point P to downstream infinity Q which divides the outer flow from the wake and is of constant strength (see Section 1.5).

$$\Gamma(s) = U_\infty \tag{7.1}$$

Such flows are thus well suited for modelling by the surface vorticity method and it is surprising that there is so little published work taking advantage of this. The author (1978) undertook such studies employing the surface vorticity model illustrated in Fig. 7.1(*b*). Making use of the mirror system outlined in Section 1.11 the model comprised

(i) a vorticity sheet of initially unknown strength $\gamma(s)$, bound to the body surface APB, and

282

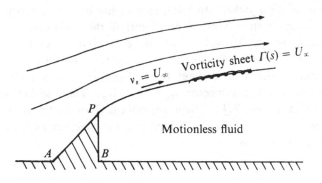

(*a*) Free-streamline separation from a ridge

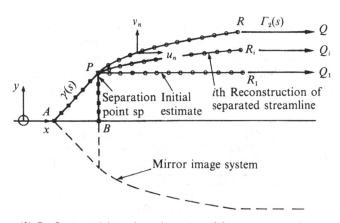

(*b*) Surface vorticity mirror image model

Fig. 7.1.   Surface and free vorticity model of 'free streamline' flow separation from a sharp edged ridge.

(ii) a shed vorticity sheet *PRQ* of prescribed strength $\Gamma(s) = U_\infty$ made up of a curved section *PR* and a straight section *RQ* extending to $x = \infty$ in order to complete the wake.

Fig. 7.1(*b*) illustrates the numerical strategy involved to arrive at the final contour of the free streamline *PR* by successive approximations. The vorticity sheet *PR* is treated as a flexible chain, initially located on some prescribed first estimate of the free streamline such as the line PR_1 parallel to the x axis. The resulting

Martensen solution yields a first estimate of the bound vorticity $\gamma(s)$ required to satisfy the boundary condition on the body surface. The velocity components (u_n, v_n) at the centre of each element $\Gamma(s_n)$ of the separation layer PR_i are then evaluated, permitting a reconstruction of the vortex sheet by realignment of each element with the local velocity vector $q_n = \mathbf{i}u_n + \mathbf{j}v_n$. The closing semi-infinite vortex sheet RQ is then repositioned to begin from the latest location of R_i. Two conditions can be checked as a test of convergence as follows.

(i) R_i approaches a limiting position.
(ii) The surface vorticity strengths of elements along PB should approach zero to satisfy the requirement of stagnation conditions in the wake region.

In practice the second of these requirements is difficult to satisfy precisely at the element sp adjacent to the separation point. As can be seen from Fig. 1.14, P is a singular point in fully attached potential flow. Smooth progression from the bound vortex sheet $\gamma(s)$ on AP to the free vortex sheet PR at P should eliminate this

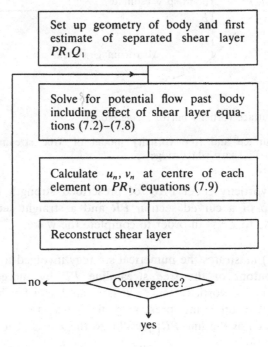

singularity entirely. In practice this is a vulnerable region where the numerical model is open to errors, but these can be reduced by closer pitching of the surface and free vorticity elements around P.

The computational sequence may be summarised as shown in the flow diagram on the previous page.

To complete the specification of this flow model, the equations referred to in the flow diagram are as follows. Potential flow past the body subject to the influence of the uniform stream U_∞ and the separation vorticity layer $\Gamma(s)$ is expressed by the Martensen equation for element m, namely

$$\sum_{n=1}^{M} K(s_m, s_n)\gamma(s_n) = -(U_\infty + U_m + U_{om})\cos\beta_m - (V_n + V_{om})\sin\beta_m$$

(7.2)

where the induced velocities due to the flexible shear layer PR represented by N elements are

$$\left. \begin{aligned} U_m &= \Gamma(s)\sum_{i=1}^{N} U(s_m, s_i) \\ V_m &= \Gamma(s)\sum_{i=1}^{N} V(s_m, s_i) \end{aligned} \right\}$$

(7.3)

The various coupling coefficients follow from the considerations of Section 1.9 (1.52) and (1.53), from which we can show that

$$\left. \begin{aligned} U(s_m, s_i) &= \frac{\Delta s_i}{2\pi}\left\{\frac{y_m - y_i}{a_{mi}^2} - \frac{y_m + y_i}{b_{mi}^2}\right\} \\ V(s_m, s_i) &= \frac{\Delta s_i}{2\pi}(x_i - x_m)\left\{\frac{1}{a_{mi}^2} - \frac{1}{b_{mi}^2}\right\} \end{aligned} \right\}$$

(7.4)

where

$$\left. \begin{aligned} a_{mi} &= \sqrt{\{(x_m - x_i)^2 + (y_m - y_i)^2\}} \\ b_{mi} &= \sqrt{\{(x_m - x_i)^2 + (y_m + y_i)^2\}} \end{aligned} \right\}$$

(7.5)

The self-inducing coupling coefficients need to include the influence of the mirror image vortex element whereupon

$$U(s_m, s_m) = \left(-\frac{1}{2} + \frac{\Delta\beta_m}{4\pi} \right) \cos \beta_m - \frac{\Delta s_m}{4\pi y_m}$$

$$V(s_m, s_m) = \left(-\frac{1}{2} + \frac{\Delta\beta_m}{4\pi} \right) \sin \beta_m \qquad\qquad \left.\right\} \qquad (7.6)$$

For all values of m the bound vorticity coupling coefficients are then given by

$$K(s_m, s_n) = U(s_m, s_n) \cos \beta_m + V(s_m, s_n) \sin \beta_m \qquad (7.7)$$

In addition, the velocities induced at any location (x_m, y_m) due to the semi-infinite wake vortex sheet $\Gamma(s)$ and its mirror image, Fig. 7.2, are given by

$$U_{om} = \frac{\Gamma(s)}{2\pi} (\phi_{2m} - \phi_{1m})$$

$$V_{om} = \frac{\Gamma(s)}{2\pi} \ln(b_{1m}/a_{1m}) \qquad\qquad \left.\right\} \qquad (7.8)$$

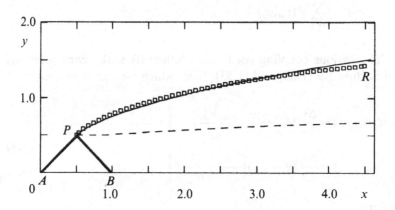

(1) —————— Free streamline exact solution
(2) – – – – Numerical method, $\Gamma(s)/U_\infty = 1.0$
(3) □ Numerical method, $\Gamma(s)/U_\infty = 1.15$

Fig. 7.2. Free streamline flow from a wedge shaped body.

286

where

$$\phi_{1m} = \pi - \arctan\{(y_m - y_R)/(x_R - x_m)\}$$

$$\phi_{2m} = \pi - \arctan\{(y_m + y_R)/(x_R - x_m)\}$$

$$a_{1m} = \sqrt{[(y_m - y_R)^2 + (x_m - x_R)^2]}$$

$$b_{1m} = \sqrt{[(y_m + y_R)^2 + (x_m - x_R)^2]}$$

$$(7.9)$$

Once the potential flow solution for $\gamma(s)$ has been obtained, convection velocities at element j of the free shear layer PR follow from

$$u_j = U_\infty + \sum_{n=1}^{M} \gamma(s_n)U(s_j, s_n) + \Gamma(s) \sum_{i=1}^{N} U(s_j, s_i) + U_{oj}$$

$$v_j = \sum_{n=1}^{M} \gamma(s_m)V(s_j, s_m) + \Gamma(s) \sum_{i=1}^{N} U(s_j, s_i) + V_{oj}$$

$$(7.10)$$

If for example the N elements of the flexible wake vortex sheet are of length $\Delta\ell$, then the change in coordinates from beginning to end of element j can be estimated by

$$\Delta x_j = \frac{u_j \Delta\ell}{\sqrt{(u_j^2 + v_j^2)}}, \qquad \Delta y_j = \frac{v_j \Delta\ell}{\sqrt{(u_j^2 + v_j^2)}} \qquad (7.11)$$

permitting integration of the free-streamline shape (x_i, y_i) through

$$x_i = x_P + \sum_{j=1}^{i} \Delta x_j, \qquad y_i = y_P + \sum_{j=1}^{i} \Delta y_i \qquad (7.12)$$

The application of this method to free streamline flow separation from a wedge shaped body is shown in Fig. 7.2, for which the flexible vortex layer PR was limited in length to about six times AP. Convergence was obtained with thirty iterations introducing a damping coefficient of 0.5. Two studies are compared in Fig. 7.2. With the vortex sheet strength set at the correct value $\Gamma(s) = U_\infty$, it is clear that the numerical model gave a serious underestimate of free-streamline deflection, curve (2). The reason for this is thought to be the problem of leakage flux due to failure to satisfy any boundary conditions on the semi-infinite vortex wake RQ. Evidence for this was the presence of significant velocity on the wake enclosed rear face *PB* of the body, signifying noticeable fluid motion within the wake region. To compensate for this the free stream vortex sheet strength $\Gamma(s)$ was increased incrementally thus providing an additional backwash velocity, until a value was found for which the surface velocity on *PB* was negligible ($v_s < 0.01\ U_\infty$). As can be seen, a value of $\Gamma(s)/U_\infty = 1.15$ produced an extremely good prediction of the free-streamline contour.

Such numerical experiments show that the predicted shape of the free streamline is extremely sensitive to the value of $\Gamma(s)$, perhaps an indication of instabilities to which vortex sheets are prone under their self-induced motion. Flow past a sharp edged plate normal to U_∞, Fig. 7.3, presents testing conditions in which the vortex sheet at the point of separation is actually normal to the uniform stream. Even worse results were obtained with $\Gamma(s)/U_\infty = 1.0$. In this case it

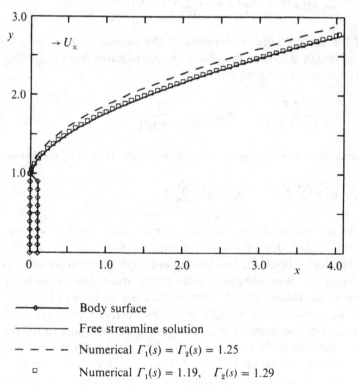

———◇——— Body surface

——————— Free streamline solution

— — — — Numerical $\Gamma_1(s) = \Gamma_2(s) = 1.25$

□   Numerical $\Gamma_1(s) = 1.19,\quad \Gamma_2(s) = 1.29$

Fig. 7.3. Free streamline separation from a sharp edged plate.

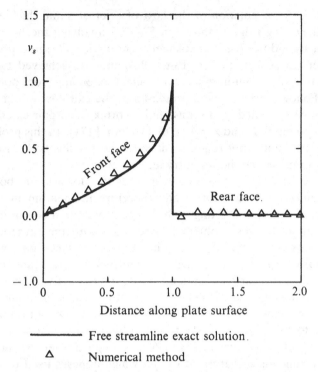

Fig. 7.4. Velocity distribution along surface of sharp edged plate in free streamline separated flow.

was found advantageous to adopt different values of $\Gamma_1(s)$ for the flexible vortex sheet and $\Gamma_2(s)$ for the downstream semi-infinite vortex sheet. After some experimentation values respectively of 1.19 and 1.29 resulted in an excellent match of the separation streamline to the exact solution and likewise close agreement between the predicted surface velocity distribution on the front surface of the plate, Fig. 7.4. The strategy adopted here was to run several cases with increasing values of $\Gamma_1(s) = \Gamma_2(s)$ until velocities on the rear face of the plate were reduced to zero (a value of 1.25 in this case). $\Gamma_1(s)$ and $\Gamma_2(s)$ were then adjusted on either side of this to provide further improvement to the streamline curvature.

7.3 Free jets

This numerical model may readily be extended to deal with jet flows by the introduction of additional vortex sheets. An obvious analogous model for simulation of the flow of a jet of initial thickness h over a deflecting ridge is shown in Fig. 7.5 involving the introduction of a second flexible vortex sheet of strength $-\Gamma(s)$ to represent the outer surface of the jet. Entry flow may be achieved by the introduction of a semi-infinite vortex sheet extending from point Q to $-\infty$. Following the previous guide-lines, the exit flow would then naturally be modelled by a semi-infinite vortex sheet pair extending between points S, R and $x = +\infty$, of strength $\pm\Gamma(s)$. In this problem a uniform stream is not required since the vortex sheets themselves are able to introduce the jet entry and exit fluxes.

However, such an arrangement imposes the downstream boundary condition of zero velocity and therefore momentum in the y direction at $x = \infty$, conflicting with the jet momentum set up in the field of interest. This problem is associated with the presence of leakage fluxes through the semi-infinite vortex sheet pair which cause major disturbances in the neighbourhood of the junctions S and R and significant backflows on the rear surface of the body. A simpler approach, which leads to much better results, is to omit the downstream semi-infinite vortex sheet pair completely, allowing the exit flux to vent freely. A solution on this basis with $h = 0.1$ and $\Gamma(s) = V_j = 1.0$, is shown in Fig. 7.6. Although the jet exit flow is causing erroneous curvature of the jet which imposes itself to some extent in the near field, the solution is generally acceptable close to the ridge, and the surface velocity prediction likewise. The velocity builds up to approximately 1.0 on the upwind face as expected and is negligible on the downwind face of the ridge.

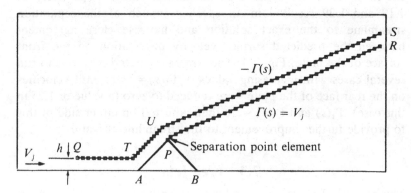

Fig. 7.5. Initial specification of vortex sheet model for flow of a jet over a ridge.

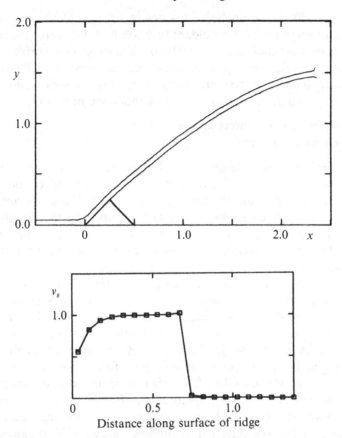

Fig. 7.6. Prediction of jet deflection and associated body surface velocity distribution.

An essential improvement above the free streamline model was the introduction of some constraint over the direction of flow separation of the lower layer. The first element leaving the separation point was set at the body slope as indicated in Fig. 7.5 and maintained at this angle throughout the computation. Such arrangements considerably improve predictions for both jet and free streamline flows.

7.4 Inverse aerofoil design

Turbomachine problems are often categorised as either direct (analysis) or inverse (synthesis). Direct methods involve the fluid

291

dynamic performance analysis of a device of prescribed geometry and such design aids are abundant in published literature. Inverse methods on the other hand, whereby the shapes of flow surfaces of engineering components are automatically selected to achieve a prescribed velocity or pressure distribution, are less well explored. Broadly speaking, singularity inverse methods are of two types.

(i) Iterative use of a direct analysis
(ii) True inverse methods

The first of these categories involves successive guesses at the final shape. For each iteration the latest estimate of device geometry (aerofoil, blade, duct, etc) is analysed and its actual surface velocity is compared with that prescribed by the designer. The error is used to contrive a correction to the geometry which will reduce the error. Wilkinson (1967a) achieved great success in the application of this technique to the design of aerofoils, cascades and slotted aerofoils and cascades, employing the surface vorticity technique and we will summarise this method later in Section 7.5. This represented a major improvement upon earlier linearised singularity theories such as that of Railly (1965) following the method of Ackeret (1942), which were of type (ii). Although the latter methods, which use internal singularities, have been widely used for axial and mixed-flow fan design, their linearisations impose constraints which limit their applicability. In view of these limitations Lewis (1982a) developed a surface vorticity true inverse method aimed at the selection of profiles for aerofoils or cascades with prescribed surface velocity distribution. This technique will be outlined in the following sub-sections and its extension to bodies of revolution in Section 7.5.

7.4.1 Basis of inverse surface vorticity design method for aerofoils and cascades

In a sense the surface singularity inverse method has already been introduced, since free streamline and jet flows represent one special case for which the surface velocity at the edge of the free streamline and its associated vorticity sheet are of prescribed constant value. The principle of deriving the free-streamline contour by successively improved reconstructions aligned with the local velocity vector, may be applied also to inverse design of two-dimensional body shapes in

plane two-dimensional flow. In direct analysis as outlined in
Chapter 1, a body of given geometry is represented by a finite
number of surface elements of fixed location but of initially
unknown vortex strengths $\gamma(s)\,\Delta s$. The problem reduces to determination of the vorticity distribution $\gamma(s)$. The inverse of this is thus
to begin by prescribing the desired surface velocity distribution
(PVD) and therefore surface vorticity $\gamma(s) = v_s$ for an initial
estimate of the required body shape. By successive approximations
the body may then be reshaped until the Dirichlet boundary
condition $v_{si} = 0$ is satisfied on the interior surface of all elements.
A suitable strategy for aerofoil design is illustrated in Fig. 7.7 which
may be explained as follows.

(i) The surface velocity v_s is chosen as a function of distance
measured along the perimeter of the proposed aerofoil.
(ii) The surface vorticity is then also known since $\gamma(s) = v_s$.
(iii) An initial shape is selected as first guess. For example, Fig.
7.7(a), an ellipse is normally a good starting point in aerofoil
design.
(iv) The guessed surface is broken into elements of suitable length
Δs and $\gamma(s)$ values are permanently assigned to each element.
To achieve this the distance along the perimeter from the
leading edge to the centre of the mth element can be
estimated by

$$s_m = \tfrac{1}{2}\Delta s_1 + \frac{1}{2}\sum_{i=1}^{m-1}(\Delta s_i + \Delta s_{i+1}) \quad \text{for} \quad m > 1 \tag{7.13}$$

(v) Velocity components u_m, v_m at each element m are calculated
including the influence of all other elements plus the uniform
stream, Fig. 7.7(c). Thus for M elements with vorticity
assumed to be concentrated at the mid-points we have,

$$\left.\begin{aligned}
u_m &= U_\infty + \frac{1}{2\pi}\sum_{\substack{n=1\\n\neq m}}^{M}\frac{(y_n - y_m)\gamma(s_n)\,\Delta s_n}{(x_n - x_m)^2 + (y_n - y_m)^2} + \Delta q_m \cos\beta_m \\
v_m &= V_\infty + \frac{1}{2\pi}\sum_{\substack{n=1\\n\neq 1}}^{M}\frac{(x_n - x_m)\gamma(s_n)\,\Delta s_n}{(x_n - x_m)^2 + (y_n - y_m)^2} + \Delta q_m \sin\beta_m
\end{aligned}\right\} \tag{7.14}$$

where the self-induced convection velocity of element m,
Section 1.7, is given in terms of profile slope β_m by

$$\Delta q_m = \frac{\gamma(s_m)}{2\pi}(\beta_{m-1} - \beta_{m+1}) \tag{7.15}$$

293

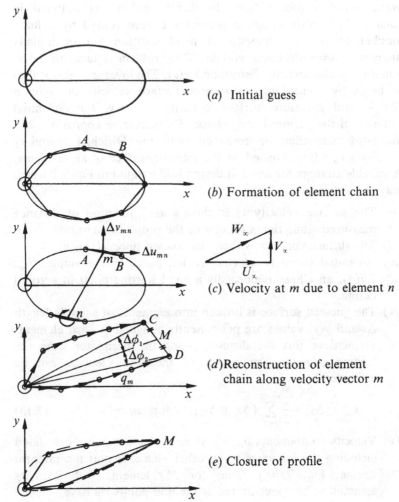

Fig. 7.7. Steps in inverse aerofoil design. Reproduced by courtesy of the American Society of Mechanical Engineers.

(vi) Upper and lower surfaces may now be realigned by treating them as flexible chains with straight line links, each of which is rearranged to lie along the local velocity vector $q_m = iu_m + jv_m$, for which equations similar to (7.11) and (7.12) apply.

(vii) Due to errors the probability arises that the trailing edge points C and D of the upper and lower surfaces will not meet, although we know that ultimately for a fully converged solution profile closure is an essential condition. A suitable strategy for handling this problem is to enforce closure by interfering with the solution at this point of each iteration in the manner illustrated in Fig. 7.7(d). This is achieved by rotating the upper and lower surfaces about the leading edge through angles $\Delta\phi_1$ and $\Delta\phi_2$ respectively until they lie on the common mean chord OM and then scaling the surfaces by the ratios OM/OC and OM/OD resulting in Fig. 7.7(e).

(viii) To conclude the iteration the velocity at the mid-point of each element is calculated from

$$q_m = \surd(u_m{}^2 + v_m{}^2) \tag{7.16}$$

Since this applies at the centre of the vortex sheet the estimated velocity stepping onto the outside of the sheet locally is

$$v_{sm}{}' = 2q_m \tag{7.17}$$

Two conditions should be met at convergence; (a) geometry should no longer change significantly and (b) $v_s{}'$ should agree with the prescribed velocity distribution v_s. Condition (a) defines convergence and condition (b) the level of accuracy which has been achieved.

Fig. 7.8 illustrates a simple test case of the design of a circular cylinder for which the PVD (prescribed velocity distribution) is given by

$$v_s = 2U_\infty \sin\theta$$

or

$$\gamma(s) = 2U_\infty \sin(s/a) \tag{7.18}$$

Adopting a fairly thin ellipse, with 20 equal length elements, as the first guess an extremely good approximation to the expected circular cylinder was obtained after only ten iterations, with close agreement also between $v_s{}'$ and the (PVD) value v_s.

7.4.2 Further refinements

Mention should be made of two further essential refinements. Firstly it is advisable and possibly essential to introduce damping in order to prevent excessive geometrical changes between iterations. Thus in the example just shown, after iteration i the velocities

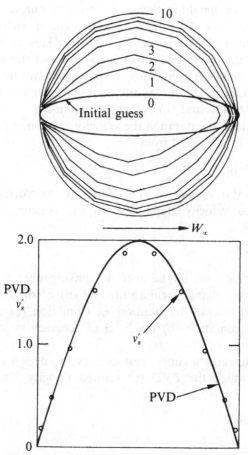

Check on final surface velocity v'_s compared
with prescribed velocity distribution

Fig. 7.8. Successive iterations in the design of a cylinder from its prescribed
velocity distribution. Reproduced by courtesy of the American Society of
Mechanical Engineers.

induced at element m were replaced by the following values before
proceeding to profile reconstruction.

$$\left.\begin{aligned}
u_{m,i} &= ku_{m,i} + (1-k)u_{m,i-1} \\
v_{m,i} &= kv_{m,i} + (1-k)v_{m,i-1}
\end{aligned}\right\} \tag{7.19}$$

A value of $k = 0.5$ is normally found to be adequate, but smaller
values may be used if divergence is encountered.

Secondly, if the PVD varies considerably over an element (e.g.

296

near a stagnation point), insufficient accuracy is obtained by lumping the vorticity at the centre of the element. To obtain greater accuracy sub-elements may be introduced as illustrated in Section 1.9. Thus the velocity components induced at the centre of element m due to element n represented by J sub-elements, are

$$
\left.
\begin{aligned}
\Delta u_{mn} &= \frac{\Delta s_n}{2\pi J} \sum_{j=1}^{J} \frac{(y_{jn} - y_m)\gamma(s_{jn})}{(x_{jn} - x_m)^2 + (y_{jn} - y_m)^2} \\[2mm]
\Delta v_{mn} &= \frac{\Delta s_n}{2\pi J} \sum_{j=1}^{J} \frac{(x_{jn} - x_m)\gamma(s_{jn})}{(x_{jn} - x_m)^2 + (y_{jn} - y_m)^2}
\end{aligned}
\right\}
\tag{7.20}
$$

where x_{jn}, y_{jn} and $\gamma(s_{jn})$ are interpolated values for sub-element j, namely

$$
\left.
\begin{aligned}
x_{jn} &= x_{An} + \left\{\frac{j - 0.5}{J}\right\}(x_{Bn} - x_{An}) \\[2mm]
y_{jn} &= y_{An} + \left\{\frac{j - 0.5}{J}\right\}(y_{Bn} - y_{An}) \\[2mm]
\gamma(s_{jn}) &= \gamma(s_{An}) + \left\{\frac{j - 0.5}{J}\right\}(\gamma(s_{Bn}) - \gamma(s_{An}))
\end{aligned}
\right\}
\tag{7.21}
$$

To apply this approach $\gamma(s_n)$ values must be prescribed not at the centres of the elements as previously recommended but at the end locations A and B, Fig. 7.7. Its introduction may be essential as illustrated by Fig. 7.9, where the improvement of the resulting solution over that without sub-elements was dramatic. However to compute velocities at element m it is normally sufficiently accurate to introduce sub-elements into the neighbouring elements $m - 1$ and $m + 1$ only. The use of six sub-elements in this manner with linear interpolation as above, will suffice. Application to the circular cylinder represented by $M = 40$ surface elements and the use of 8 sub-elements produced an extremely accurate circular body shape and close agreement between v_s and $v_s{}'$.

7.4.3 Angular constraints on leading and trailing edge elements

In the foregoing examples the elements adjacent to the leading and trailing edge stagnation points were constrained to lie at the correct angle, i.e. 72° to the vertical for 20 elements and 81° for 40

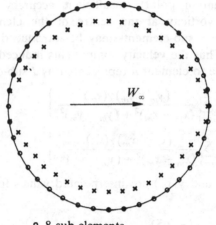

○ 8 sub-elements
× No sub-elements

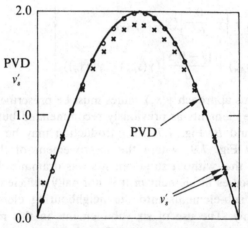

Check on final surface velocity v'_s
compared with prescribed velocity
distribution
Note: elements 1, 20, 21 and 40 fixed at
81° to horizontal axis

Fig. 7.9. Improvements resulting from use of sub-elements for circular
cylinder PVD with 40 elements. Reproduced by courtesy of the American
Society of Mechanical Engineers.

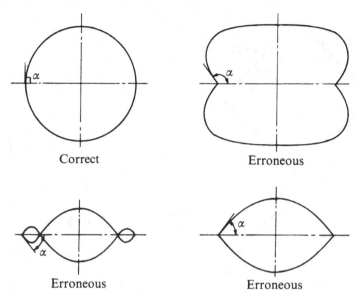

Fig. 7.10. Spurious solutions which can occur at a stagnation point in inverse design unless the element angle α is constrained. Reproduced by courtesy of the American Society of Mechanical Engineers.

elements. Because of the rapidly varying PVD in these regions where the mainstream velocity U_∞ is almost normal to the required surface, the solution is least stable if these elements remain free to adjust themselves between iterations. Freak solutions such as those sketched in Fig. 7.10 may be obtained unless the stagnation point elements are constrained as proposed here. It is of interest to observe from the circular cylinder solution of Fig. 7.9, however, that even with this constraint a poor solution was still obtained without the use of sub-elements caused by similar inaccuracies in the elements adjacent to the stagnation point elements. This is clearly a vulnerable region for such analyses.

A particular severe test for this inverse method is presented by the case of a diamond shaped body, Fig. 7.11, for which a Schwarz-Christoffel transformation is available to obtain the exact solution. Since the surface velocity is singular at the corners P and Q, the use of (7.21c) for the sub-element strengths is not possible unless some approximation of the vorticity distribution over the corner elements is made. The PVD for this body is shown in Fig.

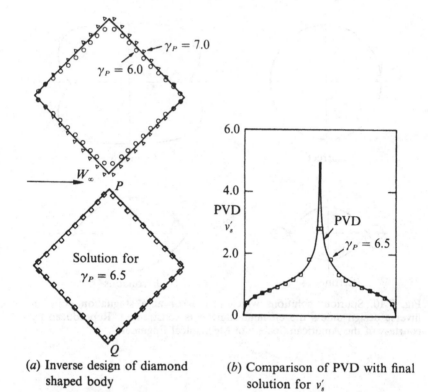

(*a*) Inverse design of diamond shaped body

(*b*) Comparison of PVD with final solution for v'_s

Fig. 7.11. Inverse design for flow past a diamond shaped body. Reproduced by courtesy of the American Society of Mechanical Engineers.

7.11(*b*). For the numerical analysis an approximation was made by replacing the values $\gamma(s) = \infty$ at the corners by a large number and retaining linear interpolation in (7.21c). As illustrated here, $\gamma(s_P)$ values of 0.6, 0.65 and 0.7 were tried, illustrating the sensitivity of the solution to this choice. Best results were obtained with the value $\gamma(s_P) = 0.65$ and comparison between the PVD and v_s' confirmed this validity. Referring back also to the plate and wedge free streamlines of Section 7.2, it was found that v_s' differed most from the correct value of U_∞ in the neighbourhood of the separation point sp, which would also be a singular point of the non-separated potential flow. Although leakage flux due to the junction between the flexible section of the free streamline and its semi-infinite closure vortex is the probable cause of most errors in that model, inaccuracies do also arise close to the sharp edge of separation.

7.4.4 Aerofoil inverse design

Two options are available for aerofoil or cascade inverse design. Option A permits the designer to specify a (PVD) on both the upper and lower surfaces, resulting in the automatic design of the entire aerofoil profile. Although this sounds attractive, the procedure is not without its set-backs. At worst it is quite possible for the designer to specify an impossible surface velocity distribution which may cause the upper and lower surfaces to cross one another at some point in 'figure of eight' manner. Even with a valid (PVD) specification profiles may be generated with impracticable thickness distributions. For this reason Option B was originally proposed by Wilkinson (1967a), whereby the (PVD) is limited to the upper surface only but profile thickness is also prescribed. In effect the inverse method is then designing the camber line shape required to achieve the desired (PVD) on the more sensitive highly loaded suction surface of the aerofoil or cascade blade. This option is decidedly the most useful and powerful design tool of the two.

The computational procedure outlined in Section 7.3.1 constitutes Option A and may be applied directly to the single compressor blade profile shown in Fig. 7.12. To ensure the adoption of a viable test case the strategy adopted here was to begin by selecting a given profile shape, namely 10C4/30C50 and subjecting the aerofoil to surface vorticity potential flow analysis. The resulting solution was then taken as the (PVD) input into an inverse surface vorticity program as described above in the manner of a 'back to back' test. Beginning with an ellipse as first guess, the profile shown in Fig. 7.12 was obtained after twelve iterations, showing very reasonable agreement with the original compressor profile and its known surface velocity distribution.

As illustrated by Fig. 7.13, greater accuracy may be obtained if the velocity is specified on the upper surface only, Option B, the lower surface being constructed by superimposing the C4 profile thickness distribution in the manner depicted in Fig. 7.14. The vorticity distribution of the upper surface is thus fixed at the outset but that of the lower surface is unknown in the first instance. For the first iteration realistic dummy values are used for $\gamma(s)$ on the lower surface but for subsequent iterations its value is determined from (7.16). Thus, since the velocity inside the profile is zero, we have the relationship

$$\gamma(s_m) = 2q_m = 2\sqrt{(u_m{}^2 + v_m{}^2)} \tag{7.22}$$

301

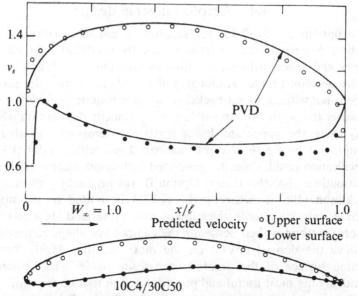

Fig. 7.12. Aerofoil design Option A, both surface PVD specified. Reproduced by courtesy of the American Society of Mechanical Engineers.

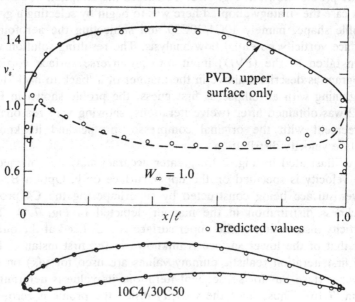

Fig. 7.13. Aerofoil design Option B, upper surface PVD and profile thickness specified. Reproduced by courtesy of the American Society of Mechanical Engineers.

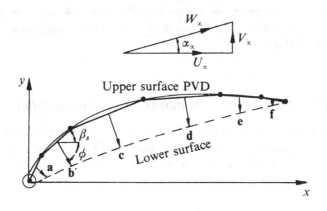

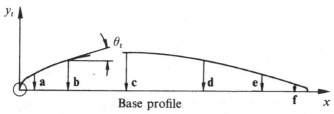

Fig. 7.14. Option B – profile reconstruction. Reproduced by courtesy of the American Society of Mechanical Engineers.

As shown by Fig. 7.14, following each reconstruction of the upper surface, the lower surface points are obtained at each location s by transferring the base profile thickness coordinates as vectors $\mathbf{a, b, c}$ ... etc inclined to the x axis at the angles ϕ, where ‚

$$\phi = \frac{\pi}{2} - \beta_s + \theta_t/2 \qquad (7.23)$$

β_s and θ_t are the upper surface and base profile slopes respectively.

7.5 Inverse design of cascades and slotted cascades

Two techniques have been proposed for the inverse design of turbomachinery cascades by the surface vorticity method, falling

303

into the two categories mentioned in Section 7.4, namely (i). Iterative use of a direct analysis and (ii) True inverse methods. We will deal with these in reverse order.

7.5.1 True inverse design method for cascades

The above aerofoil inverse method may be extended with little difficulty to cascades, by modifying equations (7.14) to include the cascade form of coupling coefficient (see Section 2.6.1). We then have

$$
\left.
\begin{array}{l}
u_m = U_\infty + \dfrac{\ell}{2t} \displaystyle\sum_{\substack{n=1 \\ n \neq m}}^{M} \dfrac{\sin\left\{\dfrac{2\pi\ell}{t}(y_m - y_n)\right\}\gamma(s_n)\,\Delta s_n}{\cosh\dfrac{2\pi\ell}{t}(x_m - x_n) - \cos\dfrac{2\pi\ell}{t}(y_m - y_n)} \\[5ex]
v_m = V_\infty - \dfrac{\ell}{2t} \displaystyle\sum_{\substack{n=1 \\ n \neq m}}^{M} \dfrac{\sinh\left\{\dfrac{2\pi\ell}{t}(y_m - y_n)\right\}\gamma(s_n)\,\Delta s_n}{\cosh\dfrac{2\pi\ell}{t}(x_m - x_n) - \cos\dfrac{2\pi\ell}{t}(y_m - y_n)}
\end{array}
\right\} \quad (7.24)
$$

These expressions give the velocity components at any point (x_m, y_m) due to an infinite array of aerofoils parallel to the y axis and with prescribed surface vorticity $\gamma(s_m)$. Although one might foresee no special problem in their direct substitution into the previous procedure for aerofoil design, there are three essential differences in design requirements for cascades which may be summarised as follows.

(i) The pitch/chord ratio t/ℓ appears in these expressions for (u_m, v_m), as an extra design parameter to be considered
(ii) Designers normally prefer to prescribe not the vector mean conditions (W_∞, β_∞) but the inlet velocity W_1 and the inlet and outlet flow angles β_1 and β_2
(iii) The PVD for aerofoils v_s/W_∞ is usually normalised by the vector mean velocity. For cascades the preferred notation may be v_s/W_1 for compressors and v_s/W_2 for turbines.

Now we have already shown* that the bound circulation of a

---

* See Section 2.6.2 (Cascade dynamics and parameters) for further details of overall cascade velocity triangle relationships.

cascade may be expressed alternatively through

$$\Gamma = U_\infty t(\tan \beta_1 - \tan \beta_2)$$

$$= \oint \gamma(s_n) \, \mathrm{d}s_n = \sum_{n=1}^{M} \gamma(s_n) \, \Delta s_n$$

$$= W_\infty \ell \oint \bar{\gamma}(\bar{s}_n) \, \mathrm{d}\bar{s}_n \qquad (7.25)$$

where normalised quantities are defined

$$\bar{s} = \frac{s}{\ell}, \qquad \bar{\gamma}(\bar{s}) = \frac{\gamma(s/\ell)}{W_\infty}, \qquad \bar{v}_s = \frac{v_s}{W_\infty} \qquad \text{etc.}$$

The pitch chord ratio may thus be expressed directly in terms of prescribed quantities through

$$\frac{t}{\ell} = \frac{\sum\limits_{n=1}^{M} \bar{\gamma}(\bar{s}_n) \, \Delta \bar{s}_n}{\cos \beta_\infty (\tan \beta_1 - \tan \beta_2)} \qquad (7.26)$$

where the vector mean angle is related to inlet and outlet angles through

$$\tan \beta_\infty = \tfrac{1}{2}(\tan \beta_1 + \tan \beta_2)$$

In the case of Option A, since $\bar{\gamma}(\bar{s})$ is completely specified at the outset, so also is t/ℓ. With Option B on the other hand, the vorticity of the lower surface is not initially known. Thus an initial estimate of t/ℓ is required (the author recommends $t/\ell = 1.0$), its real value being obtained iteratively as the solution proceeds. No convergence problems are normally encountered with Option B although for closely pitched cascades care may be needed, when selecting blade thickness, to avoid overlap of adjacent blades. With Option A it is helpful to prescribe the leading edge element geometries to avoid the instabilities in the region of the stagnation point already referred to in Section 7.4.3 in relation to Fig. 7.10.

For commercial use a high lift fan cascade was required to produce a deflection from $\beta_1 = 67.45°$ to $\beta_2 = 53.33°$. The inverse boundary layer method of Stratford (1959) was used to prescribe the upper surface velocity as shown in Fig. 7.15. Further details have been given by the author, Lewis (1982a). The velocity, in this case normalised by inlet velocity v_s/W_1, was kept constant for the first 30% of the blade surface and then diffused to a fairly high value of $v_s/W_2 = 1.2$ at the trailing edge. Adopting Option B a C4

305

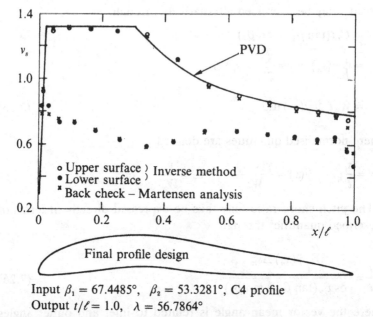

Input $\beta_1 = 67.4485°$, $\beta_2 = 53.3281°$, C4 profile
Output $t/\ell = 1.0$, $\lambda = 56.7864°$

Fig. 7.15. Inverse design of a fan cascade for prescribed velocity on upper surface and C4 blade thickness distribution – Option B. Reproduced by courtesy of the American Society of Mechanical Engineers.

profile was combined with this upper surface PVD, resulting in the acceptable lower surface PVD and profile shape shown in Fig. 7.15. This 'Stratford' type of velocity distribution tends to produce negative camber in the region of rapid diffusion due to the associated diminishing local bound circulation, but from previous application to the design of aerogenerator blade profiles by Cheng (1981), such aerofoils were known to offer high lift/drag and stall resistance properties. Good agreement was obtained between the final surface velocity $v_s{}'$, the PVD and a back check obtained by Martensen direct analysis. The design t/ℓ for this cascade was 1.03257 with a stagger λ of 56.79°.

7.5.2 Inverse cascade design by iterative use of the direct method

An alternative design method for aerofoils of small camber and moderate thickness was given by Weber (1955), based upon thin

306

aerofoil theory. Wilkinson (1967a) adapted some essential features of this, in particular her expressions for camber line distributions of vorticity, in order to produce a more advanced method of inverse design by the surface vorticity method subject to no restrictions on allowable camber and thickness. Her strategy was as follows for an Option B type design routine:

(i) Prescribe an upper surface PVD and blade thickness distribution as required
(ii) Make a first guess at the camber line and superimpose the thickness distribution normal to it
(iii) Analyse the guessed profile by the standard Martensen analysis
(iv) Convert the error of predicted upper surface velocity v_s into an equivalent camber line vortex sheet $\Gamma(\ell)$.
(v) Calculate the velocities Δv_{tm} and Δv_{nm} parallel and normal to the camber line induced by the 'error' vortex sheet $\Gamma(\ell)$
(vi) Change the local camber line slope to accommodate these velocity perturbations, with the aim to make $\Delta v_{nm} \to 0$, and the camber line a streamline
(vii) Reconstruct the profile and repeat from (iii) until convergence is achieved

Although the method for camber line reconstruction here is subject to the limitations and inaccuracies of thin aerofoil theory, this does not matter provided successive estimates of surface velocity converge towards the intended PVD. Items (iv)–(vi) are merely the means for guessing an improved estimate of the aerofoil profile. Space does not permit a full statement of the equations which have been thoroughly presented by Wilkinson. The leading equations are as follows. For a specified surface velocity distribution v_{sPVD}, the error in surface in upper surface velocity may be expressed

$$\frac{\Delta v_{sm}}{U_\infty} = \frac{v_{sPVD} - \gamma(s_m)}{U_\infty} \tag{7.27}$$

where $\gamma(s_m)$ is the latest Martensen solution at point m.

Since the thickness is to remain unchanged, this velocity increment may be considered to be caused by a continuous vortex sheet of strength $\Gamma(\ell)$ located along the camber line, Fig. 7.16. Following the arguments of Section 1.5, Δv_{sm} is then equal to approximately

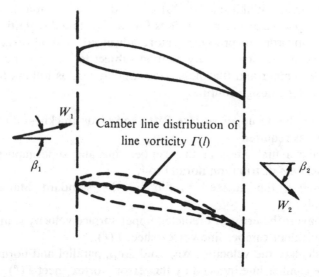

Fig. 7.16. Camber line error vorticity used in the Wilkinson inverse method for cascade design.

half of $\Gamma(\ell)$. In fact for small cambers the Riegel's factor* may be used to convert Δv_{sm} to $\Gamma(\ell)$ directly with improved accuracy through

$$\frac{\Gamma(\ell)}{2U_\infty} \approx -\frac{\mathrm{d}s}{\mathrm{d}l} \cdot \frac{\Delta v_{sm}}{U_\infty} \qquad (7.28)$$

Wilkinson recommends this for small cambers but offers more advanced formulations for highly staggered and cambered sections and for multi-section systems such as aerofoils or cascades with slots and flaps. The tangential and normal velocities on the camber line due to $\Gamma(\ell)$ may be expressed through

$$\left.\begin{array}{l}\Delta v_{tm} = \dfrac{\Gamma(\ell_m)}{2} - \dfrac{1}{2\pi}\int \Gamma(\ell)\left\{\dfrac{(X_m - X)\left(\dfrac{\mathrm{d}Y}{\mathrm{d}\ell}\right)_m - (Y_m - Y)\left(\dfrac{\mathrm{d}Y}{\mathrm{d}\ell}\right)_m}{(X_m - X)^2 + (Y_m - Y)^2}\right\}\mathrm{d}\ell \\[3em] \Delta v_{nm} = \dfrac{1}{2\pi}\int \Gamma(\ell)\left\{\dfrac{(Y_m - Y)\left(\dfrac{\mathrm{d}Y}{\mathrm{d}\ell}\right)_m - (X_m - X)\left(\dfrac{\mathrm{d}X}{\mathrm{d}\ell}\right)_m}{(X_m - X)^2 + (Y_m - Y)^2}\right\}\mathrm{d}\ell \end{array}\right\}$$

$$(7.29)$$

* See Riegels (1949), Weber (1955) or Schlichting (1955).

308

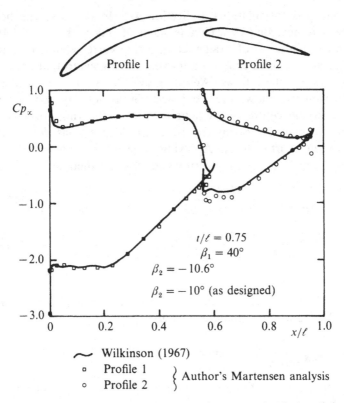

Fig. 7.17. Pressure distribution for a tandem cascade designed by the Wilkinson inverse method (1967).

where (X, Y) are the camber line coordinates. Alternative expressions are of course available for cascades and for multiple sectioned bodies. Wilkinson introduced the circular transformation of Section 2.5.2, Fig. 2.8, to improve the accuracy of evaluation of these integrals and laid down other guide lines for treatment of the special problems which may be experienced in the leading edge region. A sample tandem cascade designed by the Wilkinson inverse method is compared with the author's analysis in Fig. 7.17.

7.6 Inverse design of axisymmetric bodies

An inverse method similar to Option A of Section 7.4 may be devised for the design of axisymmetric bodies making use of the direct analysis formulations of Section 4.2. In this case the body surface is represented by a flexible sheet of ring vorticity modelled by a finite number M of straight conical ring vortex elements joined together as links of a chain in a manner analogous to that depicted in Fig. 7.7 for plane flows. Adopting cylindrical (x, r) coordinates for axisymmetric flow, the axial and radial velocity components induced at the centre of element m due to the ring vorticity on element n may be calculated from the expressions for a unit strength ring vortex (4.8). Applying these to a representation of element n by I equally spaced sub-elements, we then have

$$\left. \begin{aligned} \Delta u_{mn} &= -\frac{\Delta s_n}{2\pi I} \sum_{i=1}^{I} \frac{\gamma(s_i)}{r_i \sqrt{[x^2 + (r+1)^2]}} \left\{ K(k) - \left[1 + \frac{2(r-1)}{x^2 + (r-1)^2} \right] E(k) \right\} \\ \Delta v_{mn} &= \frac{\Delta s_n}{2\pi I} \sum_{i=1}^{I} \frac{\gamma(s_i)}{r_i \sqrt{[x^2 + (r+1)^2]}} \left\{ K(k) - \left[1 + \frac{2r}{x^2 + (r-1)^2} \right] E(k) \right\} \end{aligned} \right\}$$

(7.30)

where $K(k)$ and $E(k)$ are complete elliptic integrals of the first and second kind and the parameter k is given by

$$k = \sqrt{\left[\frac{4r}{x^2 + (r+1)^2} \right]} = \sin \phi$$

and where (x, r) are the dimensionless coordinates of sub-element i defined by

$$x = \frac{x_m - x_i}{r_i}, \qquad r = \frac{r_m}{r_i}$$

As shown in Section 4.2.1, computational speed may be increased with little sacrifice in accuracy by the use of a table look-up procedure for evaluation of the elliptic integrals, tabulated more advantageously against even intervals of the angular variable ϕ, Appendix 1. A strategy similar to that for plane aerofoil design, Section 7.4.1, may then be adopted as follows.

(i) The surface velocity U_{sPVD} is prescribed as a function of distance measured along the body perimeter s from nose to tail, corresponding to surface vorticity of equal strength $\gamma(s) = U_{sPVD}$.

(ii) An initial body shape is selected as first guess. An ellipsoid of approximately the correct aspect ratio (minor axis/major axis) will normally suffice.

310

Inverse design of axisymmetric bodies

(iii) The guessed surface is broken down into M straight line elements and local values of $\gamma(s)$ at the element end points are interpolated from the PVD data and permanently assigned to each element. Equation (7.13) is appropriate here and $\gamma(s_i)$ values may be interpolated linearly for each sub-element when evaluating equations (7.30).

(iv) Velocity components may now be calculated at the mid-point of each element m, including the influence of the uniform stream U_∞ and all other elements. On the outer surface of element m we then have

$$\left.\begin{aligned} u_m &= U_\infty + \sum_{n=1}^{M} \Delta u_{mn} + \left(\frac{1}{2} - \frac{\Delta\beta_m}{4\pi}\right) \cos\beta_m \\ v_m &= \sum_{n=1}^{M} \Delta v_{mn} + \left(\frac{1}{2} - \frac{\Delta\beta_m}{4\pi}\right) \sin\beta_m \end{aligned}\right\} \tag{7.31}$$

(v) The elements may now be re-aligned to lie along the local velocity vector $\mathbf{q}_m = \mathbf{i}u_m + \mathbf{j}u_m$ and the vortex element chain thus reconstructed to follow the latest estimate of the body surface streamline.

(vi) Due to error the estimated tail point radius $r(M+1)$ may no longer be zero. By analogy with the procedure for plane flows illustrated by Fig. 7.7, the whole body profile must now be rotated about the nose to enforce body closure. At the same time the body coordinates may be scaled to match an initially prescribed chord length if desired.

(vii) The estimated surface velocity may now be compared with the prescribed velocity since

$$q_m = \sqrt{[u_m{}^2 + v_m{}^2]} \approx \gamma(s_m) \tag{7.32}$$

which provides a check upon both convergence and the final accuracy of the design.

In stage (iv) of this process, trial and error has shown that single element representation of Δu_{mn} and Δv_{mn}, as given by equations (7.30), is sufficiently accurate for the widely separated elements when say $|m - n| > 3$. For elements in closer proximity on the other hand, i.e. $|m - n| \leqslant 3$, the use of sixteen or more sub-elements will ensure sufficient accuracy. There is then no need to employ the Lamb–Ryan formulation of Section 4.2.3 for the self-propagation velocity of a ring vortex element, as shown in Section 5.7.2. On the other hand the additional self-induced

311

velocity due to body curvature in the (x, r) plane (4.19) has been correctly included in the above expressions (u_m, v_m), equations (7.31). During stage (vi) of the computational process the body slope β_m must therefore be re-estimated at each iteration. Damping is recommended to avoid possible divergence of the solution in sensitive regions. To avoid spurious solutions in the stagnation point regions similar to those illustrated for plane flows in Fig. 7.10, the slopes of elements 1 and M should also be prescribed.

If this procedure is applied to the example illustrated by Fig. 4.6, we may begin by adopting the surface velocity solution shown there, derived from the analysis program axisym.pas (Program No. 4.2), as our PVD input. As an extreme test the initial body shape proposed, Fig. 7.18(a), was a sphere, the only merit in this choice being the consequence of equal length elements which follows from the use of equations (2.11) to define an ellipse by equal intervals of θ. To illustrate the speed of convergence intermediate estimates of the body shape are also shown for iterations Nos. 5 and 10, the final solution being obtained after 60 iterations. This example comprising a spherical nose section, cylindrical main body and conical tail cone presents a particular challenge due to the velocity disturbances at the section junctions. The inverse method has handled this extremely well, with only minor problems in the nose and tail stagnation point regions. This final velocity distribution as given by (7.32) is also in good agreement with the PVD, Fig. 7.18(b).

Probably the earliest analysis for flow past axisymmetric bodies is that attributable to Theodore Von Karman (1930). Based on an axial distribution of source/sink elements, this method has been extensively studied by Oberkampf & Watson (1974) who concluded that the system of linear equations, which is in general ill-conditioned, does not always yield reliable solutions. An extension of this type of technique by Levine (1958) to an axisymmetric body contained within a cylindrical duct was applied by the author, Lewis (1964b), to turbomachine problems but exhibited similar difficulties. Surface source panel methods developed at the Douglas Aircraft Company on the other hand present no such problems and have been extensively used to tackle both the direct and inverse problems. Using constant source density over each element, Smith & Pierce (1958) developed precise solutions for planar and axisymmetric body flows. Hess & Smith (1966) summarised numerous later extensions of the method with improved source singularity distributions and Hess (1976) published an inverse method aimed at

312

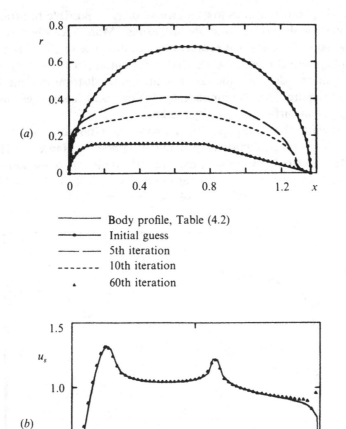

(a)

    ———— Body profile, Table (4.2)
    —•— Initial guess
    — — 5th iteration
    ------ 10th iteration
    ▲ 60th iteration

(b)

    ———— Prescribed velocity distribution
    ▲ Final solution

Fig. 7.18. Design of a body of revolution from its prescribed surface velocity distribution (c.f. example considered in Section 4.3.2).

313

producing bodies with as low surface velocity as possible to produce low drag and good cavitation properties. More recently Bristow (1974) solved the inverse problem by iterative use of the Douglas–Newmann direct method while Zedan & Dalton (1978), (1981) have continued to develop the axial source/sink distribution method which requires less computational effort and converges more quickly than surface singularity methods.

Following the technique of Parsons & Goodson (1972) for optimum shaping of axisymmetric low drag bodies, Hansen & Hoyt (1984) undertook extensive experimental studies of a submersible vehicle body designed for laminar boundary layer flow. This vehicle terminates in a cylindrical tail boom, Fig. 7.19. For present

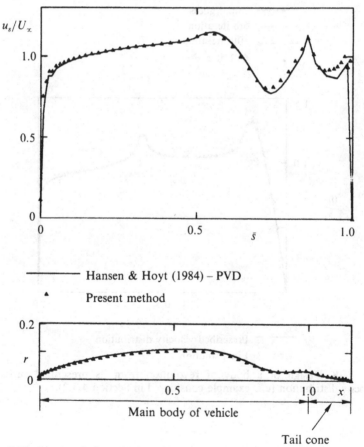

Fig. 7.19. Inverse design of a low drag body of revolution – 'Parsons' body.

analytical convenience a conical tail cone was added and the resulting 'Parsons' body was analysed by program axisym.pas. The resulting solution was then adopted as the PVD input to the author's inverse design program axipvd.pas. The initial body shape chosen was an ellipsoid of aspect ratio 0.3 and the final solution is shown in Fig. 7.19. Although there is a slight under-prediction of body diameter, in general the inverse method has been able to reproduce the complex body shape with its accelerating velocity to $x/l = 0.65$ followed by rapid diffusion towards the tail boom, remarkably well.

Although space permits us to deal here only with bodies of revolution, this method could be extended to engine annuli or intakes with little difficulty. Compared with some of the works quoted above, the surface vorticity methods as just outlined may appear to converge slowly bearing in mind that sixty iterations were undertaken for the illustrative examples. The main reason for this was the introduction of 50% damping and selection of a poor initial choice of ellipsoid. By the introduction of extrapolation between iterations and a more suitable initial ellipsoid quite reasonable solutions may be obtained with twenty iterations. However since computational speed presents no problem today, it is safer to introduce damping and closer limits on convergence.

CHAPTER 8
Vortex dynamics in inviscid flows

8.1 Introduction

The early contributors to the surface vorticity method such as Martensen (1959), Jacob & Riegels (1963) and Wilkinson (1967a) were concerned primarily with the development of a flexible numerical method for the solution of potential flows. Preceding chapters testify to the scope and power of this conceptually simple technique and to the imagination and creativity of a host of later research workers who have extended the method to deal with a wide range of engineering potential flow problems. Although the broader physical significance of the surface vorticity model, as expounded in Chapter 1, has always been realised, only recently has this been more fully explored by attempts to model the rotational fluid motion of real fluids including both boundary layer and wake simulations. The remaining chapters will lay down progressively the essential fundamentals of this work which the reader requires to proceed to practical computational schemes, employing what has come to be known as the 'vortex cloud' or 'discrete vortex' method.

All real flows involve rotational activity developed in the regions adjacent to flow surfaces or in the rear wake region in the case of bluff bodies. Some flows also exhibit spontaneous boundary layer separation or stall behaviour while in other situations flow separation occurs inevitably from sharp corners. Vortex cloud analysis attempts to model these flows by discretisation of the distributed vorticity or separated shear layers into finite numbers of small discrete vortex elements and the convective motions of such assemblies is often referred to as 'vortex dynamics'. A good deal of research into vortex dynamics has been completed linked to classical potential flow methods particularly in the field of aeronautical and off-shore engineering with no reference to surface vorticity modelling. However the physical significance of the latter, which can be thought of as a true model of infinite Reynolds Number flow of a real fluid, and its ready adaptability to deal with arbitrary body

316

shapes led the author to develop his own contributions to vortex cloud theory as a natural extension of the surface vorticity method. This is the strategy which we will adopt here.

Historically, the first serious attempt at discrete vortex modelling is probably attributable to Rosenhead (1931) who studied the Kelvin–Helmholtz instability of vortex sheets, an important fluid motion which we will explore in Section 8.3. Birkhoff & Fisher (1959) and Hama & Burke (1960) re-examined Rosenhead's work, concluding that the progressive growth of an initial disturbance, periodic along the length of a vortex sheet, tends towards the concentration of the vorticity into a series of vorticity cores or clouds. An interest in the formation of vortex street wakes behind bluff bodies has fascinated large numbers of research workers since the early experiments of Strouhal (1878) concerning the generation of 'Aeolian tones' and the famous paper by Theodore von Karman (1911). Abernathy & Kronauer (1962) extended the discrete vortex model of Rosenhead to this problem by considering the stability of a parallel pair of infinite vortex sheets of equal and opposite strength $\pm U$ subjected to sinusoidal perturbations along their length in various combinations. A sample of their solution shown in Fig. 8.1 illustrates quite clearly the progressive formation of a von Karman type vortex street as the final outcome of an initially unstable sheet vortex configuration. Apart from the initial disturbance the motion is self-induced entirely by the convective processes of all the interacting discrete vortex elements. Abernathy & Kronauer concluded from this study that the role of a bluff body is not very important in the formation of its detailed vortex street wake, excluding its function in the generation of the two separating vortex shear layers responsible for feeding vorticity into the wake. As also asserted by these authors, the constancy of drag coefficient and Strouhal number for a circular cylinder over a wide range of Reynolds Number, e.g. Re = 300–100 000, demonstrates the negligible influence of viscosity upon these flows. Being thus dominated by convective influences such fluid motions are essentially thermodynamically reversible and a good deal of progress in vortex dynamics, perhaps surprisingly, has been made on this premise.

This last remark invites further comment. Firstly it is clear that thermodynamic reversibility in this situation demands actual fluid dynamic reversibility. In other words a quality control test of any vortex dynamics model is that of its ability to return the vortex structure to its original state if run backwards in time. Although

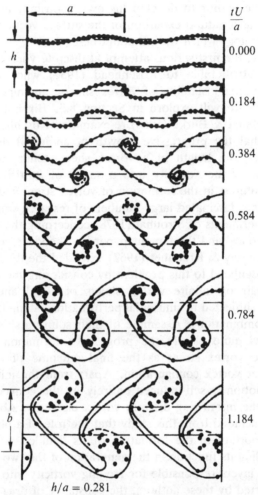

Fig. 8.1. Development of von Karman vortex street from sinusoidally perturbed vortex sheet pair. Abernathy & Kronauer (1962).

discretisation imposes some constraint on the allowable deformation of the vortex sheet being modelled, in principle the above proposition remains true and forms a useful means for testing the merit worthiness of any scheme of vortex dynamics. These constraints of discretisation lead to contortions in the trajectories of individual vortices which become less sheet-like as the rolling-up process proceeds, a feature of the model which was of some concern to Birkhoff & Fisher and Hama & Burke. Although several later authors have devised techniques such as re-discretisation to over-come these defects in modelling local sheet deformation, the main point at issue here is that of ensuring reversibility when modelling convective motions. Secondly there seems to be a clash of ideas in the proposition that these vortex instabilities, which lead ultimately to the formation of large scale eddies and are thus linked into the yet more complex matter of the generation of the turbulence, are fundamentally reversible phenomena. These are among some of the important considerations which will be discussed in this chapter within the context of inviscid and therefore reversible flows. Most practical engineering flow problems on the other hand are in-fluenced by viscous diffusion, models for which will be developed in Chapter 9.

We begin in the next section by considering the reversible convection of discrete vortex elements and by applying this to two important problems of vortex sheet stability, the roll-up of a free-ended vortex sheet, Section 8.3.1 and the Kelvin–Helmholtz instability, Section 8.3.2. This leads on to the convection of vortex elements due to interference with a body, Section 8.4 and then the important matter of vorticity production at a body surface or separation point, Section 8.5.1. The chapter is concluded with a primitive vortex cloud model for bluff bodies with pre-determined and fixed separation points, Section 8.5.2.

8.2 Vortex convection

We have already considered one class of problems involving the convection of free vorticity, namely the free-streamline and jet flow analyses of Chapter 7. As explained in Section 1.4, free vortex sheets convect themselves parallel to the line of the sheet with a velocity equal to one half of the local vorticity strength $\gamma(s)$. In this way vorticity is continually being convected along the boundaries of a jet or a separated free streamline. However, such flows, being steady, may be treated as if the vorticity were bound, since in steady flow there is a continual replacement of vorticity at the same strength arriving from upstream. The same is in fact true of the surface vorticity sheets of rigid bodies as discussed in Chapter 1 in relation to the physical significance of the surface vorticity model when compared with a real boundary layer, Fig. 1.3. Consequently there is no point in allowing the sheets to convect in steady flow

319

models for purpose of analysis, whether they are surface vorticity or free vorticity sheets. Even so it would actually be possible in principle to attempt such a model as we shall illustrate later for the case of a lifting aerofoil shedding upper and lower surface vorticity sheets as an alternative model to the traditional trailing edge Kutta Joukowski condition, Fig. 10.5.

The greatest deficiency of steady free-streamline and jet flow models is their inability to represent the true free vortex convective and viscous diffusive motions of real fluids. As asserted by Lewis & Porthouse (1983b) in a discussion of flow simulation by the vortex cloud model, all vorticity in a real flow is in fact free since the fluid velocity at the actual point of contact of a body surface is always zero. Bound vorticity is not a physical reality but merely a mathematical device of potential flow theory, albeit a very useful one. The free vorticity of a flow is created at the surface by the action of the local pressure gradient as demonstrated in Section 1.5 (1.14) and is under a continuous state of convection and diffusion within the surface boundary layer. We will now begin at the beginning of our study of these extremely complex flow mechanisms by considering first the most elementary of vortex convective processes as a foundation for the steady development of vortex dynamics theory.

8.2.1 Convection of a vortex pair

Consider a vortex pair of equal strength Γ, Fig. 8.2(a), distance d apart. If these are free vortices then they will each experience a convection velocity due to the other of magnitude $\Gamma/2\pi d$ acting in the direction normal to the line AB joining them. The self-induced convection velocity of a rectilinear vortex is zero. Since there is thus no displacement due to convection along the direction AB, it follows that the vortex pair will precess in a circular path about the mid point of AB.

Consider now a model of this simple motion in which time elapses over a succession of elementary but finite steps Δt. If we make the most simple assumption that each vortex is convected forward through the elementary distance $(\Gamma/2\pi d)\,\Delta t$, then over several time steps the vortex pair will move to $A_1 B_1$, $A_2 B_2$, $A_3 B_3$ etc. By analogy with the streamline calculations for a single point vortex discussed in Section 1.10, we see that this simple forward difference

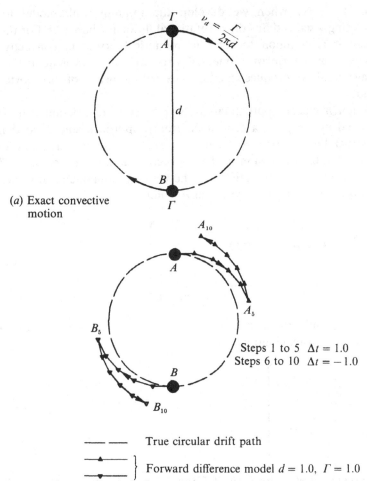

(a) Exact convective
motion

Steps 1 to 5 $\Delta t = 1.0$
Steps 6 to 10 $\Delta t = -1.0$

—— ——   True circular drift path

⎯⎯⎯⎯ ⎫
⎯⎯⎯⎯ ⎬   Forward difference model $d = 1.0$, $\Gamma = 1.0$

(b) Numerical estimate of convective motion by forward difference
method and check on reversibility

Fig. 8.2. Irreversibilities in forward differencing model of vortex
convection.

model of vortex convection leads to spiralling rather than circular
motion. Furthermore if we attempt to reverse this motion by
applying negative time steps $-\Delta t$, the vortex pair will continue to
spiral outwards, Fig. 8.2(b), and will overall rotate anticlockwise
through a reduced angle. The greater the discretisation of time Δt
the worse the degree of irreversibility. In effect this increased
separation of the vortices is analogous to viscous diffusion as will

321

become clearer when we develop the random walk model for simulating viscous diffusion of rotational flows in Chapter 9. For this reason it is common to refer to numerical errors in convective processes as a form of 'numerical viscosity', since in general they always tend to produce similar outward diffusion of the vortex cloud.

A much closer approximation to reversible convection can be achieved by adopting a central difference approach similar to that for streamline plotting introduced in Section 1.10. At this point let us broaden the discussion to the convective motion of a cloud of Z vortex elements of strength $\Delta\Gamma_n$. The velocity components at vortex location m due to a unit vortex at n would be

$$
\left.
\begin{aligned}
U_{mn} &= \frac{1}{2\pi}\left[\frac{y_m - y_n}{(x_m - x_n)^2 + (y_m - y_n)^2}\right] \\[2ex]
V_{mn} &= -\frac{1}{2\pi}\left[\frac{x_m - x_n}{(x_m - x_n)^2 + (y_m - y_m)^2}\right]
\end{aligned}
\right\}
\tag{8.1}
$$

Consequently the convection velocity components of vortex m due to the vortex cloud will be

$$
\left.
\begin{aligned}
u_m &= \sum_{\substack{n=1 \\ n \neq m}}^{Z} \Delta\Gamma_n U_{mn} \\[2ex]
v_m &= \sum_{\substack{n=1 \\ n \neq m}}^{Z} \Delta\Gamma_n V_{mn}
\end{aligned}
\right\}
\tag{8.2}
$$

Applying first a forward difference, step 1, the vortex n will convect from a to b, Fig. 8.3, where

$$
\left.
\begin{aligned}
x_{mb} &= x_{ma} + u_m\,\Delta t \\
y_{mb} &= y_{ma} + v_m\,\Delta t
\end{aligned}
\right\}
\tag{8.3}
$$

Having convected all vortices, if we now recalculate the new convection velocities at location b, (u_{mb}, v_{mb}), we may march forward again through time step 2 to point c. On the other hand if instead we average these two forward difference steps, finishing at point d, then a very much better estimate of vortex convection is achieved since curvature of the drift path is now taken into consideration to first order. In fact this is identical to using

322

Vortex convection

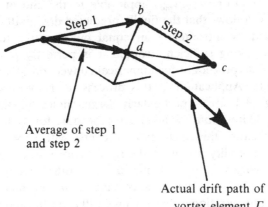

First order convection ≈ step 1

Second order convection ≈ average of step 1 + step 2

Fig. 8.3. Central difference estimate of vortex drift in convective processes.

alternatively the average drift velocity in equations (8.3) so that

$$x_{md} = x_{ma} + \tfrac{1}{2}(u_{ma} + u_{mb})\,\Delta t$$
$$y_{md} = y_{ma} + \tfrac{1}{2}(v_{ma} + v_{mb})\,\Delta t$$

$$(8.3a)$$

and can be thought of as a central difference approach. The sequence is then as indicated in the flow diagram.

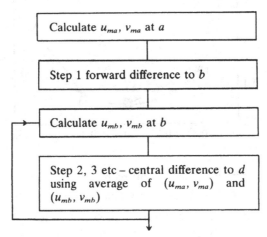

If this process were repeated many more times, always taking the average of the initial convection velocities (u_{ma}, v_{ma}) and the latest

323

estimate (u_{mb}, v_{mb}) or (u_{md}, v_{md}) applicable to the end of the x, y displacement, it follows that the finally predicted drift path could be exactly reversed by introducing an equal but negative time step $-\Delta t$. Thus by taking sufficient trouble it is perfectly possible to define a reversible procedure for numerical convection of a cloud of vortex elements. Application of this process to the vortex pair is shown in Fig. 8.4 for a particularly large choice of time step $\Delta t = 3.0$ using 50 iterations of the above procedure for each of three time steps. Repeating this calculation in reverse with $\Delta t = -3.0$ as a check upon reversibility produced the results shown in table 8.1.

In a large vortex cloud scheme on the other hand such a procedure would be extremely time consuming and normal practice is to settle for one iteration only which we will call central difference convection. In this case care is needed to choose a time step sufficiently small to maintain an acceptable level of reversibility. A related parameter more easy to visualise is the ratio of drift path

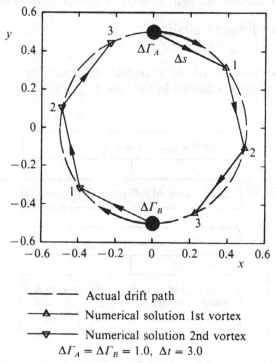

———— Actual drift path

—△—— Numerical solution 1st vortex

—▽—— Numerical solution 2nd vortex

$\Delta\Gamma_A = \Delta\Gamma_B = 1.0$, $\Delta t = 3.0$

Fig. 8.4. Predicted convection of a vortex element pair using central difference method with 50 iterations.

Table 8.1. *Reversible convection of a vortex element pair*

$(\Delta\Gamma_A = \Delta\Gamma_B = 1.0, \, d = 1.0, \, 50 \text{ iterations})$

Step no.	Δt	x 0.000 000	y 0.500 000	(Initial position of vortex A)
1	3.0	0.388 824	0.314 300	
2	3.0	0.488 908	−0.104 735	
3	3.0	0.225 929	−0.446 045	
4	−3.0	0.488 907	−0.104 736	
5	−3.0	0.388 824	−0.314 250	(Final position
6	−3.0	0.000 001	0.500 000	of vortex A)

Table 8.2. *Influence of discretisation size upon error in central difference (second order) convection of a vortex element pair*

$(\Delta\Gamma_A = \Delta\Gamma_B = 1.0, \, d_1 = 1.0)$

$\dfrac{\Delta s}{d}$	Δt	Number of steps N	Error $(d_N - d_1)/d_N$
0.5	3.141 59	3	0.142 356
0.4	2.513 27	4	0.099 757
0.3	1.884 96	5	0.052 197
0.25	1.570 79	6	0.034 191
0.2	1.256 64	8	0.027 765
0.15	0.942 48	10	0.009 000
0.1	0.628 32	16	0.003 047
0.05	0.314 16	31	0.000 383
0.025	0.157 08	64	0.000 048

length Δs to distance d of the nearest vortex. For the vortex pair we can easily show that $\Delta s/d$ is related to Δt through

$$\frac{\Delta s}{d} = \left(\frac{\Delta\Gamma}{2\pi d^2}\right)\Delta t \qquad (8.4)$$

As a fair test of the accuracy of central difference convection for various time steps, we might select instead successive values of $\Delta s/d$ and a sufficient number N of time steps for approximately 180° of precession of the vortex pair. The final error in separation distance of the vortex pair $(d_N - d_1)/d_1$ due to spiralling then gives a measure of reversibility as illustrated by Table 8.2.

We observe that to maintain the errors of second-order type convection to within 1%, the displacement ratio $\Delta s/d$ must not exceed 0.15, requiring ten time steps Δt of about 1.0 for the vortex pair to precess through 180°. The Pascal program convect.pas, given in Appendix 1 as Program 8.1, was used to complete the above study but is designed to enable the reader to experiment with more complex vortex clouds containing up to 60 vortices of varying strength. Later we will return to further study of vortex cloud convection in the presence of obstacles, Section 8.4. First we shall consider more closely some of the self-convection and stability features of vortex sheets.

8.3 Convection and stability of vortex sheets

Fig. 8.1 derived by Abernathy & Kronauer (1962) illustrates the manner in which a parallel pair of vortex sheets of equal strength but of opposite sign, when sinusoidally perturbed, will reform themselves into a vortex street under no other influence than their own self-convection. As time proceeds the initial disturbance grows while concentrating the vorticity, here represented as rows of discrete vortices, into distinct regions in the form of what these authors call 'vortex clouds'. Flow visualisation of bluff body wakes reveals that the vortex sheets roll up progressively in the manner sketched here by these authors and although the numerical discretisation leads to a more chaotic appearance at local level, the overall simulation of the eddy formation and strength is remarkable. For example the geometric ratio b/a settles down to about 0.5, which, although well in excess of the value 0.281 established by von Karman for the classical von Karman point-vortex street, is nevertheless similar to experimentally determined values. Several fundamental ideas emerge from this study which deserve further comment.

(i) Vortex sheets are apparently inherently unstable.

(ii) Vortex sheets may convect themselves into more stable configurations which are likely to be periodic.

(iii) Such rolling-up processes appear to be the prime cause for the formation of larger eddies.

(iv) These convective motions occur in the absence of viscous diffusion and are actually reversible by nature.

The most striking aspect of these observations is the notion that eddy formation resulting from the inherent instability of vortex sheets is the outcome of thermodynamically reversible convective motions. Often in our minds we associate eddy formation with loss mechanisms and certainly the presence of an eddy will result in viscous diffusion in a real fluid with consequent irreversibilities. On the other hand the *formation* of eddies due to the breakdown of unstable vortex sheets is in itself thermodynamically reversible as we shall illustrate in the next example.

8.3.1   Roll-up of a free-ended vortex sheet

Program 8.1 may be used to study the self-convection of a short length of free vortex sheet, Fig. 8.5(*a*), represented here by twenty discrete vortices of strength $\Delta\Gamma = 1.0$ located $\Delta x = 0.1$ apart along the *x* axis. As time proceeds so the sheet rotates clockwise as we would expect while at the same time undergoing a distortion due to the roll-up of its two free ends. After 50 time steps two distinct vortex cores have formed which have each already entrained seven of the vortex elements, leaving the remaining six stretched out between them as a weak connecting vortex sheet. If left to continue for a further period of time it is clear that the final outcome would be the conversion of the vortex sheet into a vortex cloud pair precessing in the manner of a point vortex pair. The solution shown in Fig. 8.5(a)–(c) was undertaken with central difference convection, selecting a time step $\Delta t = 0.005$ as determined by (8.4) and Table 8.2 to limit errors to within 0.3%.

The result of reversing this process by continuing the computation for 50 more time steps of value $\Delta t = -0.005$ is shown in Fig. 8.5(*d*). Although there is some residual error, reversibility has been achieved fairly substantially, which is remarkable bearing in mind that a total of 100 time steps were executed for what was a fairly complex fluid motion with strong convection effects especially at the ends of the line vortex. Even more surprising is the outcome of a re-run of this case aimed at the total elimination of convection errors by performing 30 iterations of the convection process at each time step, Fig. 8.5(*e*). The astonishing result was that true reversibility was limited to the four outermost vortex elements at each end of the sheet, namely the ones which experienced the maximum and most complex convective activity. For at least half of the forward

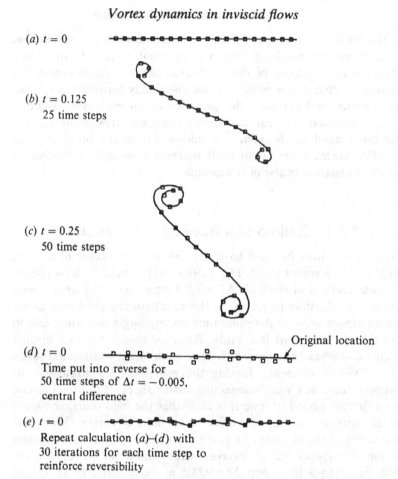

Fig. 8.5. Self-convection of a vortex sheet and check upon reversibility.

time period these vortices formed the cores of the two roll-up vortices. Equally remarkable is the behaviour of the innermost vortices. Although subject to only small displacements due to convection, the twelve remaining elements were still subject to similar error scatter but with a degree of randomness. Thus the final positional errors after time reversal were not symmetrical about the centre of the sheet.

These curious findings, which conflict with natural expectations, demand an explanation and this is to be found in the matter of vortex sheet stability. As we shall see in the next section a plane vortex sheet is inherently unstable. Whereas the end sections move rapidly towards the more stable configuration of the roll-up vortex cores, the centre twelve elements persist as a fairly flat vortex sheet for 25 time steps and likewise the innermost eight elements for all 50 time steps. During this period there is ample time for instability to be triggered by local perturbations in a manner similar to that which we shall demonstrate next.

8.3.2   Kelvin–Helmholtz instability of a vortex sheet

Before the development of electronic computers a fairly extensive study was made of vortex sheet stability by classical methods of analysis, including contributions by Rosenhead (1931) who also employed the discrete vortex or vortex cloud method. His studies involved the introduction of initial sinusoidal perturbations to the (x, y) coordinates of the vortex sheet similar to those used later by Abernathy & Kronauer (1962) for their vortex street studies, Fig. 8.1. After a number of time steps vortex roll-up develops periodically along the sheet at the crest and trough of each sine wave, culminating finally in a single row of vortex clouds not unlike half of a von Karman vortex street. The main feature of this so-called Kelvin–Helmholtz instability is the dramatic self-convective activity which follows spontaneously from only a small perturbation of the sheet. In effect the sheet breaks up into sections periodically along its length which roll themselves up into stable vortex cores in a manner similar to that which we have just considered. In fact it is by no means essential to introduce sinusoidal perturbations as shown by Porthouse (1981) (1983), who demonstrated that a small initial transverse displacement of any one of the discrete vortices modelling the sheet would result in snapping of the sheet into two halves which would then proceed to roll up about their free ends. Let us now demonstrate this calculation.

Since it is clearly impracticable to model a vortex sheet extending between $\pm\infty$ by a finite number of discrete vortices, we will first break the sheet down into a periodic array of short vortex sheets joined end to end, each of finite length T, Fig. 8.6. Now we may focus attention upon one single length located between $0 < x < T$ which we will represent by say Z vortex elements of equal strength $\Delta\Gamma_n$. Assuming a periodic flow field with pitch T, the expressions derived in Section 2.6.1 for a cascade of point vortices may be applied here to state the velocity components at the location of

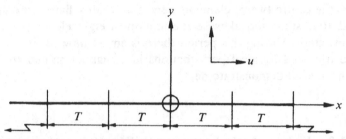

(a) Periodic array of vortex sheets of length T
forming an infinite vortex sheet

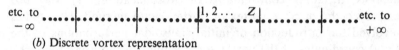

(b) Discrete vortex representation

Fig. 8.6. Modelling of an infinite vortex sheet by a periodic array of discrete vortices.

vortex element m due to a unit vortex at n and its associated periodic vortex array.

$$
\left.
\begin{aligned}
\bar{U}_{mn} &= \frac{\Gamma}{2T}\left[\frac{\sinh\dfrac{2\pi}{T}(y_m - y_n)}{\cosh\dfrac{2\pi}{T}(y_m - y_n) - \cos\dfrac{2\pi}{T}(x_m - x_n)}\right] \\[2em]
\bar{V}_{mn} &= -\frac{\Gamma}{2T}\left[\frac{\sin\dfrac{2\pi}{T}(x_m - x_n)}{\cosh\dfrac{2\pi}{T}(y_m - y_n) - \cos\dfrac{2\pi}{T}(x_m - x_n)}\right]
\end{aligned}
\right\}
\tag{8.5}
$$

The convection velocities at m due to the infinite array of discrete vortices thus become

$$
u_m = \sum_{\substack{n=1 \\ n\neq m}}^{Z} \Delta\Gamma_n \bar{U}_{mn} \qquad v_m = \sum_{\substack{n=1 \\ n\neq m}}^{Z} \Delta\Gamma_n \bar{V}_{mn}
\tag{8.6}
$$

where the self-induced velocities $\bar{U}_{mm}$, $\bar{V}_{mm}$ of vortex array $\Delta\Gamma_m$ are of course zero.

If these expressions are introduced into the previous computational procedure outlined in Section 8.2.1 to replace (8.1) and (8.2),

330

Convection and stability of vortex sheets

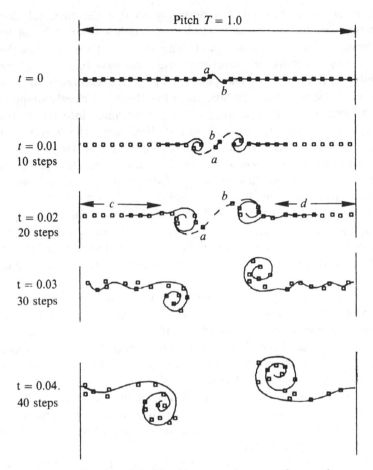

Sheet modelled by 30 vortex elements of strength
$\Delta\Gamma = 1.0$, $\Delta t = 0.001$

Fig. 8.7. Kelvin–Helmholtz instability of a vortex sheet.

Program 8.1 may be used without further changes to study the behaviour of an infinite vortex sheet subject to periodic disturbances, such as the Kelvin–Helmholtz instability.

Fig. 8.7 illustrates such a computation with 40 time steps $\Delta t = 0.001$ chosen with help from (8.4) and Table 8.1 to minimise convection errors. With a gap of 0.033 333 between the discrete vortices, the two centre vortices were displaced asymmetrically, as shown, by suitably small but significant amounts $\Delta x = \Delta y = 0.01$. The implication of (8.5) and (8.6) of course is that these distur-

331

bances are also distributed periodically so that the final solution obtained for the 30 vortices within our window $0 < x < T$ will be replicated over every other pitch length. After 10 time steps the sheet ends adjoining the perturbed vortex pair have begun to roll up tightly, leaving the vortex pair (a, b) as representative of the connecting sheet. After 20 time steps the sheet has clearly snapped as the two trigger vortices (a, b) become entrained into the vortex clouds. Proceeding through 30 and 40 time steps the vortex pair cloud formation is now well pronounced with a large gap of irrotational fluid in between. Bearing in mind the periodic nature of this solution it follows that the array of sheets finally settles down to a stable array of vortex cloud pairs similar in character to the single sheet case previously considered in Fig. 8.5, but without precession. This slight disturbance has thus triggered a dramatic instability of the vortex sheet, resulting in rapid self-convective roll-up into an array of vortex clouds which would finally settle down to a period of $T/2$.

Now although the discrete vortex patterns forming the roll-up vortices remain almost identical up to 20 time steps, there are at this stage non-symmetrical contortions appearing in the remaining straight sections of the sheet (c.f. regions c and d). After ten more time steps these small variations have suddenly blown up in scale causing finally general disintegration of the vortex sheet. As suggested at the conclusion of Section 8.3.1 this onset of instability along the entire straight section of the sheet is probably caused by local perturbations, due in this case to the mounting disturbances from the roll-up vortex eddies.

As a test of this idea the same case was re-run with twenty iterations per time step to reinforce reversibility but with no initial perturbations whatsoever to vortex elements a and b. The interesting outcome of this is shown in Fig. 8.8. For the first twenty time steps one can detect barely any disturbances to the sheet. Ten time steps later however significant convective activity is under way so that by time step 30 the vortex sheet has begun to break up along its entire length. At first this takes the form of small clusters of two or three vortex elements combining with one another locally, a feature of vortex dynamics modelling which has been studied and discussed extensively. Thus Birkhoff & Fisher (1959) repeating Rosenhead's calculations with finer discretisation and finding such contortions in the paths of individual vortices, were sceptical about the capability of discrete vortex modelling. While acknowledging these criticisms,

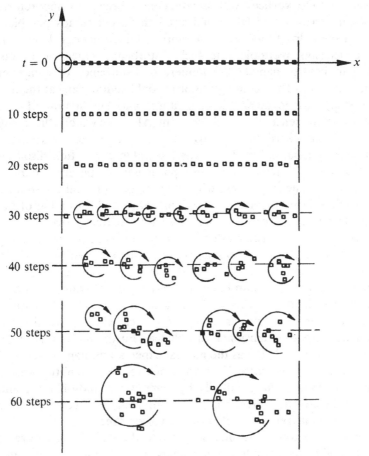

Fig. 8.8. Breakdown of a vortex sheet due to Kelvin–Helmholtz instability triggered only by computer round-off errors.

Abernathy & Kronauer felt that although fine detail was not accurately portrayed, the concentration of vorticity into clouds on the larger scale seemed acceptable as a means for simulation of vortex streets. Proceeding further to time steps 40–60 it may be observed that there is a tendency for the stronger clusters to grow by entrainment of the weaker ones. Ultimately it seems likely that there would remain just two large vortex clouds, each containing 15 vortex elements, which would be periodically pitched at intervals of $T/2$ along the original line of the vortex sheet.

Apparently then it would seem that a periodic array of equally spaced discrete vortices will spontaneously break up and reform itself into an array of two equal strength vortex clouds per pitch without the help of any externally imposed disturbances. However, this conclusion overlooks one such disturbance which we cannot afford to ignore, namely the numerical disturbance of computer round-off error. Thus although (8.6b) should sum to zero at the first time step, this involves the evaluation and summation of Z-1 convection influences for each element $\Delta\Gamma_m$. Inevitably there will be a random distribution of round-off errors spread across the vortex array resulting in the smallest possible convection displacements that can be resolved within a particular computer's accuracy. In successive time steps these will act as triggers of local instabilities at all points along the sheet, beginning with rapid formation of the small clusters of two or three elements each. As time proceeds these coalesce as the stronger clusters grow by entrainment, progressing finally to the more stable arrangement of distributed vorticity within the clouds. In effect our theoretical vortex sheet has undergone a form of transition from an initially laminar style of configuration to one in which the vorticity is repackaged by its own natural convective motions into an array of eddies which have the appearance of two-dimensional turbulence. Furthermore this study now confirms Fig. 8.5(e) as the natural expectation, namely residual disturbances of the centre 12 vortices due to irreversibilities arising from randomness fed in initially by computer round-off error and later by the growing vortex cores, but amplified over successive time steps by the inherent instability of the vortex sheet.

It is possible to gain further insight into the underlying nature of the instability by considering the analytical solution for the velocities induced by a finite length vortex sheet of strength $\gamma(x)$ per unit length, Fig. 8.9(a). For the element $\gamma(x)\,dx$ we have

$$du = \frac{\gamma(x)\,dx}{2\pi r}\sin\phi, \qquad dv = \frac{\gamma(x)\,dx}{2\pi r}\cos\phi$$

which upon integration lead to

$$\left. \begin{aligned} u &= \frac{\gamma(x)}{2\pi}(\phi_2 - \phi_1) \\[2mm] v &= \frac{\gamma(x)}{2\pi}\ln(r_1/r_2) = \frac{\gamma(x)}{2\pi}(\ln\sin\phi_1 - \ln\sin\phi_2) \end{aligned} \right\} \qquad (8.7)$$

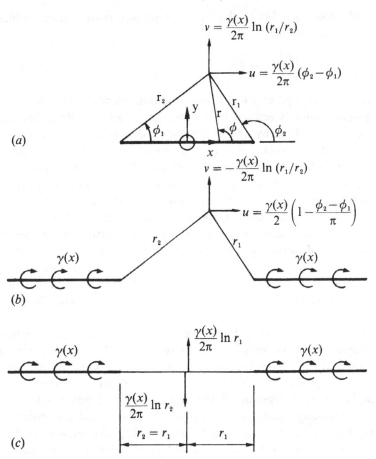

Fig. 8.9. Velocities induced by vortex sheets.

If we apply this result to an infinite vortex sheet with a central gap, Fig. 8.9(b), the velocity normal to the gap will be $-v$. On the x axis at the mid point the convection velocity is thus

$$v = \frac{\gamma(x)}{2\pi} (\ln r_1 - \ln r_1) \tag{8.8}$$

where the bracket contains the separate but cancelling contributions due to the two semi-infinite vortex sheets. If we now let the gap approach zero to obtain the self-convection velocity of a continuous vortex sheet extending between $\pm\infty$, mathematically speaking it is clear that v will also remain zero according to (8.8). Fluid dynamically speaking however, this arises from cancellation of the

335

contributions to v due to the two semi-infinite vortex sheets, which each become of value

$$v' = \pm \lim_{r_1 \to 0} \frac{\gamma(x)}{2\pi} \ln r_1 = \pm \infty \qquad (8.9)$$

Thus at any point of a doubly infinite vortex sheet the local convective velocity normal to the sheet is held in a delicate balance of zero arrived at by the subtraction of two infinite convective velocity contributions due to the two halves of the sheet, a prescription for instability. Any slight disturbance of this situation is likely to produce extremely large local values of v resulting in rapid vortex roll-up as we have demonstrated by discrete vortex modelling. Of course, once these large scale convective activities have set in (8.8) is no longer true for the distorted vortex sheet, which convects itself naturally towards some other equilibrium configuration, usually involving vortex roll-up. However, other sections of the sheet, still approximating locally to a straight vortex sheet, are open to further disturbances which lead to local instability and break-up.

Several authors have been concerned with such disturbances emanating from the very coarse discretisation of the roll-up vortex resulting directly from the simple representation used above. Thus Fink & Soh (1976) carried out a variety of numerical experiments including rediscretisation of the spiral vortex sheets resulting in smaller spacings near to the tip of the sheet and consequently smoother rolling up. Clements & Maull (1975) adopted a procedure whereby any two vortices that induced a mutual velocity in excess of a stated value were amalgamated, a strategy adopted by the present author to reduce the numbers of vortices and speed computation. Chorin & Bernard (1973) on the other hand adopted a Rankine vortex model to restrict convective velocities, again resulting in smoother roll-up of the vortex sheet. Finally Fink & Soh proposed that the error resulting in discrete vortex modelling of a vortex sheet was proportional to the logarithm of the ratio of the distances between adjacent point vortices, more detailed arguments for this being given also by Clements (1977) and by Sarpkaya & Shoaff (1979). Inevitably there will always be an element of doubt about the capability of simple discrete vortex methods to represent local flow, whereas a good and reversible convection scheme seems to be able to represent vortex motions such as periodic wakes remarkably realistically on the larger scale. Further review material has been given by Porthouse (1983), Downie (1981), Bearman & Graham (1979) and Maull (1986).

8.4 Convective interaction of free vortices with solid bodies

A sample vortex cloud solution for the flow past an obstacle is shown in Fig. 8.10, where the starting motion for flow past a wedge shaped body is portrayed. The flow is complex and is dominated by the vorticity, first generated within the boundary layers on the forward pointing faces and then projected as separated shear layers into the wake. Full vortex cloud analysis attempts to model this entire rotational flow by the use of discrete vortex elements created at the body surface over a series of small but finite time steps then shed as free vortices into the fluid to convect and diffuse naturally according to the laws of fluid motion. We have already considered the convective properties of vortex sheets and clouds. Now we focus attention upon the convective influence of the body upon the discrete vortex elements and vice versa and we will approach this in simple stages.

The circular cylinder has been used extensively for vortex dynamics studies, either in its own right as a bluff body or as a fluid dynamically more manageable shape than more complex bodies, reached through appropriate conformal transformations. Thus Gerard (1967) undertook one of the earliest known simulation calculations of the familiar von Karman vortex street by vortex

Fig. 8.10. Boundary layer and wake development for flow past a wedge shaped body predicted by vortex cloud analysis.

337

dynamics, taking advantage of the well known reflection system illustrated in Fig. 8.11(*a*). On the other hand Kuwahara (1973) and Sarpkaya (1975) both used the Joukowski transformation (2.10) to study the flow of fluid separating from the edges of a flat plate set obiquely to a uniform stream. Clements (1973) employed a

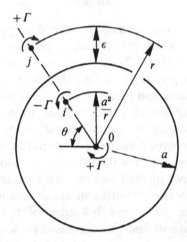

(*a*) Mirror image system for modelling the potential flow
field due to a point vortex close to a cylinder

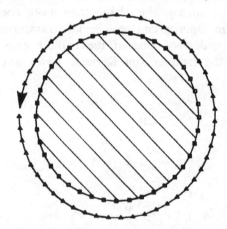

(*b*) Computer calculation by vortex cloud method for self-
induced drift path of a vortex close to a cylinder

Fig. 8.11. Interaction of a single vortex Γ with a circular cylinder and consequent drift path.

Schwartz–Christoffel transformation to study vortex street development behind a semi-infinite rectangular bluff body with a flat rear face, in this case without reference to a circle. P. Bettess & Downie (1988) developed transformations to permit modification of a true cylinder into a nearly cylindrical body with a local surface protrusion, for full vortex cloud simulation in relation to the forces experienced by off-shore structures. Conformal transformations may thus be used in certain situations to simplify treatment of more complex body shapes such as plates, polygons, ellipses or Joukowski aerofoils.

Boundary integral modelling by the surface vorticity method, on the other hand, offers the attraction of complete generality of body shape with none of the limitations or complications of conformal transformations, with consequent scope for the analysis of a wide range of engineering problems involving wake flows. There are however some numerical difficulties inherent in the modelling of free vortex singularities in close proximity to a surface represented by discrete vorticity elements. These are best explored by considering first the simplest problem of the interaction of a single vortex Γ with a circular cylinder, Fig. 8.11. This is the recommended starting point for the reader who wishes to proceed with the development of a vortex cloud computational scheme. Not only is the exact solution to this problem known, providing therefore a secure datum check, but most of the problems of convection which occur in full scale complex vortex cloud computational schemes can be studied here at close quarters in a simple and controlled manner. These problems are of two types, first the accurate representation of the potential flow field induced around the cylinder by the vortex and second the consequent convection experienced by the free vortex. A simple computational sequence for prediction of the self-induced convection experienced by the vortex depicted in Fig. 8.11(*a*) would be as shown in the flow diagram on page 340.

We will deal with these two problems in turn in the next two subsections in relation to boxes 2 and 3 of this scheme.

8.4.1 Potential flow past a cylinder due to a nearby vortex

Tackling the first problem (1.21) may be adapted, with a modified right hand side, to represent the surface vorticity boundary integral

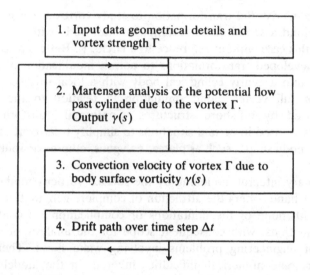

equation for a body of arbitrary shape in the (x, y) plane interacting with a vortex Γ at (x_j, y_j), namely

$$\sum_{n=1}^{M} K(s_m, s_n)\gamma(s_n) = -\Gamma(U_{mj} \cos \beta_m + V_{mj} \sin \beta_m)$$

$$= \text{rhs}_m \qquad (8.10)$$

where the velocity components (U_{mj}, V_{mj}) at element m induced by a unit vortex at j are given by expressions similar to equations (8.1)

$$U_{mj} = \frac{1}{2\pi}\left(\frac{y_m - y_j}{r_{mj}^2}\right), \qquad V_{mj} = -\frac{1}{2\pi}\left(\frac{x_m - x_j}{r_{mj}^2}\right) \qquad (8.11)$$

with

$$r_{mj} = \sqrt{[(x_m - x_j)^2 + (y_m - y_j)^2]} \qquad (8.12)$$

The mirror image system depicted in Fig. 8.11(*a*) provides an exact solution to this potential flow problem. As shown in standard texts such as Milne-Thompson (1955) and Glauert (1948), the cylinder of radius a becomes a stream surface of the flow induced by Γ if a reflection vortex of strength $-\Gamma$ is located at the inverse point, which is at radius a^2/r. A third vortex of strength Γ must also be located at the centre of the circle to ensure zero net bound circulation on the body. The exact surface velocity is then calculable by superposition of the flow fields due to the three vortices at 0, i and j, for which equations (8.11) may be conveniently made use of. Thus

$$q_{m \text{ exact}} = \sqrt{(u_m^2 + v_m^2)} \tag{8.13}$$

where

$$\left.\begin{array}{l} u_m = \Gamma(U_{m0} - U_{mi} + U_{mj}) \\ v_m = \Gamma(V_{m0} - V_{mi} + V_{mj}) \end{array}\right\} \tag{8.14}$$

A comparison of the exact and numerical predictions for a 30-element representation of a cylinder of radius 0.5 by the above methods is given in Table 8.3 for three decreasing radii of the point vortex, namely $r = 0.8$, 0.7 and 0.6. When applying the usual procedure for specification of surface elements, Fig. 1.6 and Section 1.8, the pivotal points tend to lie slightly inside the body profile. Since we wish to examine solutions for which the vortex approaches close to the pivotal points it is best for these to lie on the actual circle under consideration so that the best physical comparison with the exact solution is obtained. The 30-sided polygon representing the circle was therefore arranged to be tangential at the mid point of each element.

In this test case the vortex was located on the y axis directly opposite to the centre of surface element number 8, resulting of course in symmetry of the solution (e.g. compare elements 7 and 9, 6 and 10 etc). Extremely good numerical results were obtained for Case 1 with $r = 0.8$ and reasonably acceptable results with $r = 0.7$. With $r = 0.6$ on the other hand, disastrous numerical results were obtained. The ratio of gap $\varepsilon = r - a$ to element length Δs for Case 2 is approximately 2.0, representing the acceptable limit upon free vortex proximity imposed by surface vorticity modelling without further corrective action. Two routes towards improvement may be followed, namely zero circulation correction of the right hand side terms of the system of equations (8.10) and the use of sub-elements. We will consider these in turn.

The problem of 'numerical leakage flux' has already been discussed in Section 2.3.3 in relation to errors due to the interac-

Table 8.3. *Velocities induced on a cylinder of radius 0.5 due to a nearby vortex of strength 1.0*

Element number	Case 1 $r = 0.8$		Case 2 $r = 0.7$		Case 3 $r = 0.6$	
	num.	exact	num.	exact	num.	exact
1	0.1645	0.1644	0.2112	0.2037	1.0158	0.2543
2	0.1253	0.1252	0.1798	0.1724	0.9973	0.2358
3	0.0651	0.0650	0.1298	0.1224	0.9668	0.2054
4	−0.0315	−0.0317	0.0445	0.0370	0.9119	0.1504
5	−0.1929	−0.1930	−0.1141	−0.1215	0.7988	0.0373
6	−0.4615	−0.4616	−0.4342	−0.4417	0.5140	−0.2476
7	−0.8364	−0.8367	−1.0557	−1.0632	−0.4350	−1.1967
8	−1.0608	−1.0610	−1.5839	−1.5915	−2.4212	−3.1831
9	−0.8364	−0.8367	−1.0557	−1.0632	−0.4350	−1.1967
10	−0.4615	−0.4616	−0.4342	−0.4417	0.5140	−0.2476
11	−0.1929	−0.1930	−0.1141	−0.1215	0.7988	0.0373
12	−0.0316	−0.0317	0.0445	0.0370	0.9119	0.1504
13	0.0651	0.0650	0.1298	0.1224	0.9668	0.2054
14	0.1253	0.1252	0.1798	0.1724	0.9973	0.2358
15	0.1645	0.1644	0.2111	0.2037	1.0158	1.2543
16	0.1909	0.1908	0.2318	0.2244	1.0277	0.2663
17	0.2093	0.2091	0.2458	0.2384	1.0358	0.2743
18	0.2222	0.2221	0.2556	0.2482	1.0413	0.2798
19	0.2313	0.2312	0.2625	0.2551	1.0452	0.2837
20	0.2377	0.2376	0.2672	0.2598	1.0478	0.2863
21	0.2418	0.2417	0.2703	0.2629	1.0496	0.2881
22	0.2442	0.2441	0.2721	0.2647	1.0505	0.2891
23	0.2450	0.2449	0.2727	0.2653	1.0509	0.2894
24	0.2442	0.2441	0.2721	0.2647	1.0505	0.2891
25	0.2418	0.2417	0.2703	0.2629	1.0496	0.2881
26	0.2377	0.2376	0.2672	0.2598	1.0478	0.2863
27	0.2313	0.2312	0.2625	0.2551	1.0452	0.2837
28	0.2222	0.2221	0.2556	0.2482	1.0413	0.2798
29	0.2093	0.2091	0.2458	0.2384	1.0358	0.2743
30	0.1909	0.1908	0.2318	0.2244	1.0277	0.2663

tions of opposite surface elements of a thin body. In that context the back-diagonal coupling coefficients, which represent these opposite body profile points, were replaced by the lower mean values as recommended originally by Jacob & Riegels (1963) to ensure zero net implied circulation around the body profile interior. Following similar arguments which led to such back diagonal correction as embodied in (2.19), we may also state the necessary condition that the net circulation around the cylinder perimeter due to the externally located vortex Γ must be zero. Expressing this analytically we then have

$$\sum_{n=1}^{M} \text{rhs}_n \, \Delta s_n = 0 \qquad (8.15a)$$

from which the replacement rhs value for the nearest surface element (say p) to the vortex is given by

$$\text{rhs}_p = -\frac{1}{\Delta s_p} \sum_{\substack{n=1 \\ n \neq p}}^{M} \text{rhs}_n \, \Delta s_n \qquad (8.15b)$$

If (8.15a) were not satifisfied by the rhs values as given by (8.10) and (8.11), there would be an implied erroneous residual vorticity bound within the body profile leading to a flux through the body surface. Thus although each individual equation (8.10) is a statement of the Dirichlet boundary condition for a given surface element, it is necessary for the system of equations as a whole to satisfy the condition of internal irrotationality within the body profile. Equations (2.19) enforce this for all of the surface elements $\gamma(s_n) \, \Delta s_n$ whose influence is accounted for by the matrix of coupling coefficients. Equation (8.15b) now achieves the same end for the external vortex Γ, whose influence is accounted for entirely by the right hand sides of (8.10).

Although this may seem to be an extraordinary and artificial way to obtain a value for rhs_p, it produces quite dramatic improvements as shown by Table 8.4, which records a repeat run of the previous three cases with right hand side corrections.

Extremely good results were obtained here by the surface vorticity method for Cases 1 to 3, although element 8 in closest proximity to the vortex is still subject to significant errors for Case 3. If r is reduced further equally good predictions are obtained for all elements except element 8 for which errors mount dramatically. Case 4 illustrates this for the extreme case of $r = 0.501$ for which the

Table 8.4. *Velocities induced on a cylinder of radius 0.5 due to a nearby vortex. Solution with r.h.s. correction.*

Case 1 $r = 0.8$		Case 2 $r = 0.7$		Case 3 $r = 0.6$		Case 4 $r = 0.501$	
num.	exact	num.	exact	num.	exact	num.	exact
0.1643	0.1644	0.2037	0.2037	0.2543	0.2543	0.3176	0.3176
0.1252	0.1252	0.1724	0.1724	0.2358	0.2358	0.3173	0.3174
0.0650	0.0650	0.1224	0.1224	0.2053	0.2054	0.3170	0.3170
−0.0317	−0.0317	0.0370	0.0370	0.1504	0.1504	0.3164	0.3164
−0.1930	−0.1930	−0.1215	−0.1215	0.0373	0.0373	0.3149	0.3150
−0.4616	−0.4616	−0.4416	−0.4417	−0.2476	−0.2476	0.3109	0.3110
−0.8366	−0.8367	−1.0631	−1.0632	−1.1966	−1.1967	0.2892	0.2892
−1.0609	−1.0610	−1.5906	−1.5915	−3.1019	−3.1831	−9.1346	−8.3140
−0.8366	−0.8367	−1.0631	−1.0632	−1.1966	−1.1967	0.2892	0.2892
−0.4616	−0.4616	−0.4416	−0.4417	−0.2476	−0.2476	0.3109	0.3110
−0.1930	−0.1930	−0.1215	−0.1215	0.0373	0.0373	0.3149	0.3150
−0.0317	−0.0317	0.0370	0.0370	0.1504	0.1504	0.3164	0.3164
0.0650	0.0650	0.1224	0.1224	0.2053	0.2054	0.3170	0.3170
0.1252	0.1252	0.1724	0.1724	0.2358	0.2358	0.3174	0.3174
0.1643	0.1644	0.2037	0.2037	0.2543	0.2543	0.3176	0.3176
0.1908	0.1908	0.2243	0.2244	0.2662	0.2663	0.3177	0.3177
0.2091	0.2091	0.2384	0.2384	0.2743	0.2743	0.3178	0.3178
0.2220	0.2221	0.2482	0.2482	0.2798	0.2798	0.3179	0.3179
0.2312	0.2312	0.2550	0.2551	0.2837	0.2837	0.3179	0.3179
0.2375	0.2376	0.2598	0.2598	0.2863	0.2863	0.3179	0.3180
0.2417	0.2417	0.2629	0.2629	0.2880	0.2881	0.3179	0.3180
0.2441	0.2441	0.2646	0.2647	0.2890	0.2891	0.3180	0.3180
0.2448	0.2449	0.2652	0.2653	0.2893	0.2894	0.3180	0.3180
0.2441	0.2441	0.2647	0.2647	0.2890	0.2891	0.3180	0.3180
0.2417	0.2417	0.2629	0.2629	0.2880	0.2881	0.3179	0.3180
0.2375	0.2376	0.2598	0.2598	0.2863	0.2863	0.3179	0.3180
0.2312	0.2312	0.2551	0.2551	0.2837	0.2837	0.3179	0.3179
0.2220	0.2221	0.2482	0.2482	0.2798	0.2798	0.3179	0.3179
0.2091	0.2091	0.2384	0.2384	0.2742	0.2743	0.3178	0.3178
0.1908	0.1908	0.2243	0.2244	0.2662	0.2663	0.3177	0.3177

gap ratio $\varepsilon/\Delta s \approx 0.01$. All surface velocities were predicted with precision except for element 8 which was apparently under-predicted by an order of magnitude. The reason for this is simple. The actual velocity varies widely over element 8 ranging from about 0.3 at the element end to 318.314 at its mid point. The value of 9.1346 given by the Martensen solution represents the average vorticity strength over element 8, and the quantity $\gamma(s_8)\,\Delta s_8$ is therefore the total amount of vorticity to be found on element 8. Since this is the amount of vorticity available to be shed in a vortex cloud scheme such as we will consider later in Chapter 10, its

Table 8.5. *Velocities induced by a vortex at radius 0.52 moving past element 8 of a cylinder of radius 0.5 represented by 30 surface vorticity elements*

theta = 90°		92°		94°		96°	
num.	exact	num.	exact	num.	exact	num.	exact
0.3043	0.3044	0.3049	0.3049	0.3053	0.3054	0.3058	0.3058
0.3002	0.3003	0.3011	0.3011	0.3018	0.3019	0.3025	0.3026
0.2933	0.2934	0.2948	0.2948	0.2961	0.2961	0.2972	0.2973
0.2806	0.2807	0.2834	0.2834	0.2858	0.2859	0.2880	0.2881
0.2531	0.2532	0.2596	0.2596	0.2651	0.2651	0.2698	0.2698
0.1751	0.1751	0.1958	0.1959	0.2123	0.2123	0.2256	0.2256
−0.2337	−0.2337	−0.0915	−0.0915	0.0022	0.0022	0.0671	0.0671
−7.4014	−15.9155	−7.3056	−8.7411	−6.9685	−3.5777	−1.6804	−1.6805
−0.2337	−0.2337	−0.4650	−0.4640	−0.8707	−0.8708	−6.1806	−1.6805
0.1751	0.1751	0.1486	0.1486	0.1138	0.1139	0.0671	0.0671
0.2532	0.2532	0.2456	0.2456	0.2365	0.2365	0.2256	0.2256
0.2806	0.2807	0.2775	0.2775	0.2739	0.2739	0.2698	0.2698
0.2933	0.2934	0.2918	0.2918	0.2900	0.2900	0.2880	0.2881
0.3002	0.3003	0.2993	0.2994	0.2983	0.2984	0.2972	0.2973
0.3043	0.3044	0.3038	0.3038	0.3032	0.3032	0.3025	0.3026
0.3070	0.3070	0.3066	0.3066	0.3062	0.3063	0.3058	0.3058
0.3087	0.3088	0.3085	0.3085	0.3082	0.3083	0.3079	0.3080
0.3099	0.3100	0.3098	0.3098	0.3096	0.3096	0.3094	0.3094
0.3108	0.3108	0.3107	0.3107	0.3105	0.3106	0.3104	0.3104
0.3114	0.3114	0.3113	0.3113	0.3112	0.3112	0.3111	0.3111
0.3117	0.3118	0.3117	0.3117	0.3116	0.3117	0.3116	0.3116
0.3120	0.3120	0.3119	0.3120	0.3119	0.3119	0.3119	0.3119
0.3120	0.3121	0.3120	0.3121	0.3120	0.3121	0.3120	0.3121
0.3120	0.3120	0.3120	0.3120	0.3120	0.3120	0.3120	0.3121
0.3117	0.3118	0.3118	0.3118	0.3118	0.3119	0.3119	0.3119
0.3114	0.3114	0.3115	0.3115	0.3115	0.3116	0.3116	0.3116
0.3108	0.3108	0.3109	0.3109	0.3110	0.3111	0.3111	0.3111
0.3099	0.3100	0.3101	0.3102	0.3103	0.3103	0.3104	0.3104
0.3087	0.3088	0.3090	0.3090	0.3092	0.3092	0.3094	0.3094
0.3070	0.3070	0.3073	0.3074	0.3076	0.3077	0.3079	0.3080

prediction is actually of more practical use to us than the precise value of surface velocity at the centre of element 8.

Now so far we have considered only vortex locations opposite to the centre of element 8. Table 8.5 records results obtained with the vortex located at various angular positions moving clockwise close to element 8 at a radius of 0.52. Once again extremely good results were obtained for predicted surface velocities of all elements except the one in closest proximity to the vortex. However a problem clearly arises for location $\theta = 96°$ since pivotal points 8 and 9 are then exactly equidistant from the vortex. Since only one can be selected for the right hand side correction (8.15b) all of the circulation correction vorticity has been deposited upon element 9, whereas element 8 has the velocity value 1.6804 in agreement with the exact solution. The ideal arrangement for this particular case of $\theta = 96°$ would be as follows:

(i) The correct surface velocity at element 8 should equal that at element 9, namely 1.6804.
(ii) The value used for vortex shedding should also be equal for elements 8 and 9 and could be taken as the average, i.e. $\frac{1}{2}(1.6804 + 6.1806) = 3.9323$.

For vortex cloud schemes consideration (ii) is of more importance than (i) for an equitable distribution of vorticity shedding between elements. One way to improve this is to introduce sub-elements for improved estimation of the right hand side values for elements in close proximity to the vortex. For example we may limit the use of sub-elements for situations in which $r_{mj} < 2.0 \, \Delta s_m$ to include up to three surface elements local to the point vortex. A minimum number of sub-elements $n$subs then follows from the simple algorithm

$$n\text{subs} = 1 + \text{round}(2 * \Delta s_m / r_{mj}) \tag{8.16}$$

The velocities (U_{mj}, V_{mj}) used for evaluation of rhs$_m$ and previously given by equations (8.11) are then replaced by the average values for all sub-elements, namely

$$U_{mj} = \frac{1}{2\pi n \text{subs}} \sum_{n=1}^{n\text{subs}} \left\{ \frac{y_n - y_j}{r_{nj}^2} \right\}, \qquad V_{mj} = -\frac{1}{2\pi n \text{subs}} \sum_{n=1}^{n\text{subs}} \left\{ \frac{x_n - x_j}{r_{mj}^2} \right\}$$

$$\tag{8.17}$$

where (x_n, y_n) are coordinates of the centre of sub-element n, Fig. (8.12).

Implementation of this for the previous case is shown in Table 8.6. In comparison with Table 8.5 it will be observed that the use of sub-elements influenced results only in the vicinity of the strategic elements 8 and 9. The case just referred to with $\theta = 96°$ yielded almost equal estimates of surface vorticity for elements 8 and 9 as required, namely 3.9464 and 3.9239 respectively. Twice the minimum number of sub-elements as given by (8.16) was used here to

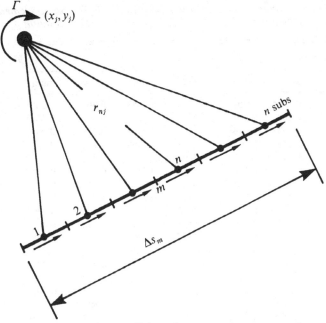

Fig. 8.12. Average velocity parallel to element m due to a nearby vortex Γ using sub-elements.

improve resolution. Four times the minimum number is recommended for general use.

8.4.2 Convection of a free vortex near a circle or an ellipse

We are now ready to consider the convective motion of the free vortex Γ due to the influence of the nearby cylindrical body. Accounting for the influence of each surface vortex element now of known vorticity strength $\gamma(s_m)$, the drift velocity components of Γ follow directly from

$$
\left.
\begin{aligned}
u_d &= -\sum_{m=1}^{M} \gamma(s_m)\, \Delta s_m U_{mj} \\
v_d &= -\sum_{m=1}^{M} \gamma(s_m)\, \Delta s_m V_{mj}
\end{aligned}
\right\}
\qquad (8.18)
$$

where the unit velocities follow directly from (8.11) or from

347

Table 8.6. *Velocities induced by a vortex at radius 0.52 close to element 8 of a cylinder of radius 0.5 with a 30-element representation*

theta = 90°		92°		94°		96°	
num.	exact	num.	exact	num.	exact	num.	exact
0.3043	0.3044	0.3049	0.3049	0.3053	0.3054	0.3058	0.3058
0.3002	0.3003	0.3011	0.3011	0.3018	0.3019	0.3025	0.3026
0.2933	0.2934	0.2948	0.2948	0.2961	0.2961	0.2972	0.2973
0.2806	0.2807	0.2834	0.2834	0.2858	0.2859	0.2880	0.2881
0.2532	0.2532	0.2596	0.2596	0.2651	0.2651	0.2698	0.2698
0.1751	0.1751	0.1958	0.1959	0.2123	0.2123	0.2256	0.2256
−0.3009	−0.2337	−0.1099	−0.0915	0.0025	0.0022	0.0745	0.0671
−7.2669	−15.9155	−7.0915	−8.7411	−6.2866	−3.5777	−3.9464	−1.6805
−0.3009	−0.2337	−0.6702	−0.4640	−1.5630	−0.8708	−3.9293	−1.6805
0.1751	0.1751	0.1592	0.1486	0.1238	0.1139	0.0745	0.0671
0.2532	0.2532	0.2456	0.2456	0.2365	0.2365	0.2256	0.2256
0.2806	0.2807	0.2775	0.2775	0.2739	0.2739	0.2698	0.2698
0.2933	0.2934	0.2918	0.2918	0.2900	0.2900	0.2880	0.2881
0.3002	0.3003	0.2993	0.2994	0.2983	0.2984	0.2972	0.2973
0.3043	0.3044	0.3038	0.3038	0.3032	0.3032	0.3025	0.3026
0.3070	0.3070	0.3066	0.3066	0.3062	0.3063	0.3058	0.3058
0.3087	0.3088	0.3085	0.3085	0.3082	0.3083	0.3079	0.3080
0.3099	0.3100	0.3098	0.3098	0.3096	0.3096	0.3094	0.3094
0.3108	0.3108	0.3107	0.3107	0.3105	0.3106	0.3104	0.3104
0.3114	0.3114	0.3113	0.3113	0.3112	0.3112	0.3111	0.3111
0.3117	0.3118	0.3117	0.3117	0.3116	0.3117	0.3116	0.3116
0.3120	0.3120	0.3119	0.3120	0.3119	0.3119	0.3119	0.3119
0.3120	0.3121	0.3120	0.3121	0.3120	0.3121	0.3120	0.3121
0.3120	0.3120	0.3120	0.3120	0.3120	0.3120	0.3120	0.3121
0.3117	0.3118	0.3118	0.3118	0.3118	0.3119	0.3119	0.3119
0.3114	0.3114	0.3115	0.3115	0.3115	0.3116	0.3116	0.3116
0.3108	0.3108	0.3109	0.3109	0.3110	0.3111	0.3111	0.3111
0.3099	0.3100	0.3101	0.3102	0.3103	0.3103	0.3104	0.3104
0.3087	0.3088	0.3090	0.3090	0.3092	0.3092	0.3094	0.3094
0.3070	0.3070	0.3073	0.3074	0.3076	0.3077	0.3079	0.3080

equations (8.17) when sub-elements are used. In view of this, computational advantage can be taken here of the (U_{mj}, V_{mj}) values already evaluated from the foregoing Martensen analysis. Thus the velocity at j induced by a unit vortex at m is equal in magnitude and opposite in direction to that at m induced by a unit vortex at j. The use of sub-elements is equally valid, representing the assumption that $\gamma(s_m)$ is uniformly spread over the body surface element Δs_m. In fact this proves to be a poor assumption as we shall see, and alternative models are required for evaluating the convective

348

motion of free vortices in very close proximity to a body. Accuracy will also always be improved by the use of second or higher order convection schemes as described in Section 8.2.1.

For the case of a free vortex close to a circular cylinder, Fig. 8.11, the exact drift velocity follows directly from the mirror image system, namely

$$q_d = \frac{\Gamma}{2\pi}\left(\frac{1}{r - a^2/r}\right) - \frac{\Gamma}{2\pi r} = \frac{\Gamma}{2\pi r}\left(\frac{a^2}{r^2 - a^2}\right) \qquad (8.19)$$

and is always normal to the radius Oj. Consequently the drift path will be circular, the vortex completing one complete rotation over the time interval

$$t = \frac{2\pi r}{q_d} = \frac{4\pi^2 r^2}{\Gamma}\left(\frac{r^2 - a^2}{a^2}\right) \qquad (8.20)$$

Fig. 8.11(b) demonstrates the precision with which this drift path may be predicted by the numerical iterative time stepping procedure described above for 50 time steps $\Delta t = 0.125068$, with $\Gamma = 1.0$, $a = 0.5$ and $r = 0.6$ for a 40-element representation of the circular cylinder. Using second order convection the radius of the vortex finally was 0.5991 and the error closure gap was 0.182618. With the order of convection raised to 30 to reinforce reversibility, the final vortex radius was 0.59886 and the closure gap error was reduced to 0.040878 which was just over 1% of the total drift path length.

The same computational procedure may be used for any other two-dimensional body shape such as the ellipse shown in Fig. 8.13. In this case, being a slender body profile, back diagonal correction was essential also. With a 60 surface element representation and a time step of $\Delta t = 0.05$, a perfectly acceptable solution was again obtained for vortex convection, which should follow a closed oval shaped contour surrounding the ellipse, Fig. 8.13(a). However, two other difficulties may arise as illustrated in Figs. 8.13(b) and (c).

Firstly, if we retain the same time step but reduce the number of surface elements by one half to $M = 30$, undulations in the drift path are obtained as the vortex drifts past each surface element. We will return to this problem shortly. Secondly, retaining 60 surface elements but increasing the time step five-fold to $\Delta t = 0.25$ errors of a different type have crept in. A credible solution was in fact obtained, but while acceptable for most of the drift path which is of modest curvature, it is clear that Δt is at the limits of acceptability for coping with the right curvature of the end sections. In practical

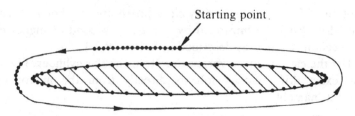

(*a*) Reversible solution with 60 surface elements, $\Delta\Gamma = 1.0$, $\Delta t = 0.5$

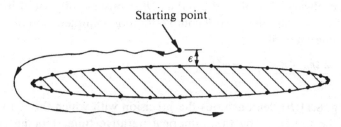

(*b*) Errors due to coarser body representation
30 surface elements, $\Delta\Gamma = 1.0$, $\Delta t = 0.05$

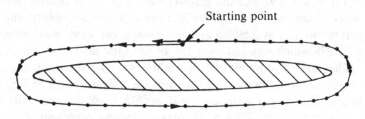

(*c*) Errors due to coarser time step
60 surface elements, $\Delta\Gamma = 1.0$, $\Delta t = 0.25$
Semi-major axis = 1.0, Semi-minor axis = 0.1

Fig. 8.13. Self-induced drift path of a free vortex close to an ellipse.

schemes the time step ought to be chosen to suit the most stringent likely convection requirements, although for the general situation of an arbitrary bluff body and its wake flow it is extremely difficult to quantify what these may be. Furthermore most of the drift path of our test case, Fig. 8.13(*c*), is well within the scope of the selected value $\Delta t = 0.25$ and an even coarser discretisation of time step would lead to computational economy with negligible loss of accuracy except in the end regions. The certainty of this could be established quite easily by focussing attention upon the local drift path curvature and selecting different Δt values to cope with this step by step. However such optimisation procedures are possible only with a single vortex and are quite impractical for a full vortex cloud modelling scheme since Δt must then be the same for all vortices in the field. In practice a fixed value of Δt must be selected arbitrarily to achieve a suitable compromise between accuracy and computational economy and which may result in some errors of significance in regions of high body curvature. The author's practice here for flow past bodies in a uniform stream W_∞ is to equate $W_\infty \Delta t$ to the average surface element length resulting in the following specification for Δt.

$$\Delta t = \frac{k}{MW_\infty} \sum_{n=1}^{M} \Delta s_n \tag{8.21}$$

where the constant k should certainly be no greater than 1.0.

8.4.3 Convection of vortices in very close proximity to a body

In the previous section reference was made to erroneous undulations in the drift path of a vortex in reasonably close proximity to an ellipse, Fig. 8.13(a). In fact these predictions were made with the normal surface vorticity model without the use of sub-elements. If sub-elements are introduced locally these undulations will be reduced to negligible proportions provided the gap ratio $\varepsilon/\Delta s_n >$ 0.4. For closer proximity it is not possible to improve the boundary element method further but fortunately recourse may be made to vortex reflection modelling instead. For example (8.19) may be used instead if the local body radius of curvature r_m is substituted for a in (8.19) provided the gap ratio $\varepsilon/r_m < 0.05$. As shown in Section 1.7 (1.31), r_m can be expressed in terms of the change in profile slope $\Delta\beta_m$ from one end of element m to the other through

$$r_m = \frac{\Delta s_m}{\Delta\beta_m} = \frac{2\Delta s_m}{\beta_{m+1} - \beta_{m-1}} \tag{8.22}$$

Alternatively we may use a plane wall mirror image approach as also illustrated in Fig. 8.14, from which the estimated convection

Vortex dynamics in inviscid flows

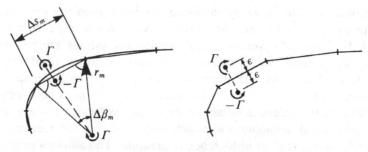

(a) Mirror image system based on vortex reflection in circular cylinder

(b) Simple image system based on plane wall reflection

Fig. 8.14. Reflection systems for calculation of self-induced convection of a vortex in close proximity to surface element m.

velocity is then

$$q_d = \frac{\Gamma}{4\pi\varepsilon} \tag{8.23}$$

The implementation of this second and more simple approach is shown in Fig. 8.15(a) for the motion of a vortex in very close proximity to a circular cylinder of radius 0.5. The starting and finishing radii of the vortex were $r = 0.51$ and $r = 0.509496$ respectively demonstrating the success of the simple mirror image reflection method used here for all time steps. Fig. 8.15(b) illustrates the more complex situation of vortex self-induced motion in close proximity to a + -shaped body, employing a computing scheme which selects whichever convection method is appropriate depending upon proximity. To summarise, the options, all of which were available during this second computation, are as follows:

(i) For gap ratios, $\varepsilon/\Delta s_m > 1.0$ normal surface vorticity modelling will suffice without the use of sub-elements. ε is defined here as the gap between the free vortex Γ and the nearest pivotal point m.

(ii) For $1.0 > \varepsilon/\Delta s_m > 0.4$ sub-elements should be used when calculating the convective velocity due to element m.

(iii) For $\varepsilon/\Delta s_m < 0.4$ mirror image modelling should be used according to (8.22) and (8.23).

It should be remarked that these studies present extremely more taxing requirements than will normally arise in a vortex cloud

352

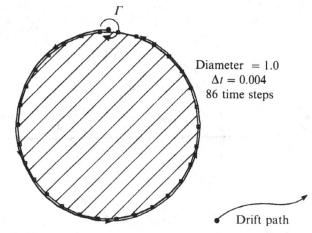

Diameter $= 1.0$
$\Delta t = 0.004$
86 time steps

Drift path

(a) Drift path of vortex in very
    close proximity to a cylinder

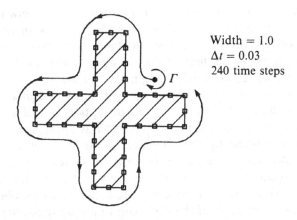

Width $= 1.0$
$\Delta t = 0.03$
240 time steps

(b) Drift path of a vortex close to a
    complex two-dimensional body

Fig. 8.15. Self-induced convection of a vortex in close proximity to a
two-dimensional body by the surface vorticity numerical model.

scheme, since the whole motion of the single vortex here has been
self-induced giving maximum play to error feedback. Two aspects
should be mentioned. Firstly, the strength of vortex cloud elements
will normally be very small and typically about $1/M$ of the value
used here where M is the number of surface elements. Secondly
convection velocity will involve a large contribution due to potential

353

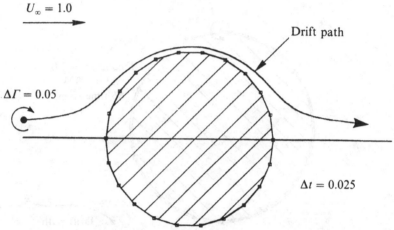

$U_\infty = 1.0$

Drift path

$\Delta\Gamma = 0.05$

$\Delta t = 0.025$

Fig. 8.16. Drift path of a vortex past a cylinder in a uniform stream.

flow caused by the uniform stream plus a contribution from each of the other free vortex elements. Thus vortex/body interaction is less significant than the present case for the majority of the vortices which soon become distant from the body as the wake develops.

As a check on the first of the above observations the flow of a typical vortex element of strength $\Delta\Gamma = 0.025$ past a cylinder of radius 0.5 in a uniform stream $U_\infty = 1.0$ has been calculated using 100 time steps $\Delta t = 0.025$, Fig. 8.16. Option (iii) above was not made use of but sub-elements were used as appropriate. Since the flow is dominated by the influence of the uniform stream, error ripples due to erroneous estimation of vortex interaction have negligible effect here. Further tests show that for such small vortex strengths good overall convection predictions may be obtained for even closer proximity of the drift path to the body but the limits set for option (iii) are still likely to lead to better predictions, especially of the actual velocity of drift. It is therefore recommended that all of options (i)–(iii) be retained in vortex cloud modelling of the type which we consider next.

8.5 Simple vortex cloud modelling for two-dimensional bodies with prescribed separation points

A true simulation of the real flow past a body must involve vorticity creation at all points of its surface accompanied by the processes of viscous diffusion and convection. Full vortex cloud modelling, as outlined in Chapter 10 and illustrated by Fig. 8.10, achieves this by the shedding of discrete vortices from all surface elements. Much simpler simulations may be attempted for bodies with known separation points such as sharp edged bluff bodies and we conclude this chapter with a consideration of such flows.

8.5.1 Vorticity shedding from a sharp edged separation point

The equilateral wedge shown in Fig. 8.17 is typical of bluff bodies for which the boundary layers formed on the forward facing surfaces are known to separate from the rear facing sharp edges A and B. The vorticity created within the boundary layers is then projected parallel to the surface with angle ϕ directly into the wake where it is free to undergo convection and diffusion. We continue for the time being to ignore viscous diffusion and to consider purely potential flows. Suppose that at some time t during such a motion the surface vorticity of the potential flow at elements A and B just upstream of the two separation points is $\gamma(s_A)$ and $\gamma(s_B)$ respectively. As shown in Section 1.5 and Fig. 1.5, the convective velocities of these vortex sheets will be $\frac{1}{2}\gamma(s_A)$ and $\frac{1}{2}\gamma(s_B)$. During the small time interval Δt the total amount of vorticity shed into the wake will thus be

$$\Delta\Gamma_A = \tfrac{1}{2}\gamma(s_A)^2 \, \Delta t, \qquad \Delta\Gamma_B = \tfrac{1}{2}\gamma(s_B)^2 \, \Delta t \qquad (8.24)$$

In the numerical scheme shown here the shed vorticity for this discrete time step is modelled by the introduction of two point

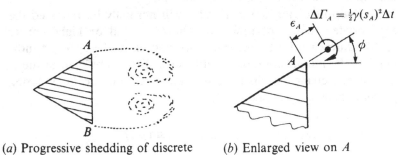

(a) Progressive shedding of discrete   (b) Enlarged view on A
   vortices Γ_A and Γ_B

Fig. 8.17. Vortex shedding from sharp edges of a wedge shaped bluff body.

Vortex dynamics in inviscid flows

vortex elements of strength $\Delta\Gamma_A$, $\Delta\Gamma_B$ as illustrated for separation point A in Fig. 8.17. The displacements, taken as the mid points of the lengths of actual vorticity sheet self-convected from the separation points, should be

$$\varepsilon_A = \frac{1}{4}\gamma(s_A)\,\Delta t, \qquad \varepsilon_B = \frac{1}{4}\gamma(s_B)\,\Delta t \qquad (8.25)$$

Since these two vortices will be in close proximity to the surface elements A and B, they will clearly exercise critical control over the values of $\gamma(s_A)$ and $\gamma(s_B)$ of the next potential flow analysis at time $t + \Delta t$. In most computational schemes the user is therefore allowed to override these values of ε and ϕ to permit subjective adjustments such that the values of $\gamma(s)$ of the four or five elements leading up to the separation points on each surface are smooth.

8.5.2 Simple vortex dynamics scheme for simulation of wake development

Beginning at time $t = 0$, the computation begins with potential flow past the wedge shaped body due to the uniform stream W_∞. After the first time step Δt two vortex elements $\Delta\Gamma_A$ and $\Delta\Gamma_B$ will be shed into the wake. For the next time step potential flow analysis is repeated including their influence in addition to that of W_∞. Convection velocities and displacements are then evaluated for the two free vortex elements using the techniques developed in Section 8.4, including also their mutual convection. Two more free vortex elements are then shed from A and B and we proceed to the next time step as indicated in the flow diagram.

The loop involving boxes 4 and 5 will normally be repeated the appropriate number of times to achieve second or higher order central difference type convection. Martensen analysis, box 4, now requires the modification of equations (8.10) to include all of the Z free vortex elements which have been shed into the fluid during previous time steps, namely

$$\sum_{n=1}^{M} K(s_m, s_n)\gamma(s_n) = -(U_\infty \cos\beta_m + V_\infty \sin\beta_m)$$

$$-\sum_{j=1}^{Z} \Delta\Gamma_{nj}(U_{mj}\cos\beta_m + V_{mj}\sin\beta_m) \quad (8.26)$$

356

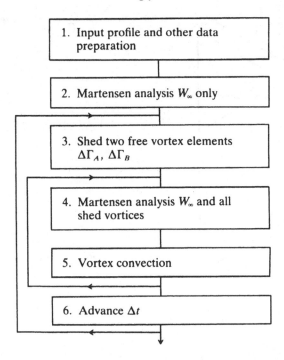

1. Input profile and other data preparation

2. Martensen analysis W_∞ only

3. Shed two free vortex elements $\Delta\Gamma_A$, $\Delta\Gamma_B$

4. Martensen analysis W_∞ and all shed vortices

5. Vortex convection

6. Advance Δt

where the unit vortex induced velocities are given by (8.11) or (8.17).

Following the strategy outlined in Section 8.2.1 for convection of a vortex cloud, extended to include the influence of the body surface vorticity as given in Section 8.4.2, the drift velocity components of vortex element m, box 5, became

$$
u_{dm} = \sum_{\substack{n=1 \\ n\neq m}}^{Z} \Delta\Gamma_n U_{mn} + \sum_{n=1}^{M} \gamma(s_n)\,\Delta s_n U_{mn}
$$

$$
v_{dm} = \sum_{\substack{n=1 \\ n\neq m}}^{Z} \Delta\Gamma_n V_{mn} + \sum_{n=1}^{M} \gamma(s_n)\,\Delta s_n V_{mn}
$$

(8.27)

Results for a typical computation are shown in Fig. 8.18 for the starting motion following time steps 10, 15 and 20 making use of (8.21) with $k = 0.5$ to select a suitable time step Δt of 0.05. In order to portray the vortex motion more clearly streak lines have been drawn along the directions of the local drift velocity as defined by (u_{dm}, v_{dm}). Some authors draw streak line length proportional to local drift velocity which can further help interpretation. On the

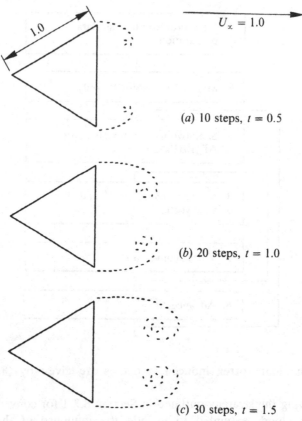

(a) 10 steps, $t = 0.5$

(b) 20 steps, $t = 1.0$

(c) 30 steps, $t = 1.5$

Fig. 8.18. Starting motion for flow past an equilateral wedge shaped body in a uniform stream $U_\infty = 1.0$. ($\Delta t = 0.05$, $\phi = 30°$, $\varepsilon = 0.025$).

other hand difficulties arise in trying to portray wide variations of drift velocity and the author's practice is to draw all streak lines of equal length. As the computation proceeds the vortex sheets continuously shed from the sharp edges immediately begin to roll up in a fashion rather similar to the behaviour of the free ends of the vortex sheet considered in Section 8.31, resulting in the development of two large eddies to form the initial symmetrical wake.

8.5.3 Vorticity shedding from a smooth surfaced bluff body

Vorticity shedding from a sharp edge is simple to model directly as we have seen. Flow separation from smooth surfaced bluff bodies such as a circular cylinder is less straightforward, since both the position of separation and the trajectory of the separated shear layer are unknown. In fact both are determined by complex fluid motions in the boundary layer involving a delicate balance between convection and diffusion, with the added important influence of local surface static pressure gradient. Despite this there have been many attempts to apply models, such as the one we have just considered, to the prediction of the vortex street wake behind a circular cylinder. Although the true position of separation will vary periodically due to upstream influence from the wake, quite reasonable results can be obtained if the simpler assumption is made of fixed separation points located at the ends A, B of the diameter normal to the mainstream flow, Fig. 8.19(a). Adopting a similar procedure to that for the wedge flow, the elements of vorticity shed during time step Δt may then be located at discrete vorticies $\Delta \Gamma_A$ and $\Delta \Gamma_B$ with polar coordinates (ε, ϕ) relative to the centres of the prescribed surface elements from which separation is assumed to occur, Fig. 8.19(b). The remaining problem is that of

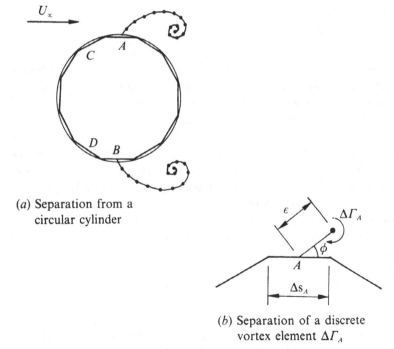

(a) Separation from a
    circular cylinder

(b) Separation of a discrete
    vortex element $\Delta \Gamma_A$

Fig. 8.19. Model for flow separation from smooth surface.

359

determining an appropriate strength representative of the real flow. Tests show that suitable values for vortex location are say $\phi = 45°$, fitting in with the typical trajectory and $\varepsilon \approx \Delta s_A/2$. However the potential flow on elements A and B is then strongly influenced by the local shed vortices, whereas in the real flow we know that the separating boundary layer vorticity is influenced predominantly by the surface flow upstream of A and B. A suitable compromise is to calculate $\Delta\Gamma_A$ and $\Delta\Gamma_B$ from the vorticity on the surface elements C and D just upstream, resulting in

$$\Delta\Gamma_A = \tfrac{1}{2}\gamma(s_C)^2\,\Delta t, \qquad \Delta\Gamma_B = \tfrac{1}{2}\gamma(s_D)^2\,\Delta t \qquad (8.28)$$

As previously explained, two new vortex elements are shed at each time step, resulting in a steady build up of data. Thus after say time step 200 there will be 400 free vortex elements in the flow field requiring the calculation of 400^2 contributions to the vortex cloud convection. Computational effort can be almost halved since the convection velocities induced at some vortex a by another vortex b may be expressed in terms of those at vortex b induced by vortex a, (u_{ba}, v_{ba}), through

$$u_{ab} = -u_{ba}\,\Delta\Gamma_b/\Delta\Gamma_a$$
$$v_{ab} = -v_{ba}\,\Delta\Gamma_b/\Delta\Gamma_a \qquad (8.29)$$

However, computational times can become excessive unless action is taken. Various techniques have been adopted including vortex merging and the redistribution of vorticity onto a grid system, some of which will be discussed in more detail in Chapter 12. Since the initial phase of the building up of a symmetrical wake, as illustrated for the wedge by Figs. 8.10, 8.17 and 8.18, can take quite a long time, a sensible strategy is to trigger asymmetry at the first time step. At least three methods have been used to achieve this, namely (i) prescribing a net bound circulation on the cylinder for step 1 only, (ii) inserting an arbitrary off-set free vortex in the wake at $t = 0$, (iii) introducing a cross-wind V_∞ for the first or subsequent few initial time steps. The result is a rapid initiation of self-propagating alternate eddy shedding and a fore-shortening of the initial symmetrical wake phase.

Fig. 8.20 illustrates such a computation for a circular cylinder of radius $a = 0.5$ represented by $m = 22$ surface elements in a uniform stream $U_\infty = 1.0$. Separation was assumed to take place from elements 6 and 17 with off-set values $\varepsilon = 0.05$, $\phi = 45°$. For time step No. 1 a transverse velocity $V_\infty = 0.5$ was used to introduce an

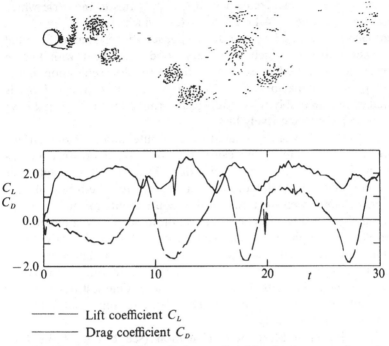

————— Lift coefficient C_L
—————— Drag coefficient C_D

Fig. 8.20. Flow past a circular cylinder predicted by simple vortex dynamics theory with two fixed separation points.

initial asymmetry and 200 time steps of magnitude $\Delta t = 0.15$, (at the limits of (8.21)), resulted in the vortex pattern and C_L, C_D curves shown in Fig. 8.20. To enhance flow visualisation, streak lines were drawn for both time steps 199 and 200 of length $0.13a$.

Eight distinct vortex clouds can be discerned including the starting vortex cloud on the far right of the streak line picture and the next vortex cloud in the final processes of formation and separation from the lower surface. Their time of shedding can be identified from the eight maxima (anticlockwise cloud) or minima (clockwise cloud) of the curve of lift coefficient C_L versus time. We may also observe that the drag coefficient C_D oscillates at twice the frequency of C_L. As a measure of vortex shedding periodicity we may adopt the reduced frequency as defined by the Strouhal number, namely

$$S_t = \frac{nU_\infty}{2a} \tag{8.30}$$

361

where n is the actual frequency. If we apply this to the cycle which occurred between $14.1 < t < 19.3$ for which both C_L and C_D produced very regular periodic response, the estimated Strouhal number is 0.1925 which is in very good agreement with known experimental tests. Despite the crudity of the flow separation model the periodic behaviour, as revealed by $C_L(t)$, $C_D(t)$ and S_t, is predicted remarkably well although the r.m.s. average lift and drag coefficients are excessively high.

Lift and drag were evaluated here by integration of the surface pressure distribution. The calculation of surface pressure in unsteady flow presents certain difficulties which we will not explore at this point but deal with more fully in Chapter 10. The principal aims of this chapter were to examine the requirements for accurate and thermodynamically reversible convection in vortex dynamics and to develop a simple computational scheme for bluff body wake simulation which implements these models. The author's computer program sepflow responsible for Figs. 8.17, 8.18 and 8.20 is based completely on the guide lines proposed here. One last point should be made before we close this chapter, and this relates to the conservation of vorticity.

Following each Martensen calculation (see box 4 of the flow diagram on page 357) the surface vorticity and previously shed vortices should satisfy the following conditions assuming that there was zero net vorticity in the flow regime at $t = 0$.

$$\sum_{n=1}^{M} \gamma(s_n)\, \Delta s_n + \sum_{j=1}^{Z} \Delta \Gamma_j = 0 \qquad (8.31)$$

This equation must be added to each of the governing equations (8.26) to ensure that vorticity is continually being conserved. This may then be written as follows.

$$\sum_{n=1}^{M} (K(s_m, s_n) + \Delta s_n)\gamma(s_n) = -(U_\infty \cos \beta_m + V_\infty \sin \beta_m)$$

$$- \sum_{j=1}^{Z} \Delta \Gamma_j (1 + U_{mj} \cos \beta_m + V_{mj} \sin \beta_m) \quad (8.32)$$

The appropriate Martensen procedure is as follows.

(i) Evaluate the coupling coefficient matrix $K(s_m, s_n)$.
(ii) Ensure zero circulation for each column by adopting a procedure similar to back diagonal correction, Section 2.3.3.

Simple vortex cloud modelling for two-dimensional bodies

In this case for column n replace the largest coupling coefficient (say in row m) with the values

$$K(s_m, s_n) = -\frac{1}{\Delta s_m} \sum_{\substack{i=1 \\ i \neq m}}^{M} K(s_i, s_m) \qquad (8.33)$$

(iii) Now add Δs_n to all coupling coefficients in each column $n = 1 \ldots M$ to complete (8.32).

CHAPTER 9

Simulation of viscous diffusion in discrete vortex modelling

9.1 Introduction

The 'random walk' model for simulation of viscous diffusion in discrete vortex clouds was first proposed by Chorin (1973) for application to high Reynolds number flows and has been widely used since. The principle involved is to subject all of the free vortex elements to small random displacements which produce a scatter equivalent to the diffusion of vorticity in the continuum which we are seeking to represent. Such flows are described by the Navier Stokes equations which may be expressed in the following vector form, highlighting the processes of convection and diffusion of the vorticity ω,

$$\frac{\partial \omega}{\partial t} + (\mathbf{q} \cdot \nabla)\omega - (\omega \cdot \nabla \mathbf{q}) = \nu \nabla^2 \omega \tag{9.1}$$

where $\mathbf{q}$ is the velocity vector and ∇^2 the Laplacian operator. The third term, applicable only in three-dimensional flows represents the concentration of vorticity due to vortex filament stretching. Otherwise in two-dimensional flows, with which we are concerned here, the vector Navier–Stokes equation reduces to

$$\frac{\partial \omega}{\partial t} + \underset{\uparrow}{(\mathbf{q} \cdot \nabla)\omega} = \underset{\uparrow}{\nu \nabla^2 \omega} \tag{9.2}$$

$$\begin{array}{cc} \text{Convection} & \text{Diffusion} \\ \text{of} & \text{of} \\ \text{vorticity} & \text{vorticity} \end{array}$$

Normalised by means of length and velocity scales ℓ and W_∞ this may be written in the alternative dimensionless form

$$\frac{\partial \bar{\omega}}{\partial \bar{t}} + (\bar{\mathbf{q}} \cdot \nabla)\bar{\omega} = \frac{1}{R_e} \nabla^2 \bar{\omega} \tag{9.3}$$

where the Reynolds number is defined by

$$R_e = \frac{W_\infty \ell}{\nu} \tag{9.4}$$

Introduction

For infinite Reynolds number (9.3) describes the convection of vorticity in inviscid flow, for which the technique of discrete vortex modelling was developed in Chapter 8. At the other end of the scale, for very low Reynolds number flow past an object of characteristic dimension ℓ, the viscous diffusion term on the right hand side (9.3) will predominate. The same is true for certain other flows such as the diffusion of a point (two-dimensional line) vortex for which the dimensional N–S equation (9.2) is more appropriate, reducing to the diffusion equation

$$\frac{\partial \omega}{\partial t} = \nu \nabla^2 \omega \qquad\qquad (9.5)$$

In Section 9.2 the classical solution of this equation for a diffusing point vortex will be used to form the basis of a practical random walk discrete vortex simulation method.

Many real flows encountered in engineering applications fall between these two extremes of Reynolds number, exhibiting both convective and diffusive motions, although often of different significance in various zones of the flow regime. Thus the flow past a circular cylinder will be strongly influenced by viscous diffusion and convection in the boundary layer along the front face prior to flow separation. For this reason a boundary layer Reynolds number based upon boundary layer or momentum thickness scale is more appropriate in (9.3) and (9.4). On the other hand the vortex dynamics of the developing vortex street wake downstream of separation is largely independent of the more appropriate body scale Reynolds number $(W_\infty D/\nu)$ provided this exceeds about 70. Indeed, as we have seen in Section 8.5.3, the periodic vortex street pattern and Strouhal number of such bluff body wake flows may be simulated with surprising reality ignoring viscous diffusion completely by simple vortex dynamics.

On the other hand many other important devices such as aerofoils and turbomachine cascades, when operated with varying angle of attack or approach flow, involve complex problems of boundary layer stability and consequent vortex shedding. In such situations it would be highly desirable to develop a consistent flow model capable of handling both boundary layer dynamics and subsequent downstream wake motion within a single computational framework. Chorin (1973), (1978) was alert to this desirable requirement at a time when grid methods for solution of the $N–S$ equations were advancing rapidly. He pointed out the conflicting requirements of grid scale for handling both the rapid vorticity rate changes within a boundary layer and the grander scale of the von Karman vortex street type motions of the downstream wake. With this in mind he sought such a consistent model in vortex dynamics including the random walk technique. By and large this generalised framework has now been achieved although care must be exercised, as with many numerical methods, in interpretation of the meaning and applicability of the models. Thus although the diffusion of a single point vortex may be modelled quite well by the random walk method as we shall see from Section 9.2, the representation of boundary layer vorticity diffusion close to a wall presents far more stringent computational requirements. As pointed out by Chorin, the aim in such regions need not be precise modelling at local eddy scale of resolution, but more of a statistical sampling process using data obtained over a period of discrete time steps. Indeed the random walk process involving single individual diffusion displacements of each vortex element at every time step, is very much like the random motions which occur at molecular level and Chorin linked his concepts to analogous theoretical studies of Brownian motion by Einstein (1956). As pointed out also by Lewis & Porthouse (1983b), some of the earliest derivations of the viscous flow continuum equations by Navier (1827) and Poisson (1831), stemmed from particulate models such as kinetic theory of gases. On reflection it seems quite natural therefore for a Lagrangian model such as vortex dynamics, where fluid rotation is pinned onto an ensemble of small free fluid elements, that the motion should resemble particle motion and obey laws of diffusion similar to those observed at molecular level.

Independently of Chorin, Porthouse & Lewis (1981) later developed an almost identical random walk technique as a natural extension of full vortex cloud modelling by the surface vorticity boundary integral method. The best starting point for this is consideration of a diffusing point vortex which will be presented in Section 9.2. This will be extended in Section 9.3 to the diffusion of a vortex sheet. The chapter is concluded in Section 9.4 with a fairly detailed application of vortex cloud modelling to boundary layers.

9.2 Diffusion of a point vortex in two-dimensional flow

The motion of a diffusing vortex of initial strength Γ centred on the origin of the (r, θ) plane is described by the diffusion equation (9.5)

Diffusion of a point vortex in two-dimensional flow

expressed in polar coordinates,

$$\frac{\partial \omega}{\partial t} = v\left\{\frac{\partial^2 \omega}{\partial r^2} + \frac{1}{r}\frac{\partial \omega}{\partial r}\right\} \tag{9.6}$$

from which we may obtain the well known solution for the subsequent vorticity distribution in space and time, Batchelor (1970),

$$\omega(r, t) = \frac{\Gamma}{4\pi v t} e^{(-r^2/4vt)} \tag{9.7}$$

Vorticity strength is a function of radius r and time t. This exact solution is shown in Fig. 9.1 for the case $\Gamma = 1.0$, $v = 1.0$, illustrating the radial diffusion of vorticity with time. The random walk method as developed by Porthouse & Lewis (1981) is based upon a numerical simulation of this diffused vorticity by an equivalent cloud of discretised vortex elements. As illustrated in Fig. 9.2, let us replace the vortex Γ by N elements each of strength

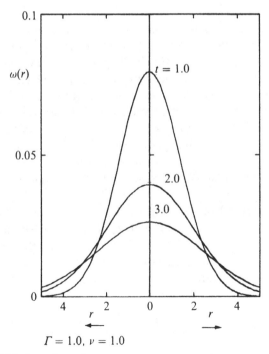

$\Gamma = 1.0$, $v = 1.0$

Fig. 9.1. Diffusion of a point vortex.

367

Viscous diffusion in discrete vortex modelling

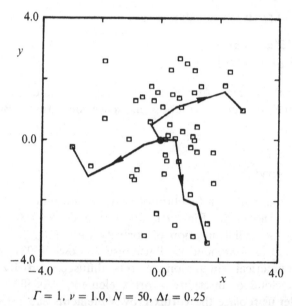

$\Gamma = 1,\ \nu = 1.0,\ N = 50,\ \Delta t = 0.25$

Fig. 9.2. Random diffusion of 50 vortex elements over four time steps.

Γ/N located at the origin at time $t = 0$ but free to move independently. Some time later the diffused vorticity may be represented by scattering the elements over the $(r,\ \theta)$ plane. The random number method outlined below achieves this in a manner which approaches the exact solution in the limit as $N \to \infty$. Figure 9.2 illustrates one particular test run with just fifty elements, taken over four time steps $\Delta t = 0.25$ each, with $\nu = 1.0$, $\Gamma = 1.0$. The paths of three selected vortex elements have been recorded to illustrate the close similarity of this numerical procedure to Brownian motion. Let us now consider how to achieve this objective by reference to the form of the exact solution.

The desired radial distribution of vorticity at time t is given by (9.7). For a vortex of unit strength split into N elements, let us assume that n vortex elements are scattered into the small area $r\,\Delta\theta\,\Delta r$ after time t. The total amount of vorticity p in this element of area then follows from

$$p = \frac{n}{N} = \left[\frac{1}{4\pi\nu t}\, e^{(-r^2/4\nu t)} \right] r\,\Delta\theta\,\Delta r \qquad (9.8)$$

Thus we could say alternatively that p is actually the probability

368

(n/N) that an element will land into the elementary area $r \Delta\theta \Delta r$ when scattered.

An appropriate strategy then is to displace each element i in the radial and angular directions by amounts r_i and θ_i over the time interval 0 to t, such that the radial probability distribution given by (9.8) is satisfied for all $r \Delta\theta \Delta r$ elements which make up a given target area, taken sufficiently large to capture all (or virtually all) of the diffused vorticity. Now it is self evident from symmetry that scattering in the θ direction should be done with equal probability. Thus we may define θ_i values independently of r_i values by the equation

$$\theta_i = 2\pi Q_i \tag{9.9}$$

where Q_i is a random number within the range $0 < Q_i < 1.0$.

As argued by Porthouse & Lewis (1981), and Porthouse (1983), the radial scattering of vortex elements may best be decided by first integrating (9.8) between $\theta = 0$ and 2π obtaining

$$p' = \left\{ \frac{1}{2vt} e^{(-r^2/4vt)} \right\} r \Delta r \tag{9.10}$$

p' is then the probability that a given element will lie between r and $r + \Delta r$. A more useful idea would be to find the probability P that an element will lie within a circle of radius r and this can be found by integrating p' from $r = 0$ to r, namely

$$P = 1 - e^{(-r^2/4vt)} \tag{9.11}$$

where the constant of integration has been set at value unity to ensure zero probability as the target area radius r approaches zero.

At this point the important observation may be made that (9.11) is in fact the normal distribution curve of statistical theory. The value of P must therefore be equally likely to lie anywhere in the range 0–1 and may be determined by selecting a random number P_i within this range. Thus for the ith vortex element (9.11) becomes

$$P_i = 1 - e^{(-r_i^2/4vt)}$$

from which we may obtain its radial random shift

$$r_i = \{4vt \ln(1/P_i)\}^{\frac{1}{2}} \tag{9.12}$$

The scatter of the N elements comprising the unit vortex is then determined by selecting N random numbers in the range 0–1 for both P_i and Q_i, the random walks (r_i, θ_i) following directly from (9.12) and (9.9).

Table 9.1. *Output from program 9.1, seed P = 0.4*

For 1000 random numbers spread in 10 bins		
Bin	Range	Number collected
1	0.0–0.1	106
2	0.1–0.2	97
3	0.2–0.3	84
4	0.3–0.4	93
5	0.4–0.5	91
6	0.5–0.6	99
7	0.6–0.7	94
8	0.7–0.8	110
9	0.8–0.9	123
10	0.9–1.0	103

9.2.1 Random number generation

Some computational facilities provide automatic random number generation otherwise of many possible techniques, that suggested by Porthouse is perfectly adequate. First we choose a real number P in the range 0.0–1.0 to seed the process, e.g. $P = 0.5$. Then we evaluate the following expression

$$P = (1.01316 + P)^5$$

$$= 7.932\,759 \tag{9.13}$$

Retaining the six figures after the decimal point as our first random number $0 = 0.932\,759$, the process may be repeated indefinitely using this as the next seed. The constant 1.01316 has been chosen arbitrarily but is of the right order to produce enough significant figures by 'real8' computation with the range $0 < P < 1.0$. Program 9.1 called Ranbox illustrates this process, sorting the resulting series of random numbers into bins to provide a cross-check on randomness. For example the results shown in Table 9.1 were obtained for 1000 random numbers collected into 10 bins.

For perfect randomness we would expect to collect 100 of the series P into each bin. The probable level of accuracy to be expected in sorting M random numbers this way is given by $1/\sqrt{[M/(\text{number of bins})]}$. Thus with 1000 numbers and 10 bins we expect an average error of 10%. The actual average error of 9.2% of the example tabulated above thus endorses the simple procedure proposed for generating random numbers.

9.2.2 Radial distribution of vorticity $\omega(r)$

As illustrated by Fig. 9.3 a similar approach may be applied to the evaluation of $\omega(r)$ for the scattered vortex elements, by defining a series of annular bins separated by equally spaced radii r_j. Suppose that n_j elements are captured by the shaded annular area lying between r_j and r_{j+1}. The vorticity at the r.m.s. radius $\sqrt{[\frac{1}{2}(r_j^2 + r_{j+1}^2)]}$ of strip j may thus be estimated from the expression

$$\omega(r) = \frac{n_j}{N\pi(r_{j+1}^2 - r_j^2)}$$

This is compared in Fig. 9.4 with the exact solution given by (9.7) for diffusion over the time period $t = 1.0$ of a unit strength vortex for a kinematic viscosity $\nu = 1.0$. In this case Γ was represented by 1000 elements of strength 0.001, each given one random walk, the radial displacement r_i being determined by (9.12). The accuracy of this result is thus a direct reflection of the quality of the random number generator. With twenty annular strips for sampling the scattered elements, a very satisfactory numerical prediction of vortex diffusion was obtained here.

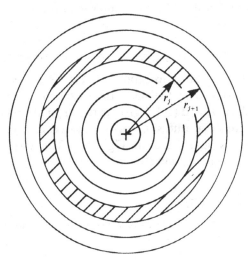

Fig. 9.3. Annular 'bins' used in vorticity calculation for random walk model of diffusing point vortex.

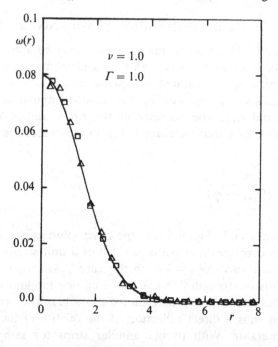

———— Exact solution

□   Numerical – single time step $\Delta t = 1.0$

△   Numerical – 10 time steps $\Delta t = 0.1$

Fig. 9.4. Prediction of the diffusion of a point vortex by the random walk method.

9.2.3   Diffusion over a series of time steps

Apart from the initial illustration of Fig. 9.2 we have so far considered diffusion over the finite time t taken in one single step only, for which (9.9) and (9.12) give us the angular and radial shifts. Let us now consider diffusion over a succession of small time increments Δt as illustrated in Fig. 9.2. The displacements of element i during time Δt will then be

$$\left.\begin{aligned} \Delta\theta_i &= 2\pi Q_i \\ \Delta r_i &= \{4\nu\,\Delta t\,\ln(1/P_i)\}^{\frac{1}{2}} \end{aligned}\right\} \tag{9.14}$$

Thus after the increment Δt the new coordinate location (x_i', y_i') of the ith element will become

$$\left.\begin{aligned} x_i' &= x_i + \Delta r_i \cos \Delta\theta_i \\ y_i' &= y_i + \Delta r_i \sin \Delta\theta_i \end{aligned}\right\} \tag{9.15}$$

Applying this to all of the N elements over several time steps Δt produces an agitated motion resembling molecular activity in which the vortex core grows in size and weakens in strength as the vortices diffuse outwards. It is not immediately obvious that this multi-stepping process is equally appropriate as a model of the vortex diffusion equation (9.7) which was based upon all of the vorticity dispersing from a concentrated vortex at the origin over the time period $0-t$. The single time step model which we have just considered in Section 9.2.2 is entirely appropriate since the theory leading to (9.12) was derived specifically on that basis. However the governing equation (9.6) is linear and its solution for a region of vorticity may be constructed by superposition of solutions for a series of zones making up this region. The scattered vortex elements after time step number 1 represent N such zones of local vorticity. Successive random walks for later time steps thus represent superimposed diffusions from the N zones. Now a much more accurate numerical representation could be obtained by splitting each of the N elements yet again into N sub-elements after time step number 1, each being diffused according to equations (9.15). However the total number of elements in the field would then escalate through N^1, N^2, N^3 ... etc. over successive time steps, and this would be quite unnecessary for adequate resolution of the solution. For example if the previous example of vortex diffusion illustrated in Fig. 9.4 is completed over ten time steps of magnitude $\Delta t = 0.1$, the numerically predicted vorticity distribution is found to be in equally good agreement with the exact solution as that for the single time step computation with $\Delta t = 1.0$.

It is evident from this study that excellent agreement with the exact solution is obtained by this simple strategy in which each of the original 1000 elements of strength $\Delta \Gamma = 0.001$ is given ten random walks for 1/10th of the time period based upon (9.14) and (9.15). There is clearly no need to increase resolution as time proceeds by splitting down elements yet further. The global effect of the 1000 random walking vortices averages out statistically to the desired representation. We may likewise anticipate similar success when applying this method directly to vortex clouds created by flow past bodies such as the vortex streets of Chapter 8.

As an aid to the reader Pascal Program 9.2 which completes this computation has been included in the appendix and sample output for the case presented in Fig. 9.2 is given in Table 9.2, providing further detail such as the radial bin limits and the numbers of

Table 9.2. *Diffusion of a point vortex – output from Program 9.2*

Number of vortex elements = 1000 Viscosity = 1.0 Step 10 $t = 1.0$

Bin no.	Number of elements in bin	Radius range	rms rad.	Vorticity distribution Numerical	Vorticity distribution Exact
1	38	0.0–0.4	0.282 843	0.075 599	0.078 002
2	113	0.4–0.8	0.632 456	0.074 935	0.072 005
3	158	0.8–1.2	1.019 804	0.062 866	0.061 358
4	170	1.2–1.6	1.414 214	0.048 315	0.048 266
5	155	1.6–2.0	1.811 077	0.034 265	0.035 048
6	134	2.0–2.4	2.209 072	0.024 235	0.023 494
7	99	2.4–2.8	2.607 681	0.015 150	0.014 537
8	47	2.8–3.2	3.006 660	0.006 234	0.008 304
9	46	3.2–3.6	3.405 878	0.005 383	0.004 379
10	17	3.6–4.0	3.805 260	0.001 780	0.002 131
11	18	4.0–4.4	4.204 760	0.001 705	0.000 958
12	4	4.4–4.8	4.604 346	0.000 345	0.000 397
13	1	4.8–5.2	5.003 999	0.000 080	0.000 152
14	0	5.2–5.6	5.403 703	0.000 000	0.000 054
15	0	5.6–6.0	5.803 448	0.000 000	0.000 018
16	0	6.0–6.4	6.203 226	0.000 000	0.000 005
17	0	6.4–6.8	6.603 031	0.000 000	0.000 001
18	0	6.8–7.2	7.002 858	0.000 000	0.000 000
19	0	7.2–7.6	7.402 703	0.000 000	0.000 000
20	0	7.6–8.0	7.802 565	0.000 000	0.000 000

elements collected in each bin. A direct comparison with the exact solution is also given.

9.3 Diffusion of a vortex sheet

As a first step towards discrete vortex modelling of boundary layers we consider next the viscous diffusion of an infinite vortex sheet, for which there is also an exact solution. For an initial sheet strength of $\gamma(s) = 2U$, Fig. 1.5, the flow regime consists of uniform streams $\pm U$ above and below the vortex sheet. As time elapses the vorticity will diffuse in the y direction resulting in a distribution $\omega(y, t)$ which satisfies the diffusion equation

$$\frac{\partial \omega}{\partial t} = v \frac{\partial^2 \omega}{\partial y^2} \qquad (9.16)$$

Diffusion of a vortex sheet

As shown by Batchelor (1970) the solution is given by

$$\omega(y, t) = \frac{U}{\sqrt{(\pi v t)}} e^{-y^2/4vt} \tag{9.17}$$

From the definition of vorticity

$$\omega = \frac{\partial u}{\partial y} \tag{9.18}$$

the velocity distribution across the diffusing shear layer may be expressed through

$$u(y, t) = \frac{U}{\sqrt{(\pi v t)}} \int_0^y e^{-\bar{y}^2/4vt} \, d\bar{y}$$

$$= U \operatorname{erf}\{y/\sqrt{(4vt)}\} \tag{9.19}$$

A discrete vortex model of this process may be constructed by considering a finite length of the sheet stretching, say, between $0 < x < \ell$. If the sheet is now broken down into N elements of strength $\gamma(s)\ell/N$, the previous procedure of random walks may be applied exactly as before. Figure 9.5 shows the vortex element scatter resulting from this for a 1000-element representation diffusing by random walks over 10 time steps $\Delta t = 0.0005$. At first glance of the scatter plot the outcome does not appear particularly significant or promising and the same could have been said of the point vortex scatter plot of Fig. 9.2. However, if a series of collection bins is defined by equally spaced contours $y = \text{constant}$

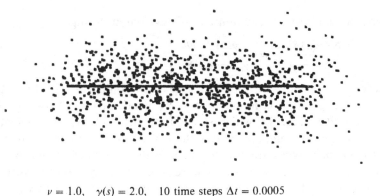

$$v = 1.0, \quad \gamma(s) = 2.0, \quad 10 \text{ time steps } \Delta t = 0.0005$$

Fig. 9.5. Diffusion of a vortex sheet simulated by the random walk method with 1000 elements.

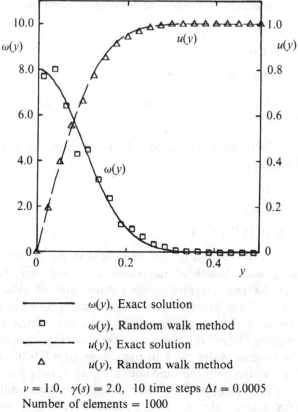

$\omega(y)$, Exact solution

□ $\omega(y)$, Random walk method

$u(y)$, Exact solution

△ $u(y)$, Random walk method

$v = 1.0$, $\gamma(s) = 2.0$,   10 time steps $\Delta t = 0.0005$

Number of elements = 1000

Fig. 9.6. Diffusion of a vortex sheet – comparison of random walk method with exact solution.

and the vorticity distribution $\omega(y)$ is evaluated through

$$\omega_i = \frac{N_i(\gamma(s)\ell/N)}{(y_{i+1} - y_i)\ell} = \frac{\gamma(s)}{(y_{i+1} - y_i)} \frac{N_i}{N} \tag{9.20}$$

where N_i elements are found in strip i, the comparison of this with the exact solution is found to be very good. A typical test run is illustrated in Fig. 9.6 which shows both the vorticity and velocity distributions in comparison with the exact solution. In this example 1000 elements were used to represent a sheet of strength $\gamma(s) = 2.0$ of length $\ell = 1.0$. Twenty bins were used to accumulate the vortex elements. Although there was some scatter in the $\omega(y)$ numerical prediction, the integration of this to obtain $u(y)$ reduced errors as one might expect, resulting in an excellent prediction.

9.4 Boundary layers by discrete vortex modelling

Convective motions were completely ignored for the diffusing point and sheet vortex flow regimes which we have just considered, an assumption which is permissible in view of symmetry in these special cases and justified for very low Reynolds numbers as shown by the dimensionless Navier Stokes equations (9.3). In both situations there was no externally imposed convective mainstream flow and in the latter case v was chosen sufficiently large to ensure random walk displacements which were small compared with the sheet length. Adoption of the simpler diffusion equations (9.6) and (9.16) was then perfectly justified. Boundary layer flows on the other hand are more complex involving two additional features:

(i) Externally imposed convection due to the main stream U, the significance of which is determined by the body scale Reynolds number (Ul/v).
(ii) Continuous creation of vorticity at the contact surface between fluid and wall, replacing the vorticity removed by diffusion and convection.

The first practical scheme for simulation of a boundary layer by discrete vortices was proposed by Chorin (1978), based on his earlier conception (1973) of the random walk model for high Reynolds number bluff body wake flows. Porthouse & Lewis (1981), (1983) later proposed similar schemes progressively developed to extend the surface vorticity method into the more

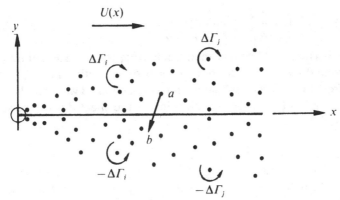

Fig. 9.7. Mirror image system for modelling vorticity creation and convection in discrete vortex boundary layer simulation.

complex field of bluff body and aerofoil flows, for which both the boundary layer and wake motions exercise important control over the global fluid motion. Both authors made use of the mirror image technique illustrated in Fig. 9.7 to simplify calculation of the convective motion and vorticity creation process referred to above. Representing the wall by M elements of length Δs, Porthouse proposed the flow sequence shown in the diagram on page 379 taken over a series of time steps Δt, the uniform stream U being introduced impulsively at $t = 0$. Let us consider this flow sequence in more detail.

9.4.1   Vorticity creation and shedding (Step 2)

Let us assume that the uniform stream U is turned on impulsively at $t = 0$, resulting in the creation of a surface vorticity sheet of strength $\gamma(s_i) = U$ at the wall surface $0 < x < \ell$. Due to viscosity this sheet will immediately begin to diffuse in the manner illustrated in Section 9.3 and also to undergo convection. Some time later the region above the plate will thus be filled with vorticity which we shall represent by a cloud of discrete vortices. At each subsequent time step a new vorticity sheet of strength $\gamma(s_i) = u_i$ will be created due to the slip flow at the wall, under the joint influence of the uniform stream U and all the discrete vortices previously shed and diffused into the flow regime. Making use of the mirror image system of Fig. 9.7 to satisfy the wall boundary condition automatically for each of the N discrete vortices in the field, the surface vorticity per unit length created at element i will be given by

$$\gamma(s_i) = u_i = U - \frac{1}{\pi} \sum_{j=1}^{N} \frac{\Delta \Gamma_j y_j}{(x_i - x_j)^2 + y_j^2} \qquad (9.21)$$

This vorticity is of course shed freely from the surface and may be represented by M new discrete vortices of strength $\Delta \Gamma_i = \gamma(s_i) \Delta s_i$, now free to undergo diffusion and convection. Two models then spring to mind to account for diffusion of the sheet during the same time step.

(i) Random walk method (Step 2)
As previously illustrated, Fig. 9.5, application of the random walk will result in the loss of half of the newly created vorticity due to diffusion across the wall and therefore out of the active flow domain. Since we are considering vorticity creation, diffusion and

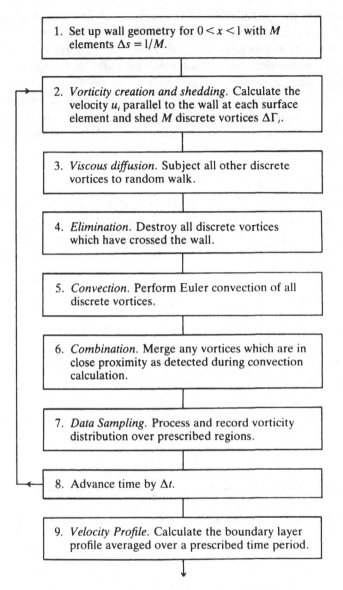

1. Set up wall geometry for $0 < x < 1$ with M elements $\Delta s = 1/M$.

2. *Vorticity creation and shedding.* Calculate the velocity u_i parallel to the wall at each surface element and shed M discrete vortices $\Delta\Gamma_i$.

3. *Viscous diffusion.* Subject all other discrete vortices to random walk.

4. *Elimination.* Destroy all discrete vortices which have crossed the wall.

5. *Convection.* Perform Euler convection of all discrete vortices.

6. *Combination.* Merge any vortices which are in close proximity as detected during convection calculation.

7. *Data Sampling.* Process and record vorticity distribution over prescribed regions.

8. Advance time by Δt.

9. *Velocity Profile.* Calculate the boundary layer profile averaged over a prescribed time period.

convection independently and in sequence in the above computational scheme, vorticity should be conserved during the diffusion and convection processes for each time step. Chorin (1978) therefore recommended the use of random walks but with a double strength surface vorticity sheet $2\gamma(s_i)\,\Delta s_i$ as employed previously in Section 2.3. Elements which cross the wall are then eliminated. Alternatively a single strength sheet $\gamma(s_i)\,\Delta s_i$ may be used if vortices which attempt to cross the wall are bounced back by assigning the value $y_i := \mathrm{abs}(y_i)$.

(ii) Offset method

As an alternative to this Porthouse (1983) recommended the representation of diffusion during the first time step after creation of each new discrete vortex by an initial offset normal to the surface of magnitude $\varepsilon = \sqrt{(4\nu\,\Delta t/3)}$, Fig. 9.8. He derived this expression by

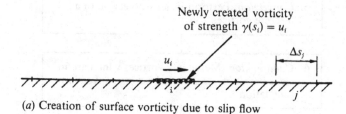

Newly created vorticity
of strength $\gamma(s_i) = u_i$

(*a*) Creation of surface vorticity due to slip flow

$$\Delta\Gamma_i = 2\gamma(s_i)\,\Delta s_i$$

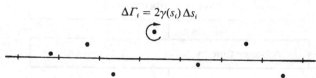

(*b*) Diffusion of double strength sheet by random walk (Chorin 1978)

$$\Delta\Gamma_i = \gamma(s_i)\,\Delta s_i \qquad \varepsilon = \sqrt{(4\nu\Delta t/3)}$$

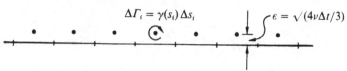

(*c*) Diffusion of single strength sheet by offset (Porthouse 1983)

Fig. 9.8. Creation and shedding of a surface vorticity sheet during boundary layer development.

considering the variance in the y direction of the depleted vorticity which would otherwise cross the wall, representing the vorticity-weighed centre position of the diffused but now conserved vorticity sheet.

On the whole the first of these methods is to be preferred since it is consistent with our previous considerations of vortex sheet diffusion by the pure random walk method which led to very successful numerical computations. However Chorin and Porthouse both obtained excellent predictions of the Blasius boundary layer despite these and other variations in their numerical schemes. In some respects however Chorin's very first model is an advance on those usually employed for greater simplicity today, since he recommended the use of discrete line rather than point vortices to represent the shed vorticity, a model better adapted to simulate strongly stratified flows. An alternative to this is the vortex blob or Rankine vortex model described by Leonard (1980) which we will discuss later in Section 9.4.4.

9.4.2   Viscous diffusion (Step 3)

Following the creation and diffusion of new vorticity, all other discrete vortices must be subjected to random walks to simulate viscous diffusion during the time step Δt. One side effect of this is of course that some of the discrete vortices follow a path such as *ab*, Fig. 9.7, crossing the wall. Porthouse recommends that these be retained for the subsequent convection process and then eliminated to enforce vorticity conservation. Chorin on the other hand recommends that these vortices should be bounced back instead. Alternatively we may simply eliminate all such vortices which cross the wall on the assumption that they will automatically be replaced by newly created vorticity shed from the wall during the next time step. These variations all lead to reasonable predictions and the latter one has the merit of a reduction in computational effort.

9.4.3   Vortex convection (Step 5)

Vortex convection is correctly modelled by the mirror image system shown in Fig. 9.7. The convection velocity components experienced by element $\Delta\Gamma_j$ due to element $\Delta\Gamma_i$ and its mirror image are then

Viscous diffusion in discrete vortex modelling

given by

$$\Delta u_j = \frac{\Delta\Gamma_i}{2\pi}\left(\frac{y_j - y_i}{r_1^2} - \frac{y_j + y_i}{r_2^2}\right) $$

$$\Delta v_j = \frac{\Delta\Gamma_i}{2\pi}(x_j - x_i)\left(\frac{1}{r_2^2} - \frac{1}{r_1^2}\right) $$

(9.22)

where

$$r_1^2 = (x_j - x_i)^2 + (y_j - y_i)^2 $$
$$r_2^2 = (x_j - x_i)^2 + (y_j + y_i)^2 $$

(9.23)

In addition to this we have the self-induced velocity of element $\Delta\Gamma_i$ due entirely to its mirror image vortex.

$$\Delta u_i = -\frac{\Delta\Gamma_i}{4\pi y_i}, \qquad \Delta v_i = 0 $$

(9.24)

Since viscous diffusion accounts for all of the irreversibilities in a boundary layer flow, it is important to ensure that the convective scheme is thermodynamically reversible. An 'Euler' scheme to achieve this is outlined in Section 8.2.1.

9.4.4 Vortices in close proximity (Step 6)

As pointed out originally by Chorin (1973) and also by Leonard (1980) in an excellent review of this subject, although the discrete vortex method faithfully solves the Euler equations, it may give only a poor representation of local vortex motion. Several investigators have therefore used vortex blobs with finite cores of various vorticity distributions to improve local convective modelling as discussed in some detail by Leonard. This technique then limits the scale of mutual convection velocities induced by nearby vortices, which for the discrete point vortex model would otherwise approach infinity should two vortices collide during the random walk. For simplicity here we will adopt the Rankine vortex model which was used very successfully in Section 4.2.3 to estimate the self-induced convection velocity of a ring vortex element in axisymmetric flows. In this case it was argued that the self convection of a ring sheet vortex element of length Δs could be represented by an equivalent smoke ring vortex with a Rankine core of radius $a = \Delta s/\pi$. Applying the same argument here we will assume that each newly created sheet vortex element of strength $\gamma(s_i)$ and length Δs_i is

382

immediately redistributed as a Rankine vortex of radius r_0 for which the induced velocity at radius r is given by

$$\left.\begin{aligned} \Delta q &= \frac{\Delta \Gamma r}{2\pi r_0{}^2} \quad \text{for} \quad r < r_0 \\ &= \frac{\Delta \Gamma}{2\pi r} \quad \text{for} \quad r > r_0 \end{aligned}\right\} \tag{9.25}$$

The convection velocities, equations (9.22), may then be suitably modified for the special cases when $r_1 < r_0$ or $r_2 < r_0$.

By the same reasoning (9.21) will lead to an over-estimate of the wall velocity Δu_i and therefore to over production of vorticity due to elements which drift close to the surface. Application of the Rankine vortex to this situation also is recommended, resulting in much improved simulation. On the other hand since the core of such a vortex element overlaps the wall, there is an implied loss of vorticity from the active domain. Thus taking the clockwise circulation integral along the entire wall we have

$$-\int_{-\infty}^{\infty} \Delta u_i \, ds_i = \Delta \Gamma_i / 2 \quad \text{for the discrete vortex}$$

$$< \Delta \Gamma_i / 2 \quad \text{for the Rankine vortex}$$

This problem was discussed in some detail in Section 8.4.1 for the case of a point vortex close to a circle modelled by the surface vorticity method. Applying the analogous corrective technique here, summarised by (8.15b), a reasonable estimate of vorticity created at surface element i closest to the discrete vortex $\Delta \Gamma$ is given by

$$\Delta \gamma(s_i) = -\frac{1}{\Delta s_i} \sum_{\substack{j=1 \\ j \neq i}}^{M} \left(\Delta u_j \, \Delta s_j + \frac{\Delta \Gamma}{2} \right) \tag{9.26}$$

where the summation is for simplicity limited to the M elements Δs_i representing the wall between $0 < x < \ell$ for which the values of Δu_j due to each discrete vortex $\Delta \Gamma$ are already available. This method due to Porthouse (1983), suggests another approach retaining the point vortex model, illustrated in Fig. 9.9. Evaluating the circulation integral along the surface element Δs_i we have

$$\int_0^{\Delta s_i} \Delta u_j \, ds_j = \frac{\phi}{2\pi} \Delta \Gamma$$

where ϕ is the angle subtended by the element. The vorticity

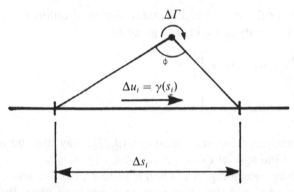

Fig. 9.9. Vorticity $\gamma(s_i)\,\Delta s_i$ created on element i due to nearby point vortex $\Delta\Gamma$.

created at element i due to the discrete vortex $\Delta\Gamma$ is then correctly given by

$$\Delta\gamma(s_i) = \frac{\phi\,\Delta\Gamma}{2\pi\,\Delta s_i} \qquad (9.27)$$

It can also be argued that if vortex elements come into such close proximity during a random walk their joint convective influence elsewhere will be adequately represented if they are merged into a single discrete vortex, resulting also in reduced computational effort. The total number of elements in the vortex cloud would still grow continuously unless further action were taken. Stabilisation of the cloud size can also be achieved by obliterating vortex elements which cross a boundary some distance downstream of the plate at say $x = 1.25l$. Little influence of this action is felt upstream but considerable reduction in computational effort is then gained.

9.4.5   Calculation of velocity profile (Step 9)

A typical vortex cloud is shown in Fig. 9.10(a) for a plate Reynolds number of 500 after 300 time steps $\Delta t = 0.015$. 60 surface elements were used to represent a plate of length $\ell = 1.0$ with a mainstream flow $U = 1.0$. By this stage of the computation the number of vortices in the cloud had stabilised at about 250 which is barely sufficient to represent the vorticity distribution. Greater accuracy may be obtained by averaging the flow field over several time steps and the velocity profile shown in Fig. 9.10(b) was derived by averaging over the period $3.0 < t < 4.5$. The vorticity distribution

Boundary layers by discrete vortex modelling

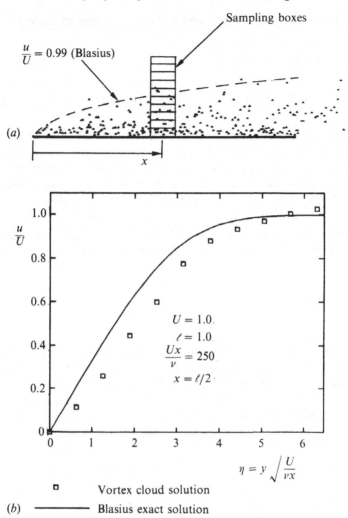

Fig. 9.10. Laminar boundary layer on a flat plate with constant mainstream velocity U.

$\omega(y)$ at some location x may be estimated by summing the total vortex strength accumulated in the sampling boxes representing the region of interest and dividing by the area of each box $\Delta x \, \Delta y$ and the number of time steps for the sampling process. Neglecting the velocity components v in the y direction, a normal assumption of laminar boundary layer theory, the velocity profile then follows from the following integral and its numerical approximation

$$u(y_N) = \int_0^{y_N} \omega(y)\, dy$$

$$\approx \sum_{n=1}^{N} \omega(y_N)\, \Delta y \tag{9.28}$$

These results were obtained by the method just outlined using the Rankine vortex model for vorticity creation and convection with merging of vortices which come closer together than the core radius. The thickness and general shape of the velocity profile were predicted reasonably well, the degree of under-estimation when compared with the Blasius exact solution being in line with the findings of Porthouse (1983). Blasius (1908)* has shown that the momentum thickness of a flat plate laminar boundary layer is given by

$$\theta = 0.664 \sqrt{\left(\frac{vx}{U}\right)}$$

This compares well with the results of Porthouse, Fig. 9.11, who undertook extensive calculations involving sampling at several x locations along the plate and with 500 or so vortex elements in the field. Porthouse also completed boundary layer calculations with varying main stream velocities of the form $U = U_1 x^m$ for which

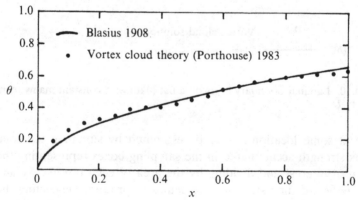

Fig. 9.11. Growth of momentum thickness θ for a flat plate boundary layer.

similarity solutions are available.* The case $m = 1$ corresponds to flow away from a stagnation point, for which the momentum thickness remains constant. Although the general shape of his predicted velocity profiles was similar to those of the similarity solutions derived by Hartree (1937), boundary layer thickness was overpredicted. The case $m = -0.0904$ corresponds to the critically decelerating boundary layer, for which the shear stress is zero along the entire wall and the boundary layer is continuously on the point of separation. Vortex cloud modelling was unable to predict this situation and its application to more complex problems such as flow past aerofoils is disappointing with regard to the important problem of predicting boundary layer separation as we will see in the next chapter.

9.4.6   Selection of element size and time step

A reasonable approach to the selection of an appropriate time step Δt is to focus attention on the average displacements of the discrete vortices due to convection and diffusion. The average convective displacement may be approximated by

$$\delta_c = \tfrac{1}{2} U \, \Delta t \tag{9.29}$$

Equation (9.14) gives the average diffusive displacement (i.e. when the random number $P_i = 0.5$), namely

$$\delta_D = \sqrt{(4v \, \Delta t \ln 2)} \tag{9.30}$$

To maintain equal discretisation of the fluid motion due to convection and diffusion we may equate δ_c and δ_D resulting in the expression

$$\Delta t = \frac{16\ell \ln 2}{URe_\ell} \tag{9.31}$$

whre $Re_\ell = U\ell/v$ is the plate Reynolds number.

It would also seem reasonable to select the surface element size at say twice δ_c leading to

$$\Delta s = U \, \Delta t = \frac{16\ell \ln 2}{Re_\ell} \tag{9.32}$$

The required number of surface elements for satisfactory dis-

* See Schlichting (1955).

cretisation of the plate is then given by

$$M = \frac{\ell}{\Delta s} = \frac{Re_\ell}{16 \ln 2} \tag{9.33}$$

It is clear from this study that enforcing equal discretisation scales δ_c and δ_D for convection and diffusion will lead to computational difficulties at large Reynolds numbers. Thus for the Blasius boundary layer considered in Section 9.4, for $Re_\ell = 500$, $M = 45$ and the choice made there of 60 elements was conservative. On the other hand for a typical engineering system value of $Re_\ell = 10^5$, (9.33) yields roughly $M = 9000$ imposing severe pressure upon computational requirements. The related time increment $\Delta t = 0.00011$ would also require 10^4 time steps to achieve one flow pass. In fact it can be shown that three flow passes involving an elapse time of $3\ell/U$ are required to reach the steady state. It is thus clear that practical computational limitations will rule out vortex cloud modelling for typical engineering system Reynolds numbers if we attempt to impose the constraint $\delta_D = \delta_c$ appropriate to the foregoing Blasius calculation.

9.4.7 Some considerations for high Reynolds number flows

One way to ease these difficulties for high Reynolds numbers would be to select different time steps for diffusion (Δt_D) and convection (Δt_c). Since convection now dominates the flow, it will be preferable to select the scale of convection displacements through

$$\frac{\delta_c}{\Delta s} = k \tag{9.34}$$

where previously we set k to 0.5. The convective time step then follows from (9.29).

$$\Delta t_c = \frac{2k \, \Delta s}{U} = \frac{2k}{M}\left(\frac{\ell}{U}\right) \tag{9.35}$$

The average random walk diffusive displacements over this same interval follow from (9.30), namely

$$\frac{\delta_D}{\Delta s} = \sqrt{\left(\frac{8Mk \ln 2}{Re_\ell}\right)} \tag{9.36}$$

Thus for the case $Re_\ell = 10^5$ let us select $M = 100$ surface elements with $k = 0.5$. Equation (9.35) then yields an appropriate convective time step $\Delta t_c = 0.01$, requiring 300 steps for three flow passes needed to achieve steady state. However (9.36) reveals a very small scale of random walk $\delta_D/\Delta s = 0.052\,655$ over this same time step. Although it would be perfectly in order to perform both the convection and random walk processes over the same time step Δt_c, a saving in computational effort could be achieved by undertaking only one random walk for every N_t convection steps with $\Delta t_D = N_t \Delta t_c$. The upper limit of N_t follows from (9.34) and (9.36) if we equate the scales δ_c and $\delta_D N_t$, namely

$$N_t = \frac{kRe_\ell}{8M \ln 2} \tag{9.37}$$

For the present example this suggests that we need only perform one random walk for every $N_t = 90$ convection steps if we wish to maintain the same maximum allowable discrete vortex displacement scales $\sigma_c/\Delta s$ and $\delta_D/\Delta s$. We must then adopt the diffusion time step

$$\Delta t_D = N_t \Delta t_c \tag{9.38}$$

At first encounter such a procedure seems worrying since it appears to over-emphasise the influence of convection. But this is of course what one should expect of a high Reynolds number flow. The degree of convective influence becomes much greater than the influence of diffusion over a given time step as Reynolds number is increased. Indeed we can perceive from this how transition of a boundary layer from laminar to turbulent motion may come about at very high Reynolds number, bearing in mind the Kelvin–Helmholtz instability considered in the previous chapter. For example the stratified shear layer of a laminar boundary layer could be considered to be made up of a series of parallel vorticity sheets, each subject to the Kelvin–Helmholtz type of instability. Viscous diffusion is known to damp down these instabilities and Porthouse (1983) has illustrated this by introducing random walks into the discrete modelling of a self-convecting vortex sheet. In the present example for $Re_\ell = 10^5$ however we were proposing at best 90 convective sub-time steps for ideal modelling in between each random walk, which might indeed provide sufficient time for the Kelvin–Helmholtz instability to take place. In fact this particular Reynolds number is known to be close to the experimental transition point. Certainly we may see from the evidence of the numerical simulations of Chapter 8 that instabilities of the shear layers analogous to the Kelvin–Helmholtz instability are likely to occur leading to large scale eddy formation.

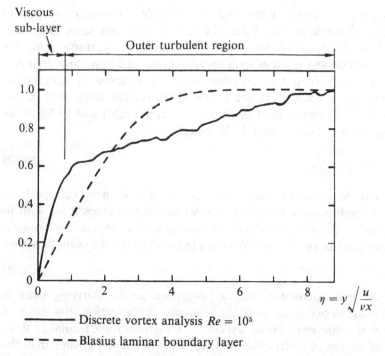

Fig. 9.12. High Reynolds number boundary layer prediction by vortex dynamics.

To conclude this chapter, vortex dynamics predictions for a flat plate boundary layer at high Reynolds number are shown in Figs. 9.12 and 9.13 for the data given in Table 9.3.

At the Reynolds number of 10^5 an actual boundary layer would still lie just within the laminar range covered by the Blasius solution but would be close to transition to turbulent flow. We might anticipate here that the presence of 'numerical turbulence' implied by the finite scale $\Delta s/\pi$ of the discrete vortices would perturb the shear layers leading to transition. Indeed we observe from Fig. 9.12 that the predicted boundary layer profile bears no resemblance to the Blasius solution for laminar flow. On the contrary it exhibits the familiar characteristics of a turbulent boundary layer. Close to the wall there is a region of concentrated vorticity generating high shear rates and dominated by viscous diffusion, the laminar sub-layer. At the outer edge of this region the profile bends sharply as would the typical turbulent boundary layer over an intermediate zone in which the damping effects imposed by both viscosity and the presence of

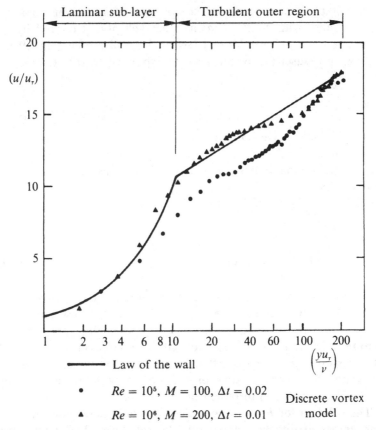

Boundary layers by discrete vortex modelling

Fig. 9.13. Discrete vortex prediction of flat plate turbulent boundary layer.

the plane wall can no longer quell the instabilities triggered by fluctuations imposed by the larger eddies of the outer layer. These outer regions of high entrainment extend over 90% of the boundary layer thickness, exhibiting the characteristics of turbulent mixing similar to those found experimentally in turbulent flow.

Table 9.3

Re	M	Δt	Number of time steps	Number of time steps averaged	Sampling zone x/ℓ
10^5	100	0.02	200	40	0.7–0.8
10^6	200	0.01	300	50	0.7–0.8

391

Viscous diffusion in discrete vortex modelling

A suitable bench mark for comparison of turbulent boundary layer profiles with constant mainstream velocity, is offered by the experimentally well established law of the wall profile, Fig. (9.13). Following Clauser (1956) this may be expressed by the curve fit

$$\frac{u}{u_\tau} = \frac{yu_\tau}{v} \qquad \text{for} \quad \frac{yu_\tau}{v} < 11$$

$$= 2.44 \ln\left(\frac{yu_\tau}{v}\right) + 4.9 \quad \text{for} \quad \frac{yu_\tau}{v} > 11 \tag{9.39}$$

where the friction velocity u_τ is defined in terms of the wall shear stress τ_0 by

$$u_\tau = \sqrt{\left(\frac{\tau_0}{\rho}\right)}$$

$$= \sqrt{\left[v\left(\frac{du}{dy}\right)_{y_0}\right]} = \sqrt{[v\omega(y_0)]} \tag{9.40}$$

Thus u_τ may be directly related to the vorticity $\omega(y_0)$ close to the wall, which may in turn be evaluated from the discrete vortex computation as described in Section 9.4.5. A sensible approach here is to take $\omega(y_0)$ as the average vorticity strength over the inner half of the sub-layer. The logarithmic scale of plotting then helps to draw out the characteristics of the boundary layer in both the laminar near wall and turbulent outer regions.

The solution for $Re = 10^6$ at which we would anticipate turbulent flow agrees extremely well with the law of the wall. For the lower Reynolds number of 10^5 the general characteristics were also exhibited although the actual profile was in less accord with the law of the wall. Bearing in mind that vortex dynamics is essentially a two-dimensional model, it is remarkable and extremely encouraging that these computations should be able to yield the characteristics of a turbulent boundary layer so competently, allaying some of the fears expressed earlier about the potential power of vortex dynamics to meet the higher Reynolds number requirements of practical engineering situations. In these two examples the number of surface elements used was a small fraction of that required for equal convection and diffusion scales δ_c and δ_D (9.33).

392

CHAPTER 10

Vortex cloud modelling by the boundary integral method

10.1 Introduction

The main objective of this chapter is to present the reader with a practical numerical approach to vortex cloud modelling of bluff body flows, drawing upon the techniques developed earlier in the book and especially the treatments of vortex dynamics and viscous diffusion considered in Chapters 8 and 9. Reporting on Euromech 17, which was entirely devoted to bluff bodies and vortex shedding, Mair & Maull (1971) remarked upon the preponderance of experimental work at that time and the need for more theoretical studies to be attempted, since there was little discussion of numerical techniques. It was felt, on the other hand, that since such flows showed marked three-dimensional characteristics (e.g. a circular cylinder von Karman street wake will not in general be correlated along its length for L/D ratios in excess of 2.0), two-dimensional computations, whilst being of interest, would not be very useful. It was admitted however that 'with an increase in the size of computers a useful three-dimensional calculation could become a reality'. By the time of the next Euromech 119 on this subject, Bearman & Graham (1979), one third of the papers focused on theoretical methods, the majority based upon the Discrete Vortex Method (DVM). Various reviews of the rapid subsequent progress with DVM were given by Clements & Maull (1975), Graham (1985a) and Roberts & Christiansen (1972) and a fairly comprehensive recent review of U.K. work was presented by Maull (1986) revealing extensive interest and progress with the application of DVM to a wide range of problems. Other useful references include a review by Sarpkaya (1979) on vortex-induced oscillations, consideration of separated flows combining vortex dynamics with boundary layer theory and schemes for economy in computation by Spalart & Leonard (1981) and useful extensions of the DVM by Spalart (1984) including the prediction of rotating stall employing Cray-1. These are representative of a fairly extensive literature which has appeared in a short period of time, revealing remarkable success in the development of the Discrete Vortex Method as a simulation technique for bluff body flows and as a predictive engineering tool. It is not the intention here to review this work but to provide a working framework which will help the reader to move forward more quickly to the preparation of computational schemes.

Flow regimes generally fall into the two categories of unseparated and separated or stalled flows. Most of this book has been devoted to the prediction of unseparated flows by surface vorticity potential flow analysis, which is a perfectly reasonable approach to many practical situations. In Chapter 8 on the other hand an outline vortex cloud method was presented for bluff body flows applicable to situations in which the flow is expected to separate in a fairly dramatic but prescribable manner such as boundary layer separation from a sharp edge. The discrete vortex method has undoubtedly made outstanding contributions to the prediction and understanding of such flows which were considered intractable prior to about 1970, and we will return to further consideration of this technique in Section 10.2. In this simplified approach, vortex shedding is limited to the prescribed separation points with potential flow modelling of the remaining body surface. The reason for its success is widely believed to lie in the dominating influence of the convective process in the wake at high Reynolds numbers and the minimal influence of the body shape (apart from its crucial function of shedding the shear layers which eventually form the von Karman street type wake).

As will be shown in Section 10.3, the method may also be applied quite simply to lifting aerofoils for which the aerofoil itself is modelled by surface vorticity in potential (but unsteady) flow and the wake is developed by natural shedding of the upper and lower surface shear layers.

Full vortex cloud modelling on the other hand attempts to simulate also the developing boundary layer on the body surface and the remainder of the chapter will be devoted to the fundamental basis of this technique. Vorticity is created over the whole body surface and diffused as a cloud of discrete vortices during a sequence of small time steps, ultimately shedding naturally from sharp edges or boundary layer natural separation points. The ultimate aim is nothing less than a full simulation of the real fluid flow with the minimum of human intervention. Schemes for shedding, diffusion and convection in relation to overall computational sequences will be outlined in Section 10.4. The calculation of surface pressure and associated problems of numerical noise will be dealt with in Section 10.5 and application of full vortex cloud modelling to flow past a circular cylinder will be presented in Section 10.6. Further developments and applications of the vortex cloud method will be presented in Chapter 11.

10.2 Vortex cloud modelling with prescribed separation points

Most early work in vortex dynamic analysis of bluff body flows was restricted to vortex shedding from distinct separation points, ignoring completely the vortex creation, diffusion and convection activity within the body boundary layers. A brief review has already been given in Section 8.4 followed by a fairly full description to which the reader is referred at this point. Typical of the two main approaches to this are the papers by Sarpkaya (1975) and R. I. Lewis (1981). Both methods employ an iterative time marching scheme such as that given in Section 8.4 in which one discrete vortex element is shed from each sharp edge for every time step, alternating with potential flow analysis and convection. The only significant difference lies in the technique used for potential flow modelling. Thus in modelling the separated flow due to a flat plate with angle of attack, Fig. 10.1, Sarpkaya made use of the Joukowski transformation to take advantage of mirror image potential flow modelling in the circle plane. In calculating the separated flow from a flat plate normal to a plane wall on the other hand, Lewis made use of the surface vorticity method to represent the plate surface, Fig. 10.2. In this case the wall boundary condition was accomplished by introducing a mirror image system as described in Section 1.11.

Such schemes are simple to define and program using the guide lines given in Section 8.4 regarding discrete vortex separation and diffusion. The only further comment needed is a cautionary one concerning two problems which can occur.

(i) Vortices may accidentally stray inside the body contour due to inaccurate convection routines.
(ii) Vortices may drift so close to the body as to generate serious potential flow errors or excessive self-convection velocities.

Vortex cloud modelling

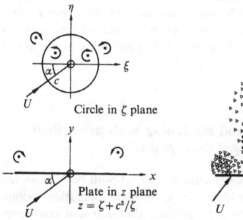

Circle in ζ plane

Plate in z plane
$z = \zeta + c^2/\zeta$

(a) Conformal transformation to ζ plane with vortex reflection system

(b) Vortex dynamics solution for $\alpha = 50°$, $U = 1.0$, $t = 62$, $\Delta t = 0.04$, Sarpkaya (1975)

Fig. 10.1. Vortex dynamics solution of the flow past a plate with angle of attack. By courtesy of the Journal of Fluid Mechanics.

The first of these problems can be detected by the circulation check given in Section 8.4.1. Thus let us reconsider the case of a unit vortex located at (x_j, y_j) close to a body represented by M surface elements (x_n, y_n), Fig. 8.11. The right hand side of the Martensen equation (8.10) then represents the flow induced tangential to the body at m by the vortex, namely

$$\text{rhs}_m = -(U_{mj} \cos \beta_m + V_{mj} \sin \beta_m) \tag{10.1}$$

where

$$U_{mj} = \frac{1}{2\pi} \left\{ \frac{y_m - y_j}{r_{mj}^2} \right\}, \qquad V_{mj} = -\frac{1}{2\pi} \left(\frac{x_m - x_j}{r_{mj}^2} \right) \tag{10.2}$$

and

$$r_{mj} = \sqrt{[(x_m - x_j)^2 + (y_m - y_j)^2]} \tag{10.3}$$

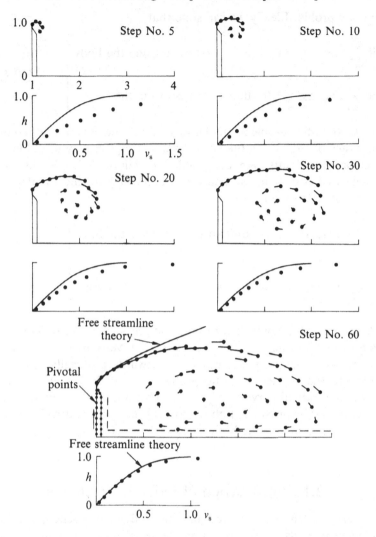

Fig. 10.2. Progressive development of flow separation from a flat plate by simple vortex dynamics with boundary integral modelling of the plate. (Reproduced from the Proceedings of the Institution of Mechanical Engineers by permission of the Council of the Institution.)

The 'right hand side' zero circulation correction given by (8.15a) and (8.15b) was based upon the assumption that the discrete vortex under consideration lay outside the body. The check may also be used to discover whether our unit vortex lies either inside or outside the body profile. Ideally we can state that

$$
\left.
\begin{aligned}
&\text{if } \sum_{n=1}^{M} \text{rhs}_n\, \Delta s_n = 0 \qquad \text{unit vortex is outside the body}\\[2ex]
&\text{if } \sum_{n=1}^{M} \text{rhs}_n\, \Delta s_n = 1.0 \quad \text{unit vortex is inside the body}
\end{aligned}
\right\} \tag{10.4}
$$

However the problem of leakage flux due to the inherent inaccuracy of rhs_m when the vortex is in close proximity would lead to erroneous results. An acceptable procedure which works extremely well is to omit the nearest element m from the circulation check, which then becomes

$$
\left.
\begin{aligned}
&\text{if } \sum_{\substack{n=1\\ n\neq m}}^{M} \text{rhs}_n\, \Delta s_n < 0.5 \quad \text{unit vortex is outside the body}\\[2ex]
&\text{if } \sum_{\substack{n=1\\ n\neq m}}^{M} \text{rhs}_n\, \Delta s_n > 0.5 \quad \text{unit vortex is inside the body}
\end{aligned}
\right\} \tag{10.5}
$$

A common strategy is then to snuff out any such vortices detected when evaluating the right hand sides of the Martensen equations. Alternatively they may be displaced outwards artificially along a normal to the surface to some prescribed boundary such as that shown as a dashed curve in Fig. 10.2. This procedure may also be applied to problem No. (ii) above to prevent vortices drawing into too close proximity to the body surface.

10.2.1   Introduction of reduced circulation

Another problem encountered in vortex dynamics schemes is a tendency towards over-production of vorticity or rather over-retention of wake vorticity leading to excessive estimates of drag coefficient, fluctuating lift coefficient and Strouhal number for circular cylinder flows as discussed by Ling (1986) and Sarpkaya & Shoaff (1979). Many researchers have encountered this by decreasing the strength of each discrete vortex continuously as time progresses and models to achieve this were suggested by Sarpkaya & Shoaff. Ling recommends the simple formula

$$
\Delta \Gamma_j\,(t + \Delta t) = (1 - \lambda)\, \Delta \Gamma_j \tag{10.6}
$$

where his numerical studies seemed to correlate best with experiment using $\lambda = 0.01$. He argued that some reduction in circulation in the real flow may be caused by turbulent dissipation, cancellation of vorticity, or even possibly by three-dimensional deformation of vortices.

On the other hand Ling's calculations have shown that reduction in circulation strength of vortices located far from a bluff body has negligible influence upon predicted characteristics. Since the problem seems linked to the near wake flow Ling & Yin (1984), Ling (1986), and Ling, Bearman & Graham (1986), have focussed attention on another possible cause of reduced circulation, namely the development of secondary separations on the rear face of a bluff body, Fig. 10.3. Flow visualisation reveals that in addition to the primary vortices A shed during start motion of flow past a circular cylinder, secondary vortices B will also be shed due to the separation of the rear face boundary layer induced under the influence of the primary vortices. Ling *et al.* attacked this problem for both a circular cylinder and a flat plate by first applying the boundary layer methods of Thwaites (1949) and Stratford (*see* White (1974)) beginning at the rear stagnation point S. For each time step the position of flow separation was estimated and the boundary layer was shed as a new discrete vortex resulting in the build up of typical predicted flow patterns as shown in Fig. 10.3. In practice Ling had to reduce the strength of this secondary vortex sheet which is so weak compared with the primary clockwise vortex that it is barely able to roll up into a well formed anticlockwise vortex core before becoming entrained into the primary vortex and totally absorbed. Thus although this secondary vorticity is of opposite sign to the primary vorticity, the net reduction in wake

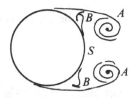

(*a*) Traced from flow visualisation
Ling *et al.* (1982)

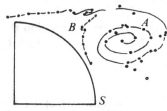

(*b*) Vortex dynamics solution
Ling *et al.* (1982)

Fig. 10.3. Starting motion for flow past a cylinder. By courtesy of the Indian Academy of Sciences.

vorticity due to this cause is perhaps less important than the other three-dimensional effects of vortex decay. Adoption of full vortex cloud modelling in any case accounts automatically for this feature which must otherwise be introduced rather artificially into the fixed-separation point models we are considering here.

10.2.2 Time growth of the vortex core

The use of a Rankine vortex for averting excessive convection velocities of any discrete vortices which come into close proximity has been recommended and explained in Section 9.4.4. Several authors including Ling *et al.* (1986) employ a more advanced technique based upon the solution for the actual velocity induced by a vortex in a viscous fluid given in Section 9.2. Integration of (9.6) with respect to r leads to

$$v = \frac{\Gamma}{2\pi r} \{1 - e^{(-r^2/4\nu r)}\} \tag{10.7}$$

The core radius r_c at time t is then approximately equal to the radial distance of the point of maximum velocity. Ling states that this can be expressed for element j in terms of its initial core radius r_{cjo} and growth age Δt_j through

$$r_{cj} = r_{cjo} + 3.17 \sqrt{\left(\frac{\Delta t_j}{Re}\right)} \tag{10.8}$$

Where $Re = U_\infty D / \nu$ is the cylinder Reynolds number. The velocity within the core then follows from (10.6), from which Ling derives the alternative expression in terms of core radius

$$v_j = \frac{\Gamma_j}{2\pi r} \left\{1 - \exp\left[-1.25643\left(\frac{r}{r_{cj}}\right)^2\right]\right\} \tag{10.9}$$

Outside the core region the induced velocity can be taken as that of a free vortex of strength Γ_i.

10.3 Application of fixed separation point analysis to a lifting aerofoil

When solving potential flow past an aerofoil by the standard Martensen method, Section 2.4, there was no need to consider

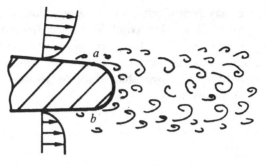

(*a*) Boundary layer vorticity shed into wake

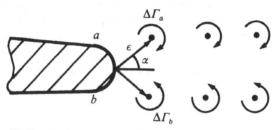

(*b*) Simulation by discrete vortex shedding

Fig. 10.4. Treatment of vorticity shedding in an aerofoil wake.

vortex shedding. Instead a trailing edge Kutta–Joukowski condition was stated equivalent to enforcing equal static pressures at the pivotal points *a* and *b*, Fig. 10.4, which was expressed by Wilkinson (1967a) through

$$\gamma(s_a) = -\gamma(s_b) \qquad\qquad [2.22]$$

The implicit physical assumption underlying this is that the two vorticity sheets which in real life are shed as boundary layers from *a* and *b* to form the wake, will coalesce and completely cancel one another in the equivalent surface vorticity potential flow model.

This is a perfectly reasonable assumption for potential flow and it is then essential to enforce a trailing edge Kutta–Joukowski condition. On the other hand we may progress to a more advanced model which attempts to simulate the real flow more realistically by actually shedding the surface vorticity sheets of the potential flow from points *a* and *b* as successive discrete vortices, Fig. 10.4(*b*). To simplify matters we will ignore viscous diffusion, although that can easily be introduced via random walks as outlined in Section 9.2. In this situation we may completely suspend any action regarding the Kutta–Joukowski trailing edge condition since the continuously shed vorticity sheets forming the near wake will fairly quickly remove the trailing edge recirculation 'singularity' which appears at $t = 0$. The strength of the two shed discrete vortices, following Lewis (1986), will be given by

$$\left. \begin{aligned} \Delta\Gamma_a = \tfrac{1}{2}\gamma(s_a)^2\,\Delta t \\ \Delta\Gamma_b = \tfrac{1}{2}\gamma(s_b)^2\,\Delta t \end{aligned} \right\} \tag{10.10}$$

The correct positioning of these discrete vortices to yield a good solution remains unclear and in application of the method some control and experimentation is desirable. The starting motion for NACA 0012 with 20° angle of attack predicted by this method is shown in Fig. 10.5 for which the shed vortex locations were set at $\varepsilon = 5\%$ of chord and $\alpha = 30°$, Fig. 10.4. As time proceeds the starting vortex rolls up characteristically and drifts downstream

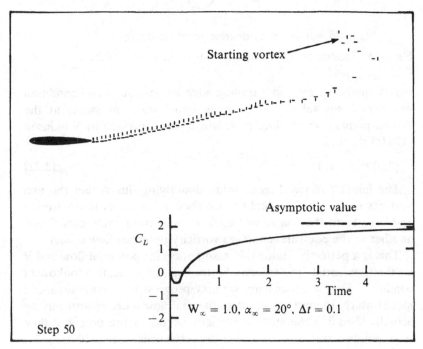

Fig. 10.5. Development of wake motion and lift coefficient for NACA 0012 aerofoil starting from rest.

402

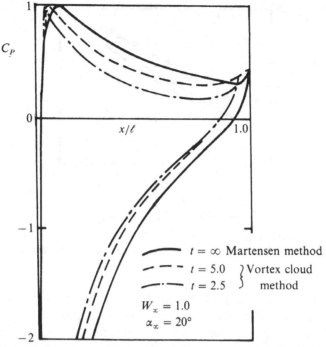

Fig. 10.6. Development of pressure distribution with time for NACA 0012 aerofoil compared with steady state solution by Martensen method.

while the lift coefficient C_L rises asymptotically towards its ultimate steady state value. Surface pressure and thereby lift were calculated here by the simple assumption of quasi-steady flow employing Bernoulli's equation at each time step. Later, in Section 10.4, we shall consider more appropriate formulations for calculation of pressure in an unsteady flow, but the present flow in any case is proceeding in the limit to steady state. Apart from the initial period of wake growth, which was highly unsteady, this simple approach has proved adequate to simulate the development of aerofoil lift by vortex dynamics modelling. Predicted surface pressure distributions also show asymptotic progression towards the steady state Martensen solution, Fig. 10.6.

In tackling the more complex problem of an aerofoil with oscillating incidence Basu & Handcock (1978) represented the wake vorticity immediately adjacent to the trailing edge by a single vorticity sheet whose length and inclination were determined as part of the solution, the sheet inclinations being found to agree with the

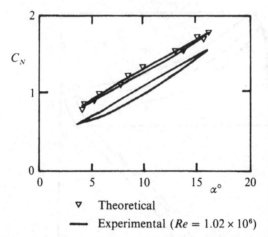

Fig. 10.7. Normal force on NACA 23012 with periodic angle of attack $\alpha = 10 + 6 \sin \omega t$ – Vessa & Galbraith (1985). By courtesy, *Int. J. Num. Meth. Fluids*.

Kutta condition proposals of Maskell (1972). Wake vorticity further downstream was discretised and the method applied to aerofoils suddenly changing incidence, oscillating or entering gusts. Using a similar representation of trailing edge flow, Vezza & Galbraith (1985) obtained reasonable agreement with experimental measurements of the 'lift loop' for an oscillating aerofoil, Fig. 10.7. This method offers considerable scope also for future modelling of turbine or compressor blade wake interactions, some reference to which will be made in Section 11.5.

10.4 Full vortex cloud modelling by the surface vorticity boundary integral method

Full vortex cloud theory attempts to simulate the real flow through solution of the Navier–Stokes equations, which may be expressed in vector form

$$\frac{\partial \mathbf{q}}{\partial t} + \mathbf{q} \cdot \nabla \mathbf{q} = -\frac{\nabla p}{\rho} + \nu \nabla^2 \mathbf{q} \qquad (10.11)$$

(A) (B)   (C)   (D)

Reading from left to right we are reminded that unsteady (A) and convective (B) fluid motions are related to pressure gradients (i.e. normal stresses) (C) and viscous shear stresses (D).

404

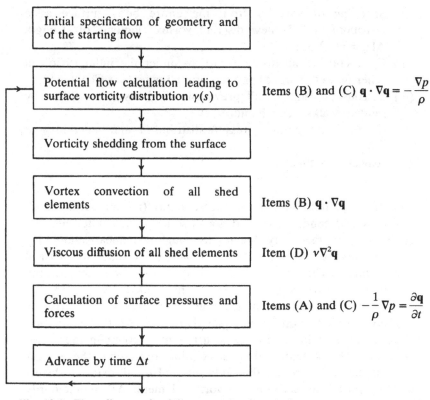

Fig. 10.8. Flow diagram for full vortex cloud analysis.

Inviscid steady potential flows incur terms (B) and (C) only. The aim of vortex cloud theory on the other hand is to simulate flows which may be unsteady, subject to viscous diffusion and possibly stalled or separated. All four terms must therefore be accounted for as illustrated by the flow diagram shown in Fig. 10.8. Although a practical computational scheme would be much more complex than this, the main underlying features of vortex cloud modelling are portrayed here. Following the initial input and data specification procedures, the computation proceeds iteratively through a series of small time increments Δt. For each time step the contributory factors which make up the Navier–Stokes equations are evaluated in sequence as follows.

(i) Analysis of potential flow past the body by Martensen's method including the influence of all uniform streams and discrete vortices.

(ii) Shedding of vorticity from the entire body surface as a distribution of M new discrete vortex elements of strength $\Delta \Gamma_n = \gamma(s_n) \Delta s_n$.

(iii) Convection of all discrete vortices under the influence of all other discrete vortices and uniform streams (see Chapter 8).

(iv) Viscous diffusion of all previously shed discrete vortices by random walks (see Chapter 9).

(v) Calculation of surface pressure distribution and hence lift and drag.

(vi) Advance by time Δt.

In the real fluid motion of course items (i)–(iv) occur simultaneously and continuously and not in series as demanded here by computational practicality. Inevitable though this 'linearisation' may be it is important to take note of it and to consider its physical interpretation. The overall effect is as if viscosity were temporarily switched to zero while convection proceeded over time Δt. Following this, convection laws are suspended while viscous diffusion is switched on for an equal elapse time Δt, the assumption being made that little error is incurred by decoupling these two activities. The potential flow calculation represents the instantaneous flow field around the body induced by all external disturbances with all the discrete free vortex elements $\Delta \Gamma_n$ momentarily frozen in position and could thus be described as a quasi-steady representation of the truly unsteady flow. All of the new surface vorticity thus created during the time step Δt is then released from the surface.

Thus the various terms in the Navier–Stokes equations are switched on or off at the appropriate moment of our numerical scheme. Consequently for the final step of calculating the surface pressure distribution, (10.11) reduces temporarily to

$$-\frac{1}{\rho} \nabla \mathbf{p} = \frac{\partial \mathbf{q}}{\partial t} \tag{10.12}$$

We will return to this equation later in Section 10.5 in relation to vorticity production to see how it may be used to calculate surface pressure distributions in unsteady flows. Now let us consider items (i)–(v) above in a little more detail.

10.4.1.   Potential flow analysis in the presence of a vortex cloud

For the author vortex cloud modelling was a natural extension of surface vorticity analysis (*see* Lewis (1978), (1981), (1986), Port-house & Lewis (1981) and Porthouse (1983)). In quasi-steady flow the potential flow past a two-dimensional body of arbitrary shape may thus be described by the boundary integral equation

$$-\tfrac{1}{2}\gamma(s_m) + \frac{1}{2\pi} \oint k(s_m, s_n)\gamma(s_n)\, \mathrm{d}s_n$$

$$+ \mathbf{W}_\infty \cdot \mathbf{ds} + \frac{1}{2\pi} \sum_{j=1}^{Z} L(m, j)\, \Delta\Gamma_j = 0 \quad (10.13)$$

The first three terms are identical to those which form Martensen's equation (1.19). The last term represents the contribution to the Dirichlet boundary condition of zero parallel velocity at surface element m due to the Z discrete vortices $\Delta\Gamma_j$ which form the vortex cloud. $L(m, j)$ is thus in form identical to the general Martensen coupling coefficient $k(s_m, s_n)$. Expressed in numerical form, and following the notation used in Section 8.5, we have the set of linear equations

$$\sum_{n=1}^{M} K(s_m, s_n)\gamma(s_n) = -(U_\infty \cos \beta_m + V_\infty \sin \beta_m)$$

$$- \sum_{j=1}^{Z} \Delta\Gamma_j(U_{mj} \cos \beta_m + V_{mj} \sin \beta_m) \quad (10.14)$$

where the unit velocities U_{mj}, V_{mj} are given by

$$U_{mj} = \frac{1}{2\pi} \left[\frac{y_m - y_j}{r_{mj}^{\,2}} \right], \qquad V_{mj} = -\frac{1}{2\pi} \left[\frac{x_m - x_j}{r_{mj}^{\,2}} \right] \quad (10.15)$$

with

$$r_{mj} = \sqrt{[(x_m - x_j)^2 + (y_m - y_j)^2]} \quad (10.16)$$

As discussed in the concluding section of Chapter 8 for fixed separation vortex cloud models, an additional statement of vorticity conservation is required which can now be expressed in the following form

$$\sum_{n=1}^{M} \gamma(s_n)\, \Delta s_n + \sum_{j=1}^{Z} \Delta\Gamma_j - \Gamma_{\mathrm{circ}} = 0 \quad (10.17)$$

407

Vortex cloud modelling

where Γ_{circ} is the cumulative strength of all vortices which are snuffed out for various reasons as the calculation proceeds. Initially Γ_{circ} must be set to zero and later increased by $\Delta\Gamma_j$ for any discrete vortex removed from the field. The two main reasons for vortex elimination are

(i) Discrete vortices which penetrate the body interior during the random walk or due to the first step in Euler convection.
(ii) Elimination of vortices some distance downstream of a body to reduce computational requirements.

Adding (10.17) to all Martensen equations ensures the creation of exactly the correct amount of new surface vorticity $\gamma(s_n) \Delta s_n$ to be shed at each time step to ensure vorticity conservation continuously as the computation proceeds. Equations (10.14) then become

$$\sum_{n=1}^{M} (K(s_m, s_n) + \Delta s_n)\gamma(s_n) = -(U_\infty \cos \beta_m + V_\infty \sin \beta_m)$$

$$- \sum_{j=1}^{Z} \Delta\Gamma_j(U_{mj} \cos \beta_m + V_{mj} \sin \beta_m + 1)$$

$$+ \Gamma_{\text{circ}} \qquad (10.18)$$

The appropriate procedure is then as follows:

(i) Evaluate the coupling coefficient matrix $K(s_m, s_n)$.
(ii) Back diagonal correction. See Section 8.5.3, (8.33).
(iii) Evaluate right hand side terms for uniform streams (U_∞, V_∞) and for the discrete vortices $\Delta\Gamma_j$.
(iv) Correction of right hand side values due to any discrete vortex $\Delta\Gamma_j$ which has drawn closer than a specified distance from its nearest surface element p. The implied circulation error and its treatment are described fully in Section 8.4.1, leading to the replacement right hand side value for the pth equation

$$\text{rhs}_p = -\frac{1}{\Delta s_p} \sum_{\substack{n=1 \\ n \neq p}}^{M} \text{rhs}_n \, \Delta s_n \qquad (8.15b)$$

where rhs_n is defined

$$\text{rhs}_n = \Delta\Gamma_j(U_{pj} \cos \beta_m + V_{pj} \sin \beta_m) \qquad (10.19)$$

(v) Add circulation conservation equation (10.17) to each equation.
(vi) Invert matrix.
(vii) Multiply right hand side of (10.18) by the inverted matrix.

10.4.2 Vortex shedding from body surface

Two models for vortex shedding are illustrated by the vortex cloud solutions for starting flow past a wedge shaped body shown in Fig. 10.9. The most commonly adopted approach assumes potential flow on the body surface but continuously shed vortex sheets from the sharp corners *A* and *B* as described already in Section 8.5, equations (8.24). Fig. 10.9(*b*) on the other hand illustrates the technique adopted for full vortex cloud analysis. Following potential flow analysis at each time step, the surface vorticity $\gamma(s_n)\,\Delta s_n$ created at every boundary element is released from the surface and shed freely into the fluid as a new discrete vortex. Following the guide-lines laid down in Section 9.4 in relation to the discrete vortex modelling of boundary layers, two methods for vortex shedding present themselves as illustrated by Fig. 9.8.

(i) Random walk method
    The newly created vorticity initially lying on the surface can be diffused by random walks. In this case double strength

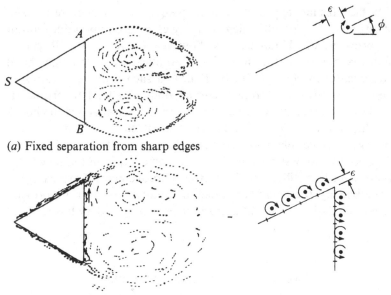

(*a*) Fixed separation from sharp edges

(*b*) Full vortex cloud simulation with vortex
      shedding from all surface elements.

Fig. 10.9. Starting motion for wedge-shaped bluff body simulated by two types of vortex cloud analysis.

discrete vortices $\Delta\Gamma_j = 2\gamma(s_n)\,\Delta s_n$ should be used as argued in Section 9.4.1.

(ii) Offset method

Each discrete vortex of strength $\Delta\Gamma_j = \gamma(s_n)\,\Delta s_n$ may be offset by a fixed distance ε, the value recommended by Porthouse (1983) being

$$\varepsilon = \sqrt{(4\nu\,\Delta t/3)} \qquad\qquad (10.20)$$

The first of these methods assumes that during the random walk half of the discrete vortices will cross the body surface where they will be lost from the main flow regime and will need to be snuffed out. This technique was justified by the end results when applied to plane wall boundary layers in Chapter 9, for both high and low Reynolds number flows. For strongly curved convex surfaces the procedure will lead to some over-production of vorticity, especially for examples close to an aerofoil leading edge. Alternatively single strength discrete vortices $\Delta\Gamma_j = \gamma(s_n)\,\Delta s_n$ may be used if vortices attempting to cross the surface during a random walk are bounced back into the flow as suggested by Chorin (1978).

In fact this last approach is directly equivalent to the use of fixed offsets given by (10.20) which is in any case based upon diffusion of a vortex sheet. Unfortunately however at very high Reynolds numbers ε is then extremely small. Application of the right hand side circulation check given by (10.5) to such a discrete vortex in close proximity to a strongly curved body can lead to erroneous solutions to Martensen's equation. The usual approach is to choose a value for the offset related to element size (say 25% of the average surface element length). Discrete vortices which stray closer to the body surface may then be removed from the flow field. There is in fact still scope for further improvement of vortex cloud modelling by the boundary integral method in order to provide better simulation of the all important surface boundary layer.

10.4.3 Convection schemes

Returning to the computational sequence for solution of the Navier–Stokes equations set out in Section 10.3, two alternative approaches to convection of the vortex cloud may be adopted as illustrated by Fig. 10.10.

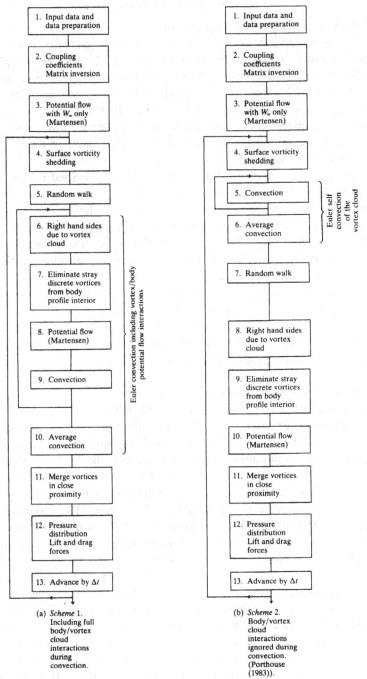

1. Input data and data preparation	1. Input data and data preparation
2. Coupling coefficients Matrix inversion	2. Coupling coefficients Matrix inversion
3. Potential flow with W_∞ only (Martensen)	3. Potential flow with W_∞ only (Martensen)
4. Surface vorticity shedding	4. Surface vorticity shedding
5. Random walk	5. Convection
6. Right hand sides due to vortex cloud	6. Average convection
7. Eliminate stray discrete vortices from body profile interior	7. Random walk
8. Potential flow (Martensen)	8. Right hand sides due to vortex cloud
9. Convection	9. Eliminate stray discrete vortices from body profile interior
10. Average convection	10. Potential flow (Martensen)
11. Merge vortices in close proximity	11. Merge vortices in close proximity
12. Pressure distribution Lift and drag forces	12. Pressure distribution Lift and drag forces
13. Advance by Δt	13. Advance by Δt

Euler convection including vortex/body potential flow interactions

Euler self convection of the vortex cloud

(a) *Scheme* 1. Including full body/vortex cloud interactions during convection.

(b) *Scheme* 2. Body/vortex cloud interactions ignored during convection. (Porthouse (1983)).

Fig. 10.10. Alternative vortex cloud computational schemes.

411

Method 1. Strict Eulerian convection

The implementation of truly Eulerian convection into a vortex cloud scheme was described in Section 8.5.2 for the simpler problem of vortex shedding from two predetermined separation points of a bluff body. A more detailed flow diagram applicable to full vortex cloud modelling is shown in Fig. 10.10(*a*). Following the initial data preparation and Martensen potential flow solution, boxes 1–3, the computation proceeds through a series of time steps Δt. First of all vorticity is created at the surface from the previous potential flow calculations, box 4, and shed as described in the last section. Following a random walk, box 5, Euler convection is completed in two steps, boxes 6–10, during the first of which discrete vortices which have accidentally entered the body profile interior are eliminated. The reason for placing vortex elimination at this point in the procedure is that such offending discrete vortices may be detected when calculating the right hand side values, box 6, as described in Section 10.12, (10.5). As a safety precaution vortex elimination (box 7) may also be repeated during step 2 of Euler convection although there should be no need for it. Finally in box 11 discrete vortices which come into close proximity may be merged following the rules proposed in Section 9.4.4.

Method 2. Simplified Eulerian convection

The above scheme offers the best attainable accuracy since all vortex/body interactions are accounted for during the convection calculations. The various rules developed in Chapter 8 may then be implemented to minimise errors and achieve the best simulation possible from the discrete vortex model. However, the repeated potential flow calculations during convection double the time required to complete the calculations. Porthouse (1983) therefore proposed the simpler scheme illustrated by Fig. 10.10(*b*) which can also be argued as perfectly justifiable from the fluid dynamic viewpoint. Thus following each potential flow calculation, box 10, the newly created surface vorticity is shed close to the body surface. A repeat potential flow analysis following box 4 ought to yield zero surface vorticity. For these reasons Porthouse argues that repeat potential flow calculations to estimate further body/vortex cloud interactions during convection are really unneccessary. The convection calculations on this basis involve only the mutual influence of the discrete vortices forming the cloud, plus the effect of the uniform stream.

It should be pointed out that the studies of vortex/body induced convections undertaken in Chapter 9 were focussed mainly on the extreme situation of a single vortex. In this case the entire motion is determined by the potential flow past the body due to the discrete vortex and its consequent reflective effect upon convection of the vortex. Strict Eulerian convection following Method 1 is essential in such situations. For full vortex cloud modelling on the other hand the self-induced convection of a particular discrete vortex element due to its own interaction with the body is insignificant compared with the sum total convective influence of the remaining discrete vortices which form the cloud combined also with the effect of the uniform stream W_∞.

10.5   Calculation of surface pressure distribution and body forces

As indicated in Fig. 10.8, once convection $\mathbf{q} \cdot \nabla \mathbf{q}$ and diffusion $\nu \nabla^2 \mathbf{q}$ have been completed in the numerical simulation, the Navier–Stokes equations (10.11) reduce to

$$-\frac{1}{\rho} \nabla p = \frac{\partial \mathbf{q}}{\partial t} \qquad (10.21)$$

At any point s_n on the body surface we have already shown (1.12) that the velocity parallel to the surface following potential flow analysis is given by $\mathbf{q} = \gamma(s_n)$. Equation (10.21) then gives the pressure gradient along the surface at s_m, namely

$$\frac{1}{\rho} \frac{\partial p}{\partial s} = -\frac{\partial \gamma(s_n)}{\partial t} \qquad (10.22)$$

from which we may derive a numerical expression for the change in surface pressure over the surface element n during the discrete time step Δt, namely

$$\Delta p_n = -\rho \frac{\gamma(s_n) \, \Delta s_n}{\Delta t} = -\rho \frac{\Delta \Gamma_n}{\Delta t} \qquad (10.23)$$

An alternative derivation of (10.22) was given in Section 1.5 where it was described as the surface vorticity production equation and discussed in particular in relation to steady and unsteady potential flows. For vortex cloud modelling vorticity is being

413

continually produced and shed as a distribution of discrete vortices $\Delta\Gamma_n$ at each time step. Equation (10.23) thus provides a surprisingly simple means for calculating the surface pressure distribution in a vortex cloud simulation of an unsteady flow.

10.5.1 Pressure distribution – full vortex cloud model

For full vortex cloud modelling in which potential flow surface vorticity is shed as discrete vortices from all surface elements during each time step, (10.23) may be integrated directly to yield the surface pressure at any point on the body,

$$p_m = p_1 + \sum_{n=1}^{m} \Delta p_n = p_1 - \frac{\rho}{\Delta t} \sum_{n=1}^{m} \Delta\Gamma_n \qquad (10.24)$$

Clearly the pressure is expressed here only relative to some datum value p_1 applicable at say station 1, Fig. 10.11 from where the integration of (10.22) begins for numerical convenience. A suitable technique is to set p_1 equal to zero initially. The p_m values in the neighbourhood of the leading edge stagnation point S may then be searched to find the highest value, say p_s. This may then be raised to the stagnation pressure of the approach flow if all p_m values are increased by the same amount $(\frac{1}{2}\rho W_\infty^2 - p_s)$. All static pressure values will then be gauge pressures (i.e. relative to p_∞).

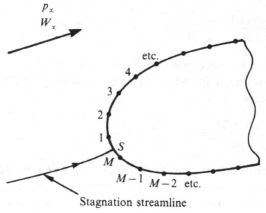

Fig. 10.11. Leading edge stagnation point.

414

10.5.2 Pressure distribution – vortex cloud modelling with fixed separation points

Curiously the simpler vortex cloud model with fixed separation points, Fig. 10.9(*a*) presents a more complex problem for computation of surface pressure distribution. Essentially we are dealing here with an unsteady 'potential' flow, the only vortex shedding occurring at the two fixed separation points *A* and *B*. Porthouse (1983) deals with this by rewriting the Navier–Stokes equations (10.11) in terms of the stagnation pressure p_0 and vorticity ω.

$$\frac{1}{\rho}\nabla p_0 = \mathbf{q} \times \omega - \nu\nabla \times \omega - \frac{\partial \mathbf{q}}{\partial t} \qquad (10.25)$$

Although we may decide to subject the vortex cloud to random walks to simulate viscous diffusion of the wake, flow at the body surface remains inviscid and irrotational, represented by an infinitely thin surface vorticity sheet which is fluctuating in strength with respect to time. Equation (10.25) then reduces to

$$\frac{1}{\rho}\Delta p_0 = \mathbf{q} \times \omega - \frac{\partial \mathbf{q}}{\partial t} \qquad (10.25a)$$

The cross product in the second term produces a vector quantity normal to both $\mathbf{q}$ and ω. But since $\omega = \nabla \times \mathbf{q}$, the vorticity ω is normal to $\mathbf{q}$. Now at the surface both $\mathbf{q}$ and ω are both parallel to the surface. $\mathbf{q} \times \omega$ thus represents the rate of change normal to the surface of a vector quantity of velocity scale (as indicated by the dimensions of the third term). Within a real shear layer this quantity would have a finite value except on the body surface itself where $\mathbf{q}$ would vanish to zero. Applying this reasoning to the surface vorticity potential flow model equivalent, Porthouse then argues that in the region beneath the sheet in unsteady flow (10.25a) reduces to

$$\frac{1}{\rho}\nabla p_0 = -\frac{\partial \mathbf{q}}{\partial t} \qquad (10.25b)$$

Thus in unsteady inviscid irrotational (potential) flow modelled by the surface vorticity method there will be a gradient of stagnation pressure along the body surface of magnitude

$$\frac{\partial p_0}{\partial s} = -\rho\frac{\partial v_s}{\partial t} \qquad (10.26a)$$

415

Expressed numerically for element m we then have

$$\Delta p_{0m} = -\rho\left(\frac{v_{sm,t+\Delta t} - v_{smt}}{\Delta t}\right)\Delta s_m \qquad (10.26b)$$

where $v_{sm,t} = \gamma(s_{m_t})$ is the surface vorticity solution for time t. The static pressure change along the surface element then becomes

$$\Delta p_m = -\rho\left(\frac{\gamma(s_m)_{t+\Delta t} - \gamma(s_m)_t}{\Delta t}\right)\Delta s_m - \tfrac{1}{2}\rho\gamma(s_m)^2 \qquad (10.27)$$

This may be applied to all surface elements and integrated around the body surface, by analogy with (10.24) for all vortex cloud modelling. However additional corrections are required to account for the discontinuities in stagnation pressure which occur across the separation vortex sheets at A and B, Fig. 10.9(a). Let us assume that the static pressure remains constant across the sheets, whose strengths $\gamma(s_a)$ and $\gamma(s_b)$ follow from the potential flow solution for the surface elements just upstream of A and B. The jumps in stagnation pressure across the sheet, moving in a clockwise direction, are then given by

$$\left.\begin{array}{l} \dfrac{1}{\rho}\Delta p_{oa} = -\tfrac{1}{2}\gamma(s_a)^2 \\[4mm] \dfrac{1}{\rho}\Delta p_{ob} = \tfrac{1}{2}\gamma(s_b)^2 \end{array}\right\} \qquad (10.28)$$

The pressure distribution relative to a datum value of zero at the leading edge S then follows from

$$\left.\begin{array}{ll} p_m = \displaystyle\sum_{n=1}^{a} \Delta p_n & \text{Front surface } SA \\[5mm] = p_a - \tfrac{1}{2}\rho\gamma(s_a)^2 + \displaystyle\sum_{n=a+1}^{b} \Delta p_n & \text{Rear surface } AB \\[5mm] = p_b + \tfrac{1}{2}\rho\gamma(s_b)^2 + \displaystyle\sum_{n=b+1}^{M} \Delta p_n & \text{Front surface } BS \end{array}\right\} \qquad (10.29)$$

For either of the above schemes the lift and drag forces may be derived by integration of $\bar{p}_n\, ds_n$ around the body profile, where $\bar{p}_n = \tfrac{1}{2}(p_n + p_{n+1})$ is the average pressure on element n. Lift and drag coefficients then follow from

$$C_L = \frac{L}{\frac{1}{2}\rho W_\infty^2 \ell} = -\frac{2}{\rho W_\infty^2} \sum_{n=1}^{M} \bar{p}_n \cos \beta_n \, \Delta s_n$$

$$C_D = \frac{D}{\frac{1}{2}\rho W_\infty^2 \ell} = \frac{2}{\rho W_\infty^2} \sum_{n=1}^{M} \bar{p}_n \sin \beta_n \, \Delta s_n \qquad (10.30)$$

10.5.3 Pressure and force fluctuations due to numerical noise

Wake patterns for the starting flow past a wedge-shaped body predicted by the two vortex cloud models outlined above are in excellent agreement, Fig. 10.9. These results were obtained after 100 time steps $\Delta t = 0.033\,333$ with 30 surface elements of equal length $\Delta s = 0.1$ and with a uniform stream $W_\infty = 1.0$. For case (a) with fixed separation points, $\varepsilon = 0.016\,667$, $\phi = 45°$ and an infinite Reynolds number was assumed. For full vortex cloud modelling, case (b), the offset ε was set at 0.01 for a Reynolds number $Re = 10^6$.

In the latter case, as time proceeds a boundary layer builds up on all three surfaces of the wedge. The front surface boundary layers are convected downstream and separate naturally from the sharp edges A and B where they merge into the outward flowing boundary layers also developing naturally on the rear face AB. To improve visualisation equal length streak lines are plotted for the three final time steps. A closer scrutiny of the two wake patterns also reveals that whereas case (a) is still almost perfectly symmetrical, the onset of asymmetry has just begun in case (b), which would soon lead to the generation of a von Karman street type of wake motion. The main reason for this lies with the differing vorticity production mechanisms. In full vortex cloud modelling the final vorticity sheet strengths, when shed at A and B, are subject to all of the random walks, vortex convections and body potential flow interactions of the front face boundary layers. The outcome of this intensive activity of the discretised vortex cloud close to the body surface can be illustrated dramatically if we compare the predicted drag coefficients for the last ten time steps, Table 10.1.

In both cases the drag coefficient fluctuated due to the global vortex wake motion. For case (a) this is more of a gentle undulation

Table 10.1. *Predicted drag coefficients for an equilateral wedge shaped body of side $\ell = 1.0$ with $W_\infty = 1.0$, over $t = 3.0$–$3.333\,33$*

Time step no.	Case (a) – fixed separation points C_D	Case (b) – Full vortex cloud model C_D
91	0.875 172	1.505 233
92	0.870 289	0.707 391
93	0.874 871	1.516 425
94	0.881 506	0.664 753
95	0.888 191	0.324 498
96	0.901 534	1.196 192
97	0.901 054	0.708 770
98	0.888 791	0.724 955
99	0.886 503	0.630 692
100	0.851 341	0.299 423
	0.88 925 (mean value)	0.827 833 (mean value)

whereas for case (b) the C_D values appear to exhibit random fluctuations sometimes of large magnitude compared with the mean value. On the other hand the mean values for the steps 91–100 are in reasonable agreement.

This chaotic behaviour of $C_D(t)$ is of course derived via (10.30) from the equally chaotic behaviour of predicted surface pressure distribution p_n. This in turn is determined by the strength of the discrete vortices $\Delta\Gamma_n$ shed during each time step and therefore of the preceding potential flow solution. In the presence of a surrounding and drifting cloud of randomly scattered discrete vortices, regularly subjected in any case to random walks, it is obvious that the resulting surface and shed vorticity and therefore pressure distribution will be equally subject to randomness. This is usually referred to as 'numerical noise', since it is linked completely to the discretisation involved in vortex cloud modelling. At high Reynolds numbers the problem will increase due to the closer proximity of the discrete vortices to the body surface and its time scale δt can be related to the average time taken for a vortex to traverse one surface element. This may be estimated crudely through

$$\frac{U_\infty}{2}\delta t = \Delta s \tag{10.31}$$

The solution to the problem is thus to average surface pressures

418

or C_L and C_D over T time steps, thereby filtering out the numerical noise, where T may be estimated as the nearest integer to

$$T = \frac{\delta t}{\Delta t} = \frac{2\,\Delta s}{U_\infty\,\Delta t} \tag{10.32}$$

A suitable strategy is to replace the first p, C_L and C_D values for time $t = 0$ by the average of the first T values and to repeat this for each successive time step. Applying this process to the wedge flow the predicted C_L and C_D values are compared for cases (a) and (b) in Fig. 10.12, where (10.32) suggests averaging over $T = 6$ time steps for the full vortex cloud solution. Encouraging agreement is then obtained between the two predictions, although fairly regular fluctuations of both C_L and C_D may still be detected. These ripples may be further reduced by doubling the number of time steps T for averaging as given by (10.32). It should be borne in mind though that averaging will also tend to filter any high frequency real events

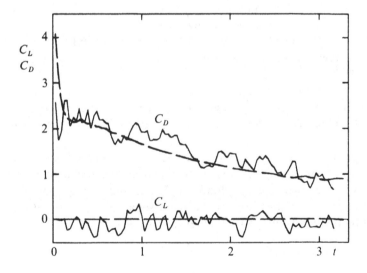

———— Fixed separation point method.

———— Full vortex cloud method averaged over 6 time steps

$\Delta t = 0.033\,333$, $U_x = 1.0$, $M = 30$, $\Delta s = 0.1$

Fig. 10.12. Predicted lift and drag coefficients for starting motion past an equilateral wedge.

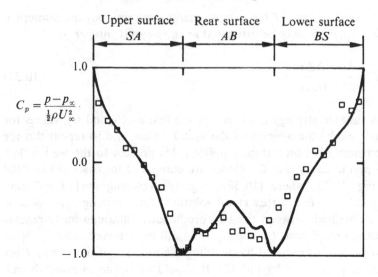

$$C_p = \frac{p - p_\infty}{\frac{1}{2}\rho U_\infty^2}.$$

$\sim$ Fixed separation point method

□   Full vortex cloud method

Solution averaged over steps 90–100 with
$\Delta t = 0.033\,333$, $U_\infty = 1.0$, $M = 30$, $\Delta s = 0.1$.

Fig. 10.13. Predicted surface pressure distributions for wedge-shaped body starting motion.

such as the flow induced by a small eddy drifting past the body. $2T$ should be regarded as the upper limit for averaging out numerical noise.

The predicted surface pressure distributions averaged over the last ten time steps, Fig. 10.13, show remarkably good agreement between these two vortex cloud methods, suggesting that the computational simplifications of the fixed separation point method should be taken advantage of whenever possible.

10.5.4 Data reduction of unsteady pressures and forces for bluff body flows

So far in this chapter we have considered only the symmetrical starting motion for a wedge-shaped body. After some time has elapsed, on the other hand, periodic wake motion will become established for any bluff body as illustrated by the vortex cloud solution for flow past a cylinder, Fig. 8.20. In addition to numerical noise there will then also be genuine surface pressure fluctuations, presenting a serious problem of how comparisons may be made between theory and experiment for the purposes of validation. A direct comparison is clearly ruled out, especially for flows influenced by an element of randomness. The best way forward is that indicated by experimental methods of statistical data reduction such as those adopted by Shim *et al.* (1988) in relation to flow through heat exchanger tube banks. The steady and fluctuating features of the flow may then be categorised as follows:

(i) An average surface pressure coefficient may be defined by

$$C_{p_{av}} = \frac{p_{av} - p_\infty}{\frac{1}{2}\rho W_\infty^2} \tag{10.33}$$

where the average surface pressure over a specified elapse time $t_2 - t_1$ is given by

$$p_{av} = \frac{1}{t_2 - t_1} \int_{t_1}^{t_2} p \, \mathrm{d}t \tag{10.34}$$

Average lift and drag forces may then be directly derived from $C_{p_{av}}$.

(ii) A r.m.s. (or standard deviation) fluctuating pressure coefficient may also be defined to categorise the scale or intensity of surface pressure fluctuations, through

$$C_p' = \frac{1}{\frac{1}{2}\rho W_\infty^2} \left\{ \frac{\int_{t_1}^{t_2} (p - p_{av})^2 \, \mathrm{d}t}{t_2 - t_1} \right\}^{\frac{1}{2}} \tag{10.35}$$

(iii) Spectral density analysis techniques may be adopted as advocated by Hill *et al.* (1986) to identify dominant frequencies f of excitation forces. This may be of particular importance in many engineering situations where dimensionless frequency may then be expressed through a Strouhal number

$$St = \frac{fd}{W_\infty} \tag{10.36}$$

d being a significant leading dimension.

A similar series of definitions can be made to categorise velocity within the wake region where the consequent data sampling and reduction procedure would follow quite closely that commonly applied to hot-wire velocity measurements. For both surface pressure and wake velocities however, it may first be necessary to reduce numerical noise by subjecting the predicted data to an averaging process such as (10.34) in the manner described in Section 10.5.3, for the short time interval given by (10.32). A technique for wake sampling for estimation of momentum losses will also be described in Section 11.5.1 in relation to turbine cascades.

10.6 Application of vortex cloud analysis to flow past a circular cylinder

Little experimental work has been completed to date on the experimental validation of bluff body flows by the above techniques.

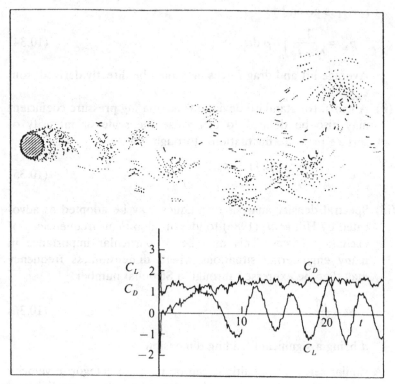

Fig. 10.14. Vortex cloud solution to flow past circular cylinder at $Re = 20\,000$.

Table 10.2. *Predicted Strouhal
number versus Reynolds
number for circular cylinder
vortex streets (Lewis & Shim
(1986))*

Re	St
20 000	0.2337
30 000	0.2230
40 000	0.2210
50 000	0.1637
60 000	0.2136

We conclude this chapter by reference to the work of Shim (1985) who applied such techniques to the flow past circular cylinders, compared later with full vortex cloud analysis by Lewis & Shim (1986). A vortex cloud simulation for $Re = 2 \times 10^4$ is shown in Fig. 10.14 for a time step $\Delta t = 0.12$ taken sufficiently large to cover several oscillations of the vortex street during 225 time steps. For reasons of computational economy discrete vortices in close proximity were recombined as follows:

for $x < 1.5d$, elements combined if closer than $0.025d$

for $x > 1.5d$, elements combined if closer than $0.05d$.

Typically this resulted in a field of about 600 vortex elements with greater concentration around the body surface and sufficient resolution elsewhere to define the general periodic wake pattern.

For this calculation discrete vortices which drifted nearer than $0.05d$ to the cylinder surface were artificially moved back to that position at each iteration. The effect of this procedure and the fairly large time step was a reduction in numerical noise. Thus according to (10.32), for a diameter $d = 1.0$ and $M = 32$ surface elements, averaging over two time steps should be sufficient to eliminate noise. On the other hand the C_L and C_D values shown in Fig. 10.14 were, in fact, not averaged and show that little numerical noise was present. Strouhal number may be derived directly from the clear oscillations of C_L, resulting in Table 10.2 of values derived for a range of Reynolds number.

Apart from the result for $Re = 5 \times 10^4$ the predicted Strouhal numbers show a regular pattern of a slight decrease with Reynolds

Vortex cloud modelling

number, the values agreeing reasonably well with experimental expectations.

For this application because of symmetry the average lift coefficient approximates to zero, whereas the actual lift coefficient varies approximately sinusoidally with an amplitude in excess of unity. An appropriate measure of the related excitation force is the r.m.s. lift coefficient defined

$$
C_{L\text{r.m.s.}} = \left\{ \frac{\int_{t_1}^{t_2} C_L{}^2 \, dt}{t_2 - t_1} \right\}^{\frac{1}{2}}
\tag{10.37}
$$

An extremely thorough experimental investigation of $C_{L\text{r.m.s.}}$ was undertaken by Kacker *et al.* (1974) for single cylinders of varying aspect ratio L/d, located between parallel side walls and at a range of Reynolds numbers, Fig. 10.15(*a*). From their findings it is apparent that $C_{L\text{r.m.s.}}$ attains a fixed value if $L/d < 2.0$, the value depending only on Reynolds number. Assuming this to represent the limiting two-dimensional flow with closely spaced side walls, Lewis & Shim (1986) compared these results with $C_{L\text{r.m.s.}}$ values predicted by vortex cloud theory, Fig. 10.15(*b*). Although the trend of decreasing r.m.s. lift with increasing Reynolds number was predicted, the theoretical $C_{L\text{r.m.s.}}$ values were a good deal greater than those measured experimentally. There is of course the added difficulty with experimental modelling that end-wall boundary layer effects would become significant at low aspect ratios, thus reducing $C_{L\text{r.m.s.}}$. However it seems probable that there are still other

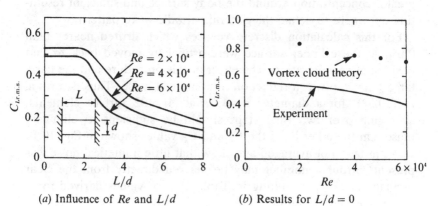

(a) Influence of Re and L/d   (b) Results for L/d = 0

Fig. 10.15. R.M.S. fluctuating lift coefficients for circular cylinders with various aspect ratios and Reynolds numbers.

424

three-dimensional effects present at $L/d < 2.0$ which would lead to deficiencies in the two-dimensional vortex cloud model. Despite these problems there is much encouragement to be gained from these comparisons with experimental test, especially regarding the prediction of the Strouhal number and general characteristics of the flow.

Further insights have been gained by comparison with experiment of the predicted average and fluctuating pressure distributions. Three such calculations given by Lewis & Shim (1986) are shown in Fig. 10.16 for various discretisations of the body surface and time steps and for fairly similar Reynolds numbers. Quite reasonable predictions of $C_{p_{av}}$ have been obtained by these and other authors, although there is a tendency for flow separation to occur slightly further downstream according to the full vortex cloud simulation. Upstream of separation the accelerating potential flow is predicted quite well over the front facing 120° arc of the cylinder. Downstream of separation vortex cloud modelling tends to over-predict the wake average suction pressures resulting in 10–15% overestimate of the r.m.s. drag coefficient.

Comparison with experiment of the predicted fluctuating surface pressure coefficients C_p', on the other hand, are less good and indicative of the probable presence of the three-dimensional effects referred to in the last but one paragraph, Fig. 10.16(b). The general form of C_p' is quite good over the ranges $\theta < 70°$ and $\theta > 150°$. However in the intermediate range, which represents the region adjacent to the cylinder in which the von Karman vortices build up and separate, theory considerably over-predicts the actual pressure fluctuations. In fact the nature of the experimental results is rather surprising and not according to expectations whereas the theoretical results follow the trend one would expect of maximum fluctuations in the region $\theta = 120°$. Thus experimentally it was found that C_p' increased rapidly up to the separation point $\theta = 70°$ and then began to decrease slightly over the region in which the separation eddies build up, $70° < \theta < 150°$. Well into the wake region $150° < \theta < 180°$ theory and experiment again follow the same trend of decreasing fluctuations and the predicted level is not dissimilar from that measured experimentally.

One possible conclusion which may be drawn from these very interesting observations is that in reality the vortices, as they roll up on alternate sides of the cylinder downstream of separation to form the ultimate periodic vortex wake, become unstable and break

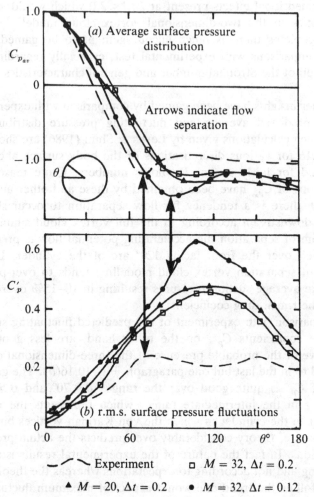

Fig. 10.16. Average surface pressure $C_{p_{av}}$ and r.m.s. fluctuating pressure $C_p{}'$ for a circular cylinder at $Re = 3 \times 10^4$.

down locally into a three-dimensional flow. The associated re-distribution of energy would then result in a reduced level of fluctuations. Bearing in mind these observed substantial differences between measured and predicted fluctuations, it is perhaps surprising that vortex cloud modelling is so capable of simulating the general and periodic features of cylinder wakes and that such real wakes remain so nearly two-dimensional. Despite the reservations expressed by Mair & Maull (1971) during the early days of discrete vortex modelling, referred to in the introduction to this chapter, it is clear that two-dimensional modelling copes surprisingly well and remains a relatively simple but important analytical tool for predicting bluff body wakes.

CHAPTER 11

Further development and applications of vortex cloud modelling to lifting bodies and cascades

11.1 Introduction

A basic outline of full vortex cloud modelling was presented in Chapter 10 but with limited application primarily to bluff body flows for which separation occurs spontaneously and dramatically at reasonably cetain separation points, resulting normally in the development of a broad periodic wake. The main aim of this chapter is to apply the full vortex cloud method to lifting bodies such as aerofoils and cascades for which the aerodynamic aim usually is to avoid flow separations, maintaining low losses. Full vortex cloud modelling represents an attempt to solve the Navier–Stokes equations including both the surface boundary layer near field and the vortex wake far field flows. Boundary layer separations are then self-determining. In practice however, as discussed by Porthouse & Lewis (1981), Spalart & Leonard (1981) and Lewis (1986) vortex cloud modelling in its present state of development seems unable quite to cope with the general problem of boundary layer stability and various techniques are proposed by these authors to avert premature stall as often experienced during vortex cloud analysis of aerofoils or cascades. These problems will be considered in Sections 11.2–11.4. Extension of vortex cloud modelling to cascades will be given in Section 11.5 and studies of acoustic excitation due to wake vortex streets from bluff bodies in ducts are briefly discussed in Section 11.6.

11.2 Flow past a lifting aerofoil by vortex cloud analysis

The full vortex cloud numerical scheme described in relation to bluff body wake flows in Chapter 10 may in fact be applied without modification to simulate the flow past bodies of arbitrary shape including two-dimensional lifting aerofoils. Predictions of flow

428

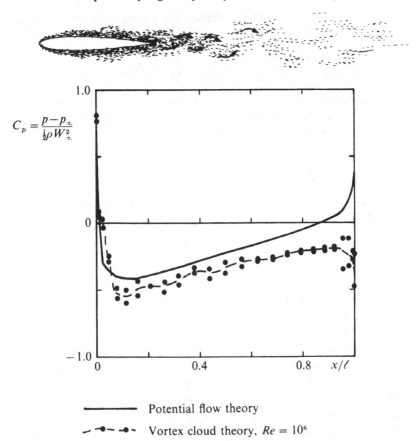

Flow past a lifting aerofoil by vortex cloud analysis

$$C_p = \frac{p - p_\infty}{\frac{1}{2}\rho W_\infty^2}$$

———————— Potential flow theory

– •– –•– Vortex cloud theory, $Re = 10^6$

Fig. 11.1. Surface pressure distribution for NACA 0012 at zero incidence by vortex cloud theory.

pattern and surface pressure distribution for the simplest situation of a symmetrical aerofoil, NACA 0012, at zero incidence are shown in Fig. 11.1 after fifty time steps $\Delta t = 0.05$ and with $M = 50$ surface elements. One discrete vortex was shed from each surface element at each time step with an off-set ε from the surface (see Fig. 10.9) of 1% of chord using computational scheme No. 2 of Fig. 10.10. Streak lines have been plotted for the last two time steps. In these calculations point vortices rather than Rankine vortices were used, control over excessive convection velocities being exercised by merging vortices which came into close proximity. A simple approach to this permits the operator to choose the maximum allowable convection velocity between interacting vortices, a suit-

429

Further development of vortex cloud modelling

able value being W_∞. At the outset an estimate of the average element strength $\Delta\Gamma = \gamma(s)\,\Delta s$ may be obtained from

$$\Delta\Gamma = \frac{\text{typical velocity} \times \text{perimeter}}{\text{number of surface elements}} = \frac{W_\infty}{M}\sum_{n=1}^{M}\Delta s_n$$

If the minimum allowable gap Δr is that for which the convection velocity close to a vortex $\Delta\Gamma$ equals the maximum allowable convection velocity, say W_∞, then

$$\Delta r = \frac{\Delta\Gamma}{2\pi W_\infty} = \frac{1}{2\pi M}\sum_{n=1}^{M}\Delta s_n \qquad (11.1)$$

In the present calculations the following merging arrangements were then made

(i) For $x < 1.5\ell$ (main target area) mergers if gap $<\Delta r$.
(ii) For $1.5\ell < x < 4.0\ell$, mergers if gap $<3\,\Delta r$.
(iii) For $x > 4.0\ell$, mergers if gap $<5\,\Delta r$.

Items (ii) and (iii) here were introduced purely to reduce the total number of vortices in the wake region in order to minimise computational effort.

The surface pressure distribution, averaged over time steps 25–50, is reasonably symmetrical and is generally, as might be expected, lower than that predicted by potential flow theory due to the influence of boundary layer displacement thickness. However there are disturbances in the trailing edge region and a lowering of static pressure which require comment. The predicted flow pattern provides a partial explanation. Despite the aerofoil symmetry the wake flow is in fact periodic. This may be attributed to convective interactions of the two vortex sheets of opposite sign shed into the wake from the upper and lower surface boundary layers. In addition to this the low average value of static pressure in the trailing edge region is largely due to the non-imposition of any trailing edge Kutta–Joukowski condition in this scheme. Although circulation conservation is rightly enforced by the addition of the circulation equation (10.17) to all Martensen equations as described in Section 10.4.1, the quantity Γ_{circ} tends to fluctuate in response to numerical noise and occasional absorption of discrete vortices into the body profile. Furthermore the oscillating wake will undoubtedly be coupled to consequent induced asymmetries in the trailing edge region. This phenomenon is known to exist in real aerofoil and turbine wake flows and has indeed resulted in fatigue failures of the

430

blade trailing edges of water turbines. The static depression predicted here is caused by overproduction of vorticity from surface elements close to the trailing edge (see the vorticity production equation (10.22)), associated with alternating recirculations around the trailing edge, first clockwise then anticlockwise, as the simulated flow attempts to attain on average to the true trailing edge Kutta Joukowski condition. Generally speaking however the vortex cloud simulation here is satisfactory.

Less satisfactory is the predicted behaviour of this aerofoil with 5° angle of attack Fig. 11.2, where results are shown for 100 time steps $\Delta t = 0.05$. To reduce numerical noise the lift and drag coefficients have been averaged over 5 local time steps as recommended in Section 10.5.3. Although this aerofoil is known to be well below its stall limit at $\alpha_\infty = 5°$, it is clear from the C_L data that the simulated aerofoil entered intermittent stall after two main stream flow passes $t = 2.0$. The flow pattern shown here for the last time step at $t = 5.0$, reveals that the aerofoil was then badly stalled on the upper surface. Immediately downstream of the leading edge,

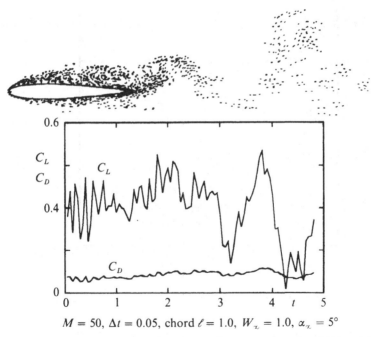

$$M = 50, \Delta t = 0.05, \text{chord } \ell = 1.0, W_x = 1.0, \alpha_x = 5°$$

Fig. 11.2. Vortex cloud prediction of flow past NACA 0012 with 5° incidence.

431

boundary layer separation can be discerned, followed by a reattachment. Further downstream at $x/\ell = 0.5$ a large separation and recirculation zone indicates major stall in agreement with the collapse of the lift coefficient over the period $4.0 < t < 5.0$.

During the initial period $t < 3.0$, quite reasonable lift coefficients were developed, rising to an average value $C_L \approx 0.5$ for time steps 30–60 which compares well with the experimental steady state value $C_L = 0.53$, Miley (1982). The predicted drag coefficient of 0.094 for this period was on the other hand well in excess of the published experimental value $C_D = 0.0122$ for $Re = 1.31 \times 10^6$ (c.f. $Re = 10^6$ used for these calculations). Examination of the predicted pressure distribution, Fig. 11.3, sheds further light on the matter. For this small angle of attack we would expect the potential flow model to behave quite well. By comparison however the vortex cloud simulation has performed particularly badly in the region of the expected upper surface leading edge suction peak which was

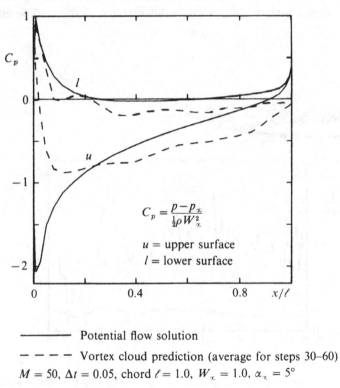

$$C_p = \frac{p - p_\infty}{\frac{1}{2}\rho W_\infty^2}$$

u = upper surface
l = lower surface

———— Potential flow solution

– – – – Vortex cloud prediction (average for steps 30–60)

$M = 50$, $\Delta t = 0.05$, chord $\ell = 1.0$, $W_\infty = 1.0$, $\alpha_\infty = 5°$

Fig. 11.3. Surface pressure distribution for NACA 0012 with 5° incidence.

seriously under-estimated due largely to the leading edge flow separation. The resulting errors are largely responsible for the excessive form (pressure) drag. On the other hand quite reasonable results were obtained for the lower surface pressure distribution for which, according to potential flow theory, fairly constant static pressure is anticipated. Furthermore the static pressure rise downstream of $x/\ell = 0.4$ on the upper surface and also the loading $(C_{p_\ell} - C_{p_u})$ downstream of $x/\ell = 0.2$ were reasonably well predicted. Static pressure towards the trailing edge is in fact an improvement upon potential flow modelling, settling down to a value just below p_∞. Troublesome though the problem of 'numerical' stall is in this pure and unsophisticated vortex cloud simulation, there is clear potential given scope for future inprovements and adaptations, some of which we will shortly consider. The time step chosen here was in excess of that suggested by the rules set out in Chapter 10. Reduction by about one half to the recommended value improves the results only marginally, the flow still being vulnerable to intermittent stall but of a higher frequency.

Despite these problems of representing unstalled flows, Lewis & Porthouse (1981) and Porthouse (1983) have demonstrated the

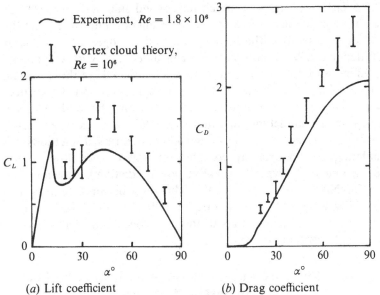

(a) Lift coefficient   (b) Drag coefficient

Fig. 11.4. Lift and drag coefficients for NACA 0012 in the stalled range – Vortex cloud theory compared with experimental results due to Critzos *et al.* (1954).

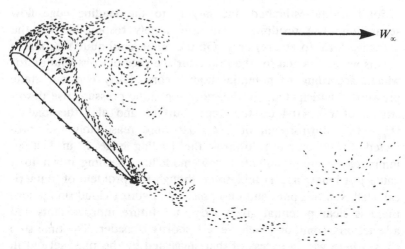

$M = 50$, $\epsilon = 0.01$, after 60 time steps $\Delta t = 0.025$

Fig. 11.5. Onset of deep stall for NACA 0012 with 45° incidence.

capability of vortex cloud theory to model fully stalled aerofoil flows extremely well. Predicted C_L and C_D values are compared in Fig. 11.4 with experimental tests published by Critzos *et al.* (1954) for NACA 0012 for angles of attack α_∞ beyond stall and up to 90°. In this situation the aerofoil is of course in effect a bluff body and the flow separations from the leading and trailing edges are reasonably well defined. Several authors have concluded that vortex dynamics modelling is extremely well suited to such flows where the upstream boundary layer is not disposed to flow separation and the vortices shed into the wake are mainly remote from the body surface. Vortex cloud modelling undoubtedly offers the most powerful, economic and accurate of all available theoretical methods for simulating these extremely complex separated wake flows at the present time and a typical streakline flow prediction is shown in Fig. 11.5 for NACA 0012 at $\alpha_\infty = 45°$. The rewards for extending this capabililty to unstalled or naturally stalling aerofoil flows are so great that various approaches to this have been considered, two of which will now be reviewed.

11.3 Alternative vortex cloud modelling techniques by Spalart & Leonard

Many of the above problems have been discussed by Spalart & Leonard (1981) who adopted a quite similar approach based on surface vorticity shedding but using a stream function method to satisfy the potential flow step of the time marching procedure. Newly created discrete vortices were placed at a distance ε from the local wall boundary condition point equal to one core radius r_0. These authors found that solutions were insensitive to r_0 provided it was within the range of about 0.5% of the body dimension, the same core radius being adopted for all shed vortices. The vorticity profile chosen within the core was a second degree polynomial

$$\left.\begin{aligned} \omega &= \frac{2\Delta\Gamma}{\pi r_0^{2}} \left(1 - \left(\frac{r}{r_0}\right)^{2}\right) & r < r_0 \\ \omega &= 0 & r > r_0 \end{aligned}\right\} \qquad (11.2)$$

higher degree formulations having been tried but showing no significant difference. Discrete vortices whose centres come back to within r_0 of the surface are suppressed. Naturally, as in the present author's scheme, circulation thus lost will be recreated before the next time step and vortices are not allowed to remain in the body profile if they have accidentally strayed there during a random walk.

The potential flow solution was derived by considering the stream function at a distribution of wall points (+, Fig. 11.6) due to contributions from all discrete vortices in the field plus the uniform stream and stating the Neumann boundary condition of zero normal

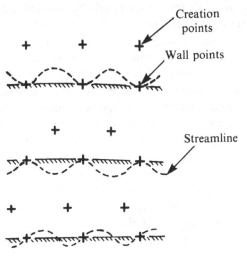

Fig. 11.6. Disposition of wall points and creation points by P. R. Spalart and A .Leonard (1981). (Reprinted with permission of the American Institute of Aeronautics and Astronautics.)

Further development of vortex cloud modelling

velocity. Thus with one equation applicable to each surface element the equation on line j of the matrix may be written $\psi_{j+1} - \psi_j = 0$ where ψ_i is the stream function at wall location i. In many ways similar to the Martensen method, solution of the equation results in M new surface vorticity values which are then deposited as new discrete vortices at the creation points, Fig. 11.6. Spalart & Leonard comment that errors derive from the fact that although the wall points remain on the same streamline once the vortices have been offset by displacements ε to their creation points, this streamline is wavy. This feature was encountered in Section 8.4.2, manifesting itself in undulating self-convection of a single vortex in close proximity to a discretised wall. Configurations (a) and (b), they argue, thus lead to greater error than (c) in which the newly created vortices are placed at one-fourth of the interval between adjacent body points. This argument does not apply to the use of Martensen's equation however, employing the Dirichlet boundary condition since, as shown in Section 8.4.1, the influence of a discrete vortex in close proximity to a surface element can best be estimated as the negative complement to its circulation around the remainder of the body surface.

Another difference in the recommendations of Spalart & Leonard lies in the choice of core radius r_0 which for their method should be not less than the interval (i.e. surface element size Δs) because of these undulations. On the other hand the present author recommends in Section 9.4.4 the use of a Rankine vortex (constant core vorticity ω) of core radius $r_0 = \Delta s / \pi$ based upon an entirely different and more logical criterion; namely that its core edge maximum induced velocity should equal the surface velocity induced by the sheet from which it is created. However Spalart & Leonard have used an order of magnitude more surface elements for their calculations, typically 600, compared with the mere 50 adopted for the calculations shown in Figs. 11.2 and 11.3. Despite this they also seem to have experienced premature stall. As illustrated by the vortex cloud plot shown in Fig. (11.7), these authors have developed a procedure to overcome this which deliberately prevents flow separation by re-assembling the vortex cloud. At each time step all of the vorticity in the boundary layer region is removed and recreated at the wall, thus artificially reducing the thickening of the boundary layer by a large amount. This of course over-rides the natural influences of convection and diffusion in the boundary layer and prevents the vortex cloud model

436

(*a*) Computation without separation control

Boundary layer discrete vortices relocated
to the vortex creation points

(*b*) Computation with separation control

Fig. 11.7. Introduction of boundary layer numerical control over separation, Spalart & Leonard (1981).

from providing a true simulation in any sense except that of a fully attached potential flow. The authors combined this technique with boundary layer integral methods to determine anticipated regions of fully attached flow and also to locate separation points. Such a technique, artificial though it may seem, can thus be justified where boundary layer integral methods are able to predict the surface flow with greater certainty than the full vortex cloud method.

The application of this quasi-potential flow analysis to a Karmann–Trefftz aerofoil at 5° incidence is compared with the exact potential flow solution in Fig. 11.8, showing remarkably good agreement except for forgiveable errors in the suction peak region.

11.3.1 NACA 0012 aerofoil in dynamic stall

Spalart & Leonard (1981) also extended their vortex dynamics method to the case of an aerofoil in dynamic stall, applied to NACA 0012 oscillating in pitch about the quarter-chord position between $\alpha_\infty = 5°–25°$ with a reduced frequency of 0.25 and $Re = 2.5 \times 10^6$. Up to this point only stationary bodies have been considered but vortex dynamics methods can be readily adapted to bodies in motion or responding to elastic constraints. This particular

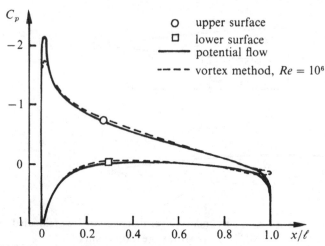

Fig. 11.8. Pressure distribution on a Karmann–Trefftz aerofoil at 5°
incidence. Comparison of exact potential flow solution with discrete vortex
method of Spalart & Leonard (1981). (Reprinted with permission of the
American Institute of Aeronautics and Astronautics.)

problem involves major boundary layer motions which influence
stability and the solution presented by these authors represents a
courageous and impressive attempt to model an extremely complex
seemingly intractible flow regime. Simple steady state boundary
layer integral methods were used, Thwaites (1949) for laminar flow
and M. R. Head's method* for turbulent flow with the Schlichting–
Granville criterion to predict the position of transition. Knowing the
pressure distribution from the previous time step the attached part
of the boundary layer was computed by boundary layer theory to
obtain the primary separation location. It was then possible to
prevent separation of the discrete vortices upstream of this location
in the manner just described and illustrated by Fig. 11.7. Downstream of separation the vortex cloud was allowed to develop
without restraint.

Space permits the presentation of only a limited sample of their
predicted vortex cloud plots during the first cycle of oscillation, Fig.
11.9, illustrating dynamic stall on the upper surface. As α rose to
24° leading edge separation was followed by reattachment and later
separation at mid-chord of a large eddy, which drifted down the
upper surface and was ultimately deposited into the wake. During

* See Cebeci & Bradshaw (1977)

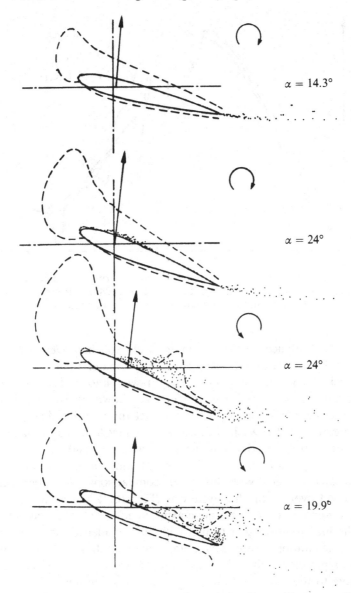

$\alpha = 14.3°$

$\alpha = 24°$

$\alpha = 24°$

$\alpha = 19.9°$

Fig. 11.9. NACA 0012 airfoil in dynamic stall, oscillating cyclically between $0° < \alpha < 25°$. Spalart & Leonard (1981). (Reprinted with permission of the American Institute of Aeronautics and Astronautics).

439

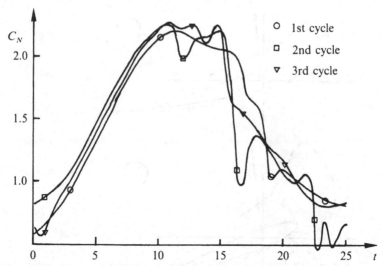

Fig. 11.10. Time evolution of normal force during dynamic stall of oscillating NACA 0012. Spalart & Leonard (1981). (Reprinted with permission of the Institute of Aeronautics and Astronautics.)

the rise in incidence the flow remained fully attached over $5° < \alpha < 23°$, the lift coefficients being significantly greater than the steady state values for the same incidence, a well known experimental phenomenon, e.g. Carr *et al.* (1977), as shown by the predicted normal force coefficients C_N over the first three cycles, Fig. 11.10. Although the stall is washed away and each cycle makes a virtually clean start, trailing edge buffetting towards the end of the cycle leads to some variation. The authors recommend averaging over several cycles to deal with this. Their comparisons with experiment were good overall, the lift hysteresis (Fig. 10.7) and moment stall being observed. However, the maximum lift predicted was lower and the lift evolution near to the maximum incidence, where upper surface separations were developing, were qualitatively different. The computations, like the experiments, were found to be very sensitive to the effect of boundary layer transition and free-stream turbulence, an allowance for the latter being incorporated with the Granville criterion used to predict the location of transition.

For these calculations each cycle involved 838 time steps of value $\Delta t = 0.03$, with an aerofoil chord of $\ell = 2.0$ and vortex core radius $r_0 = 0.006$. The numbers of discrete vortices tended to range from about 400 to 900 depending upon the size of the vortex cloud

440

regions during the cycles. A powerful scheme for speeding computations by grouping vortices into a rectilinear network of cells was given by Spalart & Leonard and will be described in Section 12.2.

11.4 Mixed vortex cloud and potential flow modelling

Enforcement of boundary layer attachment by the method of Spalart & Leonard described above is tantamount to potential flow modelling in the regions concerned. An alternative approach, to which the surface vorticity boundary integral method lends itself quite naturally, is to go directly for potential flow modelling in regions where we wish to enforce attached flow but to use full vortex cloud modelling elsewhere on the body surface. Lewis and Lo (1986) developed such a technique for application to lifting aerofoils with the deployment of airbrake spoilers for the aerodynamic control of wind turbines. A suitable starting point for such studies is the comparison of starting motions for flow past a wedge shaped body shown in Fig. 10.9 for the two basic methods of discrete vortex modelling. The first of these, which involves vortex shedding only from the two prescribed rear sharp corners A and B, assumes that the entire body surface is in potential flow, which is evaluated by the normal Martensen surface vorticity analysis. Perhaps surprisingly, the wake pattern, for what seems at first a rather crude model, agrees remarkably well with the full vortex cloud simulation, Fig. 10.9(b). Furthermore the predicted surface pressure distributions are in equally excellent agreement, Fig. 10.13. In effect the surface vorticity solution along the front faces SA and SB of the wedge with potential flow modelling are equivalent to squeezing the actual boundary layer of the vortex cloud simulation back onto the body surface in a manner similar to that undertaken by Spalart & Leonard. For this particular problem a favourable pressure gradient ensured stable numerical modelling of the front face boundary layers by vortex cloud theory, making it a good datum case for checking out the quality of the simpler potential flow model with shear layer fixed point separations.

A hybrid model may now be constructed quite simply by combining these two cases as follows, Fig. 11.11:

(i) Front face SA in potential flow with vortex sheet separation for sharp corner A of strength

$$\Delta\Gamma_A = \tfrac{1}{2}\gamma(s_A)^2 \, \Delta t \tag{11.3}$$

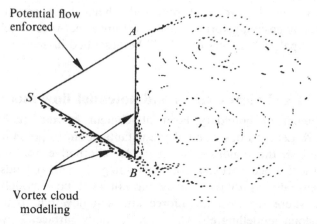

Flow visualisation after 90 time steps $\Delta t = 0.33333$
with $W_\infty = 1.0$, $\alpha_\infty = 0.0°$

Fig. 11.11. Hybrid potential flow/vortex cloud model simulating wedge wake starting motion.

(ii) **Front face *SB* and rear face *AB* full vortex cloud shedding of local strength**

$$\Delta\Gamma = \gamma(s)\,\Delta s$$

A sample flow pattern for an equilateral wedge with side lengths 1.0, taken after 90 time steps $\Delta t = 0.033\,333$ with $U_\infty = 1.0$ is shown in Fig. 11.11, with offsets $\varepsilon = 0.02$ for the vortex cloud shedding and with $\varepsilon = 0.016667$, $\phi = 30°$ for the upper surface vortex sheet shed from *A*. The most important observation from this study is the reasonable preservation of symmetry as expected from Fig. 10.9, despite the quite different vortex production models used to simulate the upper and lower front surfaces. This was borne out by the predicted lift coefficients which average over steps 50–100 to the insignificant value $-0.068\,749$.

Application of this to a long term run for the wedge is shown in Fig. 11.12. Periodicity of the wake is now firmly established as shown by the graph of lift coefficient versus time, from which the estimated Strouhal number was $St = 0.189$. As also shown by Lewis (1987c), the predicted surface pressure distribution for the hybrid model was on average symmetrical, as it should be, establishing that no unpleasant coupling side effects were created by mixing these two rather different simulation techniques.

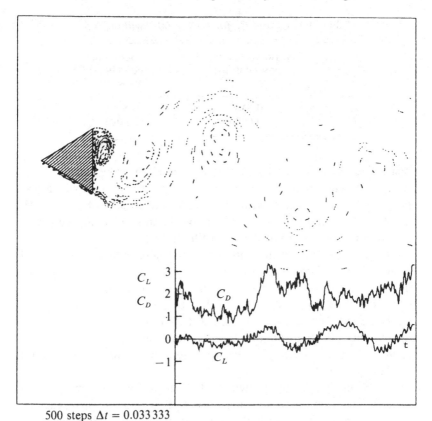

500 steps $\Delta t = 0.033\,333$

Fig. 11.12. Wake flow for a wedge shaped body with C_L and C_D curves predicted by hybrid potential flow/vortex cloud model.

11.4.1 Lifting aerofoil by the hybrid potential flow/vortex cloud method

Now we are in a position to apply the same computer code to a lifting aerofoil for which the upper surface is stabilised by enforcing potential flow while the lower surface flow is represented by full vortex cloud modelling. The predicted flow pattern and lift–drag data are shown in Fig. 11.13 for NACA 0012 with $W_\infty = 1.0$, $\alpha_\infty = 5°$, $\ell = 1.0$ taken over 200 time steps $\Delta t = 0.033\,333$, for $Re = 10^6$. Discrete vortex shedding offsets were $\varepsilon = 0.02$ for the vortex cloud and $\varepsilon = 0.016\,667$, $\phi = 30°$ for the upper surface vortex sheet. C_L and C_D data were averaged over five time steps to minimise numerical noise, although clearly there were still some

443

Table 11.1. C_L and C_D for NACA 0012 with $\alpha_\infty = 5°$

	Hybrid vortex cloud/ potential flow $Re = 10^6$	Experiment Miley (1982) $Re = 10^6$
C_L	0.5109	0.5400
C_D	0.0030	0.0090

residual undulations of C_L, probably linked to wake vortex street formation. The lift coefficient rose progressively at a rate which was generally in agreement with the published results of Spalart & Leonard (1981). In Table 11.1 average values are compared with experimental results published by Miley (1982).

The low drag coefficient value predicted here indicates the virtual elimination of form drag and associated separations as expected. The predicted lift coefficient is in reasonable agreement with experiment, the potential flow value being slightly higher, $C_L = 0.5909$. Predicted surface pressure distribution is now in very good agreement with that obtained by potential flow analysis, Fig. 11.14, representing a considerable improvement upon full vortex cloud modelling, Fig. 11.3.

Attention should be drawn to the need for care in defining trailing edge geometry, to which this hybrid method proves sensitive, since this reflects strongly upon the strength of vortex shedding from the upper surface and ultimate satisfaction of the Kutta–Joukowski condition. The data in Table 11.2 used for this calcula-

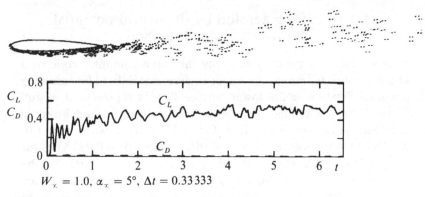

$W_x = 1.0, \ \alpha_x = 5°, \ \Delta t = 0.33333$

Fig. 11.13. Simulation of flow past NACA 0012 aerofoil by hybrid potential flow/vortex cloud model.

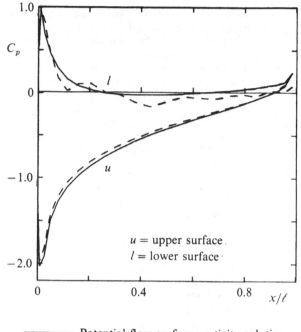

———— Potential flow surface vorticity solution

– – – – Hybrid potential flow/vortex cloud model

Fig. 11.14. Predicted pressure distributions for NACA 0012 aerofoil with $\alpha_\infty = 5°$.

tion were based upon the circular chordwise distribution of data points recommended in Section 2.5.2, Fig. 2.8. However, with 50 elements the consequent trailing edge elements were then only of length $\Delta s/\ell = 0.004$. Better results are obtained if the last three elements next to the trailing edge on each surface are merged into a single element, which is then of length 0.0356, a scale similar to that of the mainstream drift per time step.

11.4.2 Aerofoil with air brake spoiler by the hybrid potential flow/vortex cloud method

Using the same computer code Lewis (1986), (1987c) undertook investigations of the lift/drag behaviour of NACA 0025 with the presence of an air brake spoiler on one surface, in relation to requirements for the aerodynamic control of wind turbines, Fig.

Table 11.2. *NACA 0012 data points used for the hybrid potential flow/vortex cloud computations*

x	y upper $(+)$ y lower $(-)$
0.000 000	0.000 000
0.003 943	0.010 637
0.015 708	0.021 910
0.035 112	0.030 758
0.061 847	0.038 974
0.095 492	0.046 079
0.135 516	0.051 949
0.181 288	0.056 219
0.232 087	0.058 895
0.287 110	0.060 000
0.345 492	0.059 645
0.406 309	0.057 800
0.468 605	0.054 872
0.531 395	0.050 884
0.593 691	0.046 157
0.654 508	0.040 938
0.712 890	0.035 378
0.767 913	0.029 725
0.818 712	0.024 133
0.864 484	0.018 807
0.904 509	0.013 917
0.938 153	0.009 621
0.964 888	0.005 841
1.000 000	0.000 000

11.15. With full vortex cloud modelling the upper surface was subject to severe numerical stall and no means to alleviate this has yet been achieved except the present one of imposing potential flow. The upper surface vortex sheet was shed at the trailing edge according to (11.3) and the lower surface including the spoiler was modelled by full vortex cloud theory. Vortex shedding offsets were as before. A full exposition of these investigations is given in the above mentioned references. Sample solutions are shown in Figs. 11.15 and 11.16.

As may be seen from the streak line plot, the pulsating vortex flow shed from the flap induces a periodic vortex roll-up and shedding from the upper surface to form ultimately a broad oscillating wake flow. In response to this the lift coefficient builds up

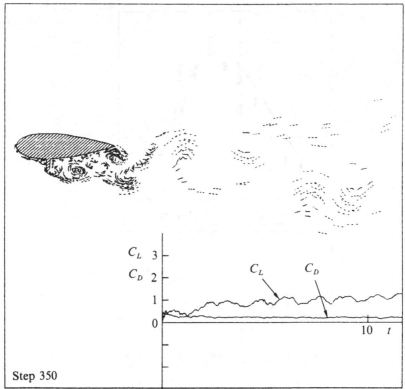

$\Delta t = 0.033\,333,\ W_\infty = 1.0,\ \alpha_\infty = 5°$

Fig. 11.15. Flow past NACA 0025 with flap by the hybrid vortex cloud method.

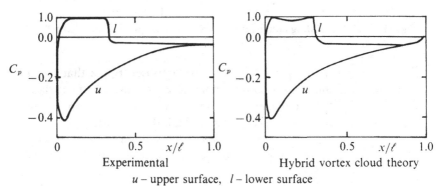

Experimental   Hybrid vortex cloud theory

u – upper surface, l – lower surface

Fig. 11.16. Surface pressure distribution for NACA 0025 with air brake flap for $\alpha_\infty = 10°$.

447

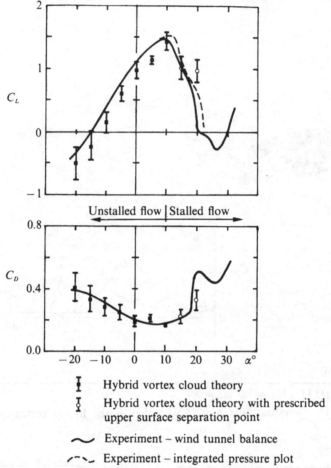

Fig. 11.17. Lift and drag coefficients for NACA 0025 with an air brake flap.

to a mean value of $C_L = 1.1$ with a superimposed ripple that has a Strouhal frequency based on total body thickness including the flap, of 0.26. The drag coefficient, dominated of course by separation from the flap, is as intended extremely high due to form drag. This is further illustrated by the pressure distribution shown in Fig. 11.16, demonstrating excellent comparison with experimental tests for an angle of attack $\alpha_\infty = 10°$.

Remarkable agreement with experiment was exhibited by this hybrid model for C_L and C_D over a wide range of incidence angles below stall, $-20° < \alpha_\infty < 10°$, Fig. 11.17. As also shown by Lewis

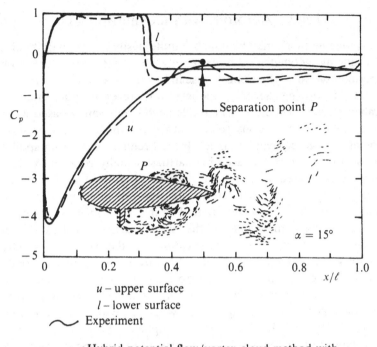

u – upper surface
l – lower surface
⌣ Experiment

,⌐⌐⌐⌐ Hybrid potential flow/vortex cloud method with
prescribed separation point P

Fig. 11.18. Pressure distribution for NACA 0025 with an air brake flap at a stalling angle of attack.

(1987c), quite good simulations may even be obtained within the stall region if the potential flow zone on the upper surface is simply limited to the region for which the boundary layer is known to remain attached. As shown by Fig. 11.17, the drag coefficient, which was the most important item required of this computation, was predicted adequately and quite a reasonable prediction of the surface pressure distribution was obtained as illustrated for $\alpha_\infty = 15°$ in Fig. 11.18. These analytical procedures were deliberately developed for evaluation at the 16-bit microcomputer scale of computation, resulting of course in limits to resolution. Thus no more than 50 surface elements were used in any of these simulations, explaining the much larger numerical noise in comparison with the work of Spalart & Leonard. These tactics lead to more rapid development of schemes but recourse to mainframe computation at tighter resolution is to be recommended for improved accuracy.

449

11.4.3   Aerofoils with moving spoilers

The above methods may be adapted quite easily to deal with bodies
of changing geometry. An extremely thorough survey of both
experimental and theoretical research into the vortex dynamic
effects of moving aerofoil spoilers was recently published by
Graham (1988). Theoretical methods have been mainly based upon
conformal transformations selected to refer the complex aerofoil
geometry to an equivalent circle plane. Frequently the thin spoiler
will be modelled by surface singularities, usually sources. Vortex
shedding is limited to the separation of discretised vortex sheets
from the spoiler tip and from the aerofoil trailing edge, the
remainder of the body being treated as if it were in potential flow.
This is precisely the same as the simple discrete vortex model
presented in Section 8.5. Application of the surface vorticity
boundary integral method however removes restrictions on body
shapes imposed by conformal transformation modelling to solve the
potential flow part of the sequence. On the other hand special care
must be taken with rotating bodies when applying the Martensen
method, rules for dealing with the implied relative eddy having
been explained in Section 3.5.2 in relation to rotating cascades. A
sample of Graham's results is shown in Fig. 11.19 for starting
motion for a symmetric Joukowski aerofoil at zero angle of attack,
with a moving spoiler on the upper surface.

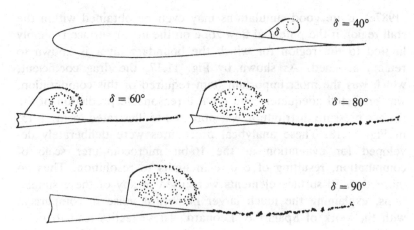

Fig. 11.19. Separation from a moving spoiler at $u_t/U_\infty = 0.19$ from a
starting angle of $\delta = 30°$. By courtesy of Dr J. M. R. Graham, Imperial
College of Science and Technology.

11.5 Application of vortex cloud modelling to turbomachinery blade rows

Vortex cloud modelling offers tremendous potential for theoretical analysis of a number of important problems in turbomachines including the prediction of rotating stall in compressors and vibrations induced by blade row wake interaction. A basic scheme for vortex cloud modelling of cascades has been developed by Lewis (1988) and the background theory will be presented in the next section. However the first prediction of the complex problem of cascade rotating stall was published by Lewis & Porthouse (1981) and Porthouse (1983) followed by some impressive fairly comprehensive studies by Spalart (1984). These will be briefly covered in Section 11.5.2.

11.5.1 Vortex cloud analysis for periodic flow through linear cascades

The flow through a stalled cascade will in general vary from blade to blade due to the randomness associated with turbulence and eddy growth. Each blade would then need to be modelled individually using the multiple body surface vorticity method described in Section 2.7. Adaption of the vortex cloud program used above is then an extremely simple matter requiring only minor adjustments to the profile data preparation procedure. For unstalled blade rows under bombardment from the wakes of other upstream blade rows of differing pitch, a large number of blades N would likewise need to be modelled individually by M surface elements each, resulting in NM boundary elements, each shedding one discrete vortex per time step, a vast computational undertaking. On the other hand the flow through an unstalled linear cascade with a uniform stream at inlet is truly periodic, leading to some simplification which we will now consider as a sensible starting point.

As shown in Section 2.6 Martensen's integral equation (1.21) for single aerofoils can be applied directly to rectilinear multiple body problems provided the coupling coefficient is replaced by the 'cascade coupling coefficient' (2.53). This can be formed from the following expressions derived in Section 2.6.1 for the velocities induced by a periodic array of vortices of strength Γ equally spaced along the y-axis with pitch t, Fig. 2.13.

$$u = \frac{\Gamma}{2t} \left(\frac{\sin \frac{2\pi y}{t}}{\cosh \frac{2\pi x}{t} - \cos \frac{2\pi y}{t}} \right) \\[2em] v = -\frac{\Gamma}{2t} \left(\frac{\sin \frac{2\pi x}{t}}{\cosh \frac{2\pi x}{t} - \cos \frac{2\pi y}{t}} \right) \right\}$$

(11.4)

The coupling coefficient giving the boundary condition of flow parallel to the surface at m due to a unit strength vortex at n then becomes

$$k_{mn} = \frac{1}{2t} \left[\frac{\sinh \frac{2\pi}{t}(x_m - x_n) \sin \beta_m - \sin \frac{2\pi}{t}(y_m - y_n) \cos \beta_m}{\cosh \frac{2\pi}{t}(x_m - x_n) - \cos \frac{2\pi}{t}(y_m - y_n)} \right]$$

(11.5)

Martensen's equations for this problem may then be written, suitably adapting the single body equations (8.26) to read

$$\sum_{n=1}^{M} k_{mn} \gamma(s_n) \, \Delta s_n = -(U_\infty \cos \beta_m + V_\infty \sin \beta_m)$$

$$-\sum_{j=1}^{Z} \Delta \Gamma_j k_{mj}$$

(11.6)

where the same unit vortex cascade coupling coefficient k_{mj} can be used for the velocities induced at the body surface by the discrete vortices $\Delta \Gamma_j$ forming the cloud. For reasons explained in Section 10.4.1 we must also assert vorticity conservation through

$$\sum_{n=1}^{M} \gamma(s_n) \, \Delta s_n + \sum_{j=1}^{Z} \Delta \Gamma_j - \Gamma_{\text{circ}} = 0$$

(11.7)

where Γ_{circ} is the cumulative strength of all vortices which are snuffed out if they accidentally enter the body contour. Adding this to each Martensen equation *after back diagonal correction of the*

452

Vortex cloud modelling to turbomachinery blade rows

coupling coefficients, we have, finally

$$\sum_{n=1}^{M} (k_{mn} + 1)\gamma(s_n) \, \Delta s_n = -(U_\infty \cos \beta_m + V_\infty \sin \beta_m)$$

$$-\sum_{j=1}^{Z} (k_{mj} + 1) \, \Delta\Gamma_j + \Gamma_{\text{circ}} \qquad (11.8)$$

which is identical to the vortex cloud equations for single bodies, equations (10.18), apart from the coupling coefficient formulation.

Although the single body and cascade numerical forms are almost the same, special consideration must be given to fluid deflection and velocity triangles in dealing with cascades. As illustrated in Fig. 11.20, and proved in Section 2.6.2 which deals with cascade dynamics, W_∞ is the vector mean of the inlet and exit velocities W_1 and W_2. The above formulations are then correct for potential flow modelling involving W_∞ as thus defined and with $\Delta\Gamma_j$ and Γ_{circ} put to zero.

If we wish to extend the vortex cloud method to deal with cascades on the other hand, it is necessary to replace W_∞ by the inlet velocity $W_1 = U_1$ in (11.8). The reason for this can be seen from Fig.

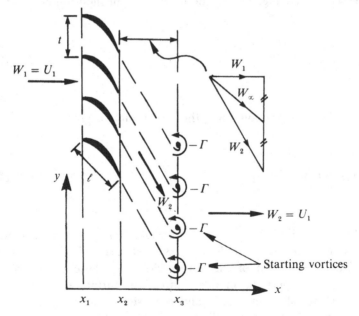

Fig. 11.20. Flow from rest through a turbine cascade including starting vortices.

453

11.20 which reminds us that the whole of the vorticity including the starting vortex system is retained in the flow field from $T = 0$ onwards. Thus some distance downstream of the starting vortices $x > x_3$ the velocity returns to the inlet value W_1. In effect W_1 becomes the vector mean velocity for the complete vortex cloud system. This observation is consonant with the need for an imposed constrain upon the Martensen equations to ensure the conservation of vorticity, namely (11.7).

Lift and form drag are directly calculable from the integrated surface pressure distribution, the latter being obtained by the method described in Section 10.5. Alternatively the blade circulation Γ may be estimated from the summation

$$\Gamma = - \sum_{n=1}^{N} \Delta \Gamma_n \quad \text{for} \quad x_1 < x < x_2 \tag{11.9}$$

which represents the sum total of all discrete vortices captured within one blade passage between the leading and trailing edge planes x_1 and x_2, Fig. 11.20. The lift coefficient then follows from

$$C_L = \frac{2\Gamma}{\ell W_\infty} \tag{11.10}$$

and, as shown in Section 2.6.1, the outlet angle is given by

$$\tan \beta_2 = \tan \beta_1 + \frac{\Gamma}{2tU_1} \tag{11.11}$$

Scheme 2, Fig. 10.10, is appropriate for the vortex cloud procedure, introducing random walks to simulate viscous diffusion and Euler convection of the vortex cloud. Because the coupling coefficients are periodic in the y direction, surface elements and discrete vortex shedding need only be considered for a single blade. Convection velocities then follow from (11.4). Thus the convection velocities experienced by a vortex at (x_m, y_m) due to a discrete vortex $\Delta\Gamma_n$ at (x_n, y_n) and its periodic array, with pitch t, will be

$$
\left.
\begin{aligned}
U_{mn} &= \frac{\Delta\Gamma_n}{2t} \left[\frac{\sin\dfrac{2\pi}{t}(y_m - y_n)}{\cosh\dfrac{2\pi}{t}(x_m - x_n) - \cos\dfrac{2\pi}{t}(y_m - y_n)} \right] \\[2em]
V_{mn} &= -\frac{\Delta\Gamma_n}{2t} \left[\frac{\sinh\dfrac{2\pi}{t}(x_m - x_n)}{\cosh\dfrac{2\pi}{t}(x_m - x_n) - \cos\dfrac{2\pi}{t}(y_m - y_n)} \right]
\end{aligned}
\right\} \quad (11.12)
$$

Experimental investigations were made by J. M. Hill (1971), (1974) of the flow through a turbine nozzle cascade with the following geometry

Stagger angle λ	$-42.7°$
Camber angle θ	$-47.7°$
Camber line	$-$parabolic
x/l for maximum camber	-0.4
Pitch chord ratio t/l	-0.752
Inlet flow angle β_1	$-0°$
Base profile	$-$NGTE (Smith & Johnson (1966))

A comparison of the surface pressure distribution with predictions by vortex cloud theory is shown in Fig. 11.21, after 80 time steps of magnitude $\Delta t = 0.025$ with inlet velocity $W_1 = U_1 = 1.0$ and for a Reynolds number based on W_∞ of 1.5×10^5 applicable to the experimental tests. The blade profile was modelled by 40 elements and arrangements were made for the merging of discrete vortices in close proximity as follows

for all x	mergers if gap $<0.005\,75\ell$
$x > 1.10237\ell$	mergers if gap $<0.017\,23\ell$
$x > 2.93966\ell$	mergers if gap $<0.028\,72\ell$

In this way most of the discrete vortex elements were retained in the blade boundary layers, the above lowest constraint limiting maximum convection velocities to a value no greater than W_∞. The predicted surface pressure distribution using the Martensen surface vorticity program No. 2.4 (see Appendix) is also shown in Fig. 11.21. For this case with decreasing pressure gradients over most of the surface, the potential flow method should be and was good.

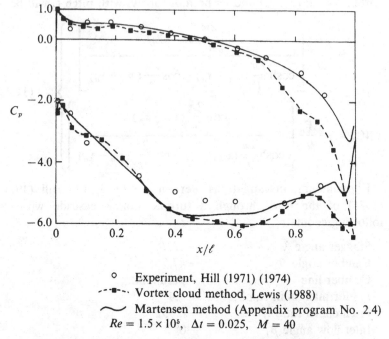

○   Experiment, Hill (1971) (1974)

--■--· Vortex cloud method, Lewis (1988)

⌒   Martensen method (Appendix program No. 2.4)

$Re = 1.5 \times 10^5$, $\Delta t = 0.025$, $M = 40$

Fig. 11.21. Prediction of surface pressure distribution for a N.G.T.E. turbine cascade. By courtesy of International Journal of Turbo and Jet Engines.

Vortex cloud analysis averaged over the last twenty time steps was likewise reasonable except in the trailing edge region where static pressures were depressed, probably due to periodic wake formation as encountered for single aerofoils and discussed in Section 11.2. The predicted outlet angle β_2 was 61.0° which compared with 64.12° according to Martensen's analysis and an experimental value of 64.6°.

The main reason for this error is thought to be the strong suction peak over the trailing edge leading to excessive form drag. For this reason a wake survey was undertaken in the manner illustrated in Fig. 11.22 which also shows the predicted streak line plot. A traverse plane $\bar{Y}\bar{Y}$ is defined downstream of the trailing edge, comprising a row of P small boxes each of area $\Delta A = \Delta y\, \Delta x$. If solutions are surveyed over S time steps, summing the discrete vortices which fall into each box, then for the ith box the local

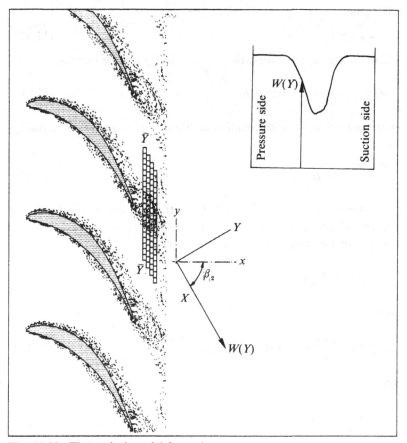

Fig. 11.22. Theoretical model for wake survey.

vorticity strength on average will be

$$\omega(y_i) = \frac{1}{S\,\Delta A}\sum_n \Delta\Gamma_{ni} \qquad (11.13)$$

Assuming that we are sufficiently downstream of the trailing edge for static pressure to be uniform, then the wake vorticity is given by

$$\omega = \frac{dW}{dY} \qquad (11.14)$$

where W is the local wake velocity and (X, Y) are coordinates along and normal to the wake. Thus y and Y are connected through the

457

wake angle β_2 by

$$Y = y \cos \beta_2 \tag{11.15}$$

Combining these results, the wake velocity profile becomes

$$W(y) = \cos \beta_2 \int_0^y \omega \, dy + K \tag{11.16}$$

where the constant of integration follows from mass flow continuity since

$$\int_0^t W(y) \cos \beta_2 \, dy = U_1 t \tag{11.17}$$

Combining (11.16) and (11.17), further reduction and use of trapezoidal integration leads to an expression for wake velocity suitable for numerical computation, namely

$$W(i) = \cos \beta_2 \left[I(i) + \frac{U_1}{\cos^2 \beta_2} - \frac{1}{t} I_t \right] \tag{11.18}$$

where

$$I(i) = \Delta y \sum_{j=1}^{i} \omega(j) \quad i = 1, 2, 3, \text{ etc.} \tag{11.19}$$

and

$$I_t = \Delta y \sum_{j=1}^{P} I(j) \tag{11.20}$$

To increase accuracy a series of staggered boxes may be introduced, lined up with the exit flow direction, permitting averaging over a larger area of the wake. The turbine cascade loss coefficient based upon exit dynamic head then follows from the definition

$$\zeta_2 = \frac{\Delta p_{0_{\text{loss}}}}{\frac{1}{2}\rho W_2^2} = \frac{1}{t} \int_0^t \frac{W(y)}{W_2} \left\{ 1 - \left(\frac{W(y)}{W_2} \right)^2 \right\} dy$$

$$= \frac{\Delta y}{t} \sum_{i=1}^{P} \frac{W(i)}{W_2} \left\{ 1 - \left(\frac{W(i)}{W_2} \right)^2 \right\} \tag{11.21}$$

An example survey for the blade row under consideration is shown in Fig. 11.21 resulting in a predicted loss coefficient of 0.105. This value is in fact badly in excess of the known loss coefficient which is within the range 0.035 to 0.05, a result which confirms earlier concern over the excessive form drag likely to be caused by poor resolution of the trailing edge flow.

Lewis (1988) also applied this method to a fan cascade, results being ruined for this diffusing flow by premature numerical stall. Use of the hybrid model described in Section 11.4 to stabilise the suction surface however led to excellent predictions of surface pressure. This vortex cloud method of cascade simulation also offers great future scope for studies of blade wake interactions in multi-blade row situations, including acoustic radiation which has already been attempted for bluff bodies by Hourigan *et al.* (1986). These authors have successfully coupled vortex cloud analysis to solution of the wave equation for studies of acoustically excited resonances caused by vortex shedding from the trailing edge of a thick plate aligned parallel to the flow within a duct. This work will be reviewed in Section 11.6.

11.5.2 Rotating stall in compressors

Lewis & Porthouse (1981), Porthouse (1983), Lewis & Porthouse (1983b), published the first simulation of compressor rotating stall shown in Fig. 11.23 for a three bladed fan with symmetrical NACA 0012 aerofoils set at a stagger angle of 60° and with an inlet angle $\beta_1 = 85°$ and a pitch/chord ratio of 1.5. Representation of the unwrapped blade into a rectilinear cascade in this case requires a periodic solution over three blade pitches, the outcome of which at this sharp angle of attack is the development of one stall cell for every three blades. As time progresses the cell runs along the cascade in the direction opposite to rotation. For very high stagger fan cascades a cell may remain upstream for some considerable time, although closer inspection of these plots indicates reverse flow in the nearest blade passage. Vertical and horizontal lines indicate clockwise and anticlockwise vortex rotation, and the length of each discrete vortex plotted is proportional to strength, indicating that many vortices were recombined in this computation to reduce processing time. The estimated speed of propagation of the stall cell along the cascade from this simulation was about 0.57 which is in very good agreement with experimental results published by Stenning & Kriebel (1957) and Horlock (1958).

Much more detailed and extensive vortex cloud simulations were published by Spalart (1984) for NACA 009 aerofoil cascades with

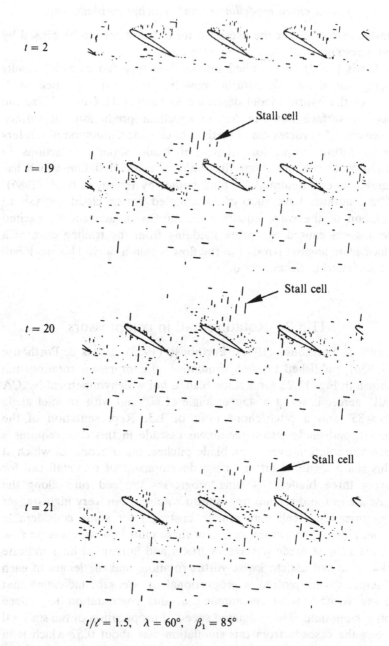

$t/\ell = 1.5, \quad \lambda = 60°, \quad \beta_1 = 85°$

Fig. 11.23. Deep rotating stall through a three bladed fan rotor of NACA 0012 profiles. Lewis & Porthouse (1983). (Reproduced from the Proceedings of the Institution of Mechanical Engineers by permission of the Council of the Institution.)

460

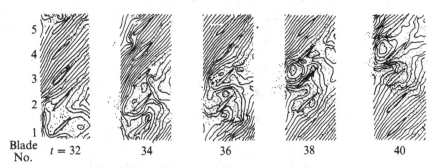

Blade No. $t = 32$ 34   36   38   40

Fig. 11.24. Stall call propagation, P. R. Spalart (1984). Reprinted with permission of the Institute of Aeronautics and Astronautics.

stagger angles of 45°, 50°, 55° and 60°, $t/\ell = 1.0$ and for angles of attack in the range 13° to 35°. A sample solution for 45° stagger, typical of an axial compressor blade row and with a typical angle of attack of 15° (i.e. $\beta_1 = 60°$) is shown in Fig. 11.24 where the interpolated streamline contours bring out more clearly the local presence and propagation of the stall cell along the line of the cascade. In this case calculations were undertaken on a Cray-1 computer using 80 blade locations (pivotal points) and 300 discrete vortices per blade, for time steps $\Delta t = 0.02$, inflow velocity $W_1 = 1.0$ and blade chord $\ell = 1.0$. The numerical model for single aerofoils described in Section 11.3 and based upon solution for the stream function was used here including the introduction of a finite core radius for each discrete vortex to effectively smooth out the rotational flow field. Following Lamb (1945) the stream function for a periodic array of vortices with pitch p in the y direction was given as

$$\psi = \frac{\Gamma}{2\pi} \log \left[\left| \sin\left(\frac{2\pi i z}{p} \right) \right| \right] \tag{11.22}$$

In order to economise, the flow pattern was assumed to repeat every fifth blade by setting $p = 5t$ and providing 5×80 equations, one for each body pivotal point. To economise further, discrete vortices were shed only in separated flow regions, boundary layer theory being used where possible to detect separation points. The smoothing of the velocity field by the use of a 'core', both to control convection of nearby vortices and for integration of streamline

461

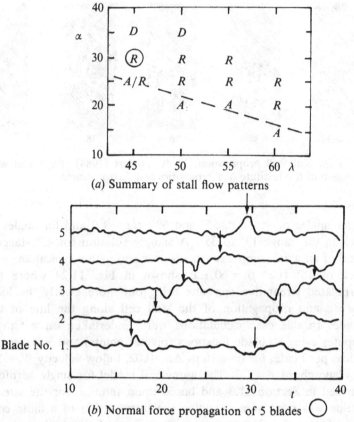

(a) Summary of stall flow patterns

(b) Normal force propagation of 5 blades

Fig. 11.25. Propagating stall in a compressor blade row, Spalart (1984). (Reprinted with permission of the Institute of Aeronautics and Astronautics.)

patterns, was accomplished by rewriting the above expression

$$\psi = \frac{\Gamma}{4\pi}\log\left[\left|\sin\left(\frac{2\pi i z}{p}\right)\right|^2 + r_0^2\right] \tag{11.23}$$

from which derivatives yield the velocity components. For these calculations a value $r_0 = 0.006$ was used.

Shown in Fig. 11.25(a) is a summary of flow regimes for a range of staggers λ and angles of attack α (angle between W_1 and the chord line); '*A*' means attached unstalled flow, '*R*' rotating stall and '*D*' deep stall in which all blades are simultaneously stalled. For the case considered in Fig. 11.24, plots of blade normal force versus

462

time are shown in Fig. 11.25(*b*). From the latter, which were filtered to remove numerical noise, one can readily discern the propagation of stall along the blade row reflected by a surge in the normal force indicated by an arrow. The propagation velocity according to this was roughly 0.42 W_1 which is a little low but quite reasonable in comparison with the experimental results of Stenning & Kriebel (1957). The flow patterns reveal, in accord with the findings of Porthouse, that the stall cell may for some period of the traverse remain upstream of the cascade as visualised for $t = 38$ and 40, Fig. 11.24.

11.6 Flow induced acoustic resonances for a bluff body in a duct

Hourigan *et al.* (1986) (1987) and Welsh *et al.* (1984) have investigated the problem of acoustic resonance caused by the vortex street downstream of a thick plate located parallel to the axis of a wind tunnel. These authors employed the method of Lewis (1981) for an isolated body to simulate the vortex shedding process by shedding discrete vortices from two fixed separation points. Vortex convection was carried out by a second-order Euler scheme assuming discrete Rankine vortices with core radius $r_0 = 0.055H$, H being the plate thickness. A time step of $0.05(H/U_\infty)$ was used, new vortices being shed every fourth time step. The plate and duct geometry are recorded in Fig. 11.26(*a*) together with a typical wake flow pattern.

The acoustic mode to be modelled is a standing wave 'organ pipe' resonance normal to the wind tunnel section. For low Mach number flows the acoustic pressure then satisfies the wave equation

$$\frac{\partial^2 p}{\partial t^2} = c^2 \nabla^2 p \tag{11.24}$$

where c is the speed of sound. The time variable may be removed from the wave equation by substituting the standing wave solution $p = \phi e^{i2\pi ft}$ where ϕ is the amplitude and f the frequency. ϕ then satisfies the Helmholtz equation

$$\nabla^2 \phi + \left(\frac{2\pi f}{c}\right)^2 \phi = 0 \tag{11.25}$$

463

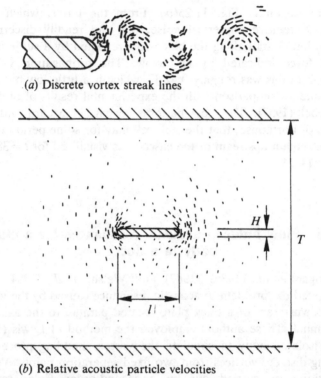

(a) Discrete vortex streak lines

(b) Relative acoustic particle velocities

Fig. 11.26. Duct acoustic resonance due to a thick plate shedding a vortex street wake, Hourigan *et al.* (1986). By courtesy of C.S.I.R., Australia.

The boundary conditions for the transverse modes are

$$n \cdot \nabla \phi = 0 \tag{11.26}$$

on rigid surfaces and

$$\phi = 0 \tag{11.27}$$

on the duct midline, except for that region occupied by the plate. Triangular finite elements were used to solve this elliptic boundary value problem resulting in the relative acoustic particle velocity field illustrated in Fig. 11.26(b). This represents the simplest or β-mode as defined by Parker (1966) for transverse acoustic resonance. The predicted resonant frequency was found to be 2.4% lower than the corresponding mode with the plate removed, in line with experimental investigations. As a wake centre-line boundary condition the amplitude of the acoustic particle velocity at the trailing edge was

464

set to $0.25U_\infty$ in line with experimentally observed values and the acoustic velocities were added into the flow field when undertaking the next vortex cloud convection step.

The release of large-scale vortex structures from the plate trailing edge was found to be 'locked-on' to the acoustic frequency, although, according to Hourigan *et al.* (1986), the duct velocity U_∞ was chosen to give a Strouhal number of 0.21 typical for natural vortex street generation. Following Howe (1975) the rate at which a vortex of vorticity strength ω moving with velocity **v** does work on a sound field with local acoustic particle velocity **u** is given by

$$P = |\omega|\,|\mathbf{v}|\,|\mathbf{u}|\sin\alpha \qquad\qquad (11.28)$$

where ω is assumed normal to **u** and **v** and α is the angle between **u** and **v**. For net positive power to be transferred to the acoustic field to feed the resonance, an imbalance in P must occur over an acoustic cycle. Since a constant acoustic amplitude is known to occur on the centre-line as assumed also in these calculations, the imbalance must arise from the variations in strength, velocity amplitude or varying directions of the developing vortices which form the vortex street wake. The acoustic power output for a large-scale vortex structure during its formation over one acoustic

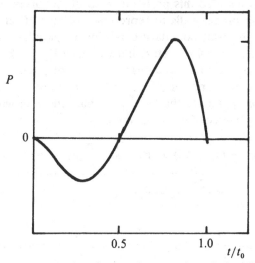

Fig. 11.27. Power generation due to the growth and shedding of a large scale vortex structure over acoustic cycle period t_0, Hourigan *et al.* (1986). By courtesy of C.S.I.R., Australia.

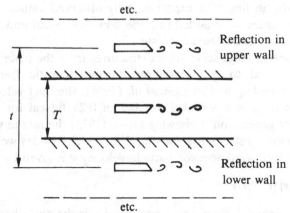

etc.

Reflection in
upper wall

Reflection in
lower wall

etc.

Fig. 11.28. Use of cascaded reflections to achieve side wall boundary conditions for vortex flow through a duct.

cycle is shown in Fig. 11.27, revealing a net positive imbalance as expected.

Although these authors accounted for wind tunnel blockage in deriving their finite element solution for the acoustic fields, with fully stated boundary conditions, they ignored blockage effects for the vortex cloud analysis using only isolated single body theory. Spalart (1984) handled this problem by locating vortex singularities along the wind tunnel walls to represent blockage effects correctly. Alternatively the wall boundary condition of parallel flow may be obtained by the reflection system illustrated in Fig. 11.28 which can be modelled as two interlaced cascades each of pitch $t = 2T$. Since the solution is the same for all wall reflections but inverted, (11.6) may simply be applied to the body plus one reflection with the new combined coupling coefficient written as

$$
k_{mn} = \frac{1}{2t} \left[\frac{\sinh \dfrac{2\pi}{t} (x_m - x_n) \sin \beta_m - \sin \dfrac{2\pi}{t} (y_m - y_n) \cos \beta_m}{\cosh \dfrac{2\pi}{t} (x_m - x_n) - \cos \dfrac{2\pi}{t} (y_m - y_n)} \right]
$$

$$
\hspace{4cm} (11.29)
$$

$$
- \frac{1}{2t} \left[\frac{\sinh \dfrac{2\pi}{t} (x_m - x_n) \sin \beta_m - \sin \dfrac{2\pi}{t} (y_m + y_n) \cos \beta_m}{\cosh \dfrac{2\pi}{t} (x_m - x_n) - \cos \dfrac{2\pi}{t} (y_m + y_n)} \right]
$$

466

Such blockage effects can exercise a significant influence upon vortex shedding from struts, partly due to the effective increase in local velocity and partly due to interference from the reflected periodic wakes. In view of the thinness of the plate in Hourigan's application, the effect of blockage upon vortex shedding was in fact probably negligible but could nevertheless be allowed for with relatively simple modifications to the code.

11.7 Potential for future development of vortex cloud analysis

It is difficult to forecast with certainty the future directions which vortex cloud modelling will follow but two factors provide the main driving forces. Firstly the Lagrangian form is attractive and economic for solutions of the Navier–Stokes equations for flows dominated by localised rotational regimes. Secondly the coupling of vortex cloud analysis to other phenomena is now receiving consideration. For example parallel solution of the wave equation for studies of acoustic resonance discussed in the previous section has proved possible and more attractive than the use of other techniques for solution of wake induced sound. Recently Smith & Stansby (1989a) have successfully extended the discrete vortex method to incorporate the influence of forced convective heat transfer. For such problems the analogous equations for viscous diffusion and heat transfer may be stated

$$
\left.
\begin{aligned}
\frac{\partial \omega}{\partial t} &= \nu \frac{\partial^2 \omega}{\partial t^2} \\
\frac{\partial T}{\partial t} &= \alpha \frac{\partial^2 T}{\partial t^2}
\end{aligned}
\right\}
\qquad (11.30)
$$

where T is temperature and α is the thermal diffusivity. Discrete vortex and temperature particles are handled together throughout each time step, temperature particles being created at the body surface to satisfy the condition of prescribed constant surface temperature. Heat thus released into the fluid is then subjected in discrete packets to the same convection and random walk motions as the discrete vortex blobs to which it is attached.

Other areas of future concern must be the extension of this work to compressible flows and also to rotational meridional flows in turbomachines, in which field there is still a pressing need for good models which incorporate annulus boundary shear layer development including tip leakage disturbances. The use of grids to manage meridional flows with distributed vorticity has been covered in Chapter 6. Other grid techniques for more efficient management of vortex dynamic computations will be described in the next chapter indicating also improved schemes for boundary layer calculations. These methods are likely to offer ways forward for handling other problems such as those of heat transfer in which key field variables must be calculated and tracked throughout the flow during time dependent motions.

CHAPTER 12

Use of grid systems in vortex dynamics and meridional flows

12.1 Introduction

Numerical schemes for the simulation of viscous rotational flows usually adopt one of two well known frameworks of reference, Eulerian or Lagrangian. Attention is focussed upon the whole of the relevent flow regime in Euler methods, usually by means of a spatially distributed fixed grid or cellular structure upon which to hang such data as the local velocity and fluid properties, updated at each stage of a time stepping procedure. Vortex dynamics on the other hand generally follows the alternative route of Lagrangian modelling in which attention is focussed upon individual particles as they move through the fluid. According to vortex cloud theory all disturbances in incompressible viscous flow can be linked to vorticity creation at solid boundaries, followed by continuous convection and diffusion. A cloud of discrete vortices may thus in principle be able to represent any rotational viscous fluid motion, accuracy depending upon the degree of discretisation and the quality of the convection and diffusion schemes.

The special attraction of this approach for external aerodynamic flows in particular is the removal of any need to consider the rest of the flow regime which of course extends to infinity. In such problems Euler models require the establishment of suitable grids extending sufficiently far out into space to define acceptable peripheral boundary conditions around the target flow regime. For simple body shapes such as cylinders or plates this may be straightforward enough. For bodies of arbitrary shape such as the aerofoil with flap considered in Section 11.4.2, the problem of selecting an optimum grid distribution can be extremely complex. In either event the Eulerian scheme may demand an extensive grid of some complexity to improve resolution in regions of key interest, resulting in excessive computational requirements. In view of this, selection of optimum grids has become a study in its own right and sophisticated 'adaptive grid' techniques have been developed for finite element analysis to optimise local grid resolution step by step in relation to local predicted variations of a chosen property such as, say, density in a high speed flow, Peraire *et al.* (1986).

Vortex cloud Lagrangian modelling on the other hand concentrates the solution derivation at the points of vital fluid action, namely the discrete vortices, and therefore usually provides essential flow information just within the actual area of interest. Such an approach thus offers the computational bonus that only the essential driving action of the fluid motion need be calculated. At any time the velocity distribution elsewhere can easily be evaluated by a Biot–Savart law integral as required, cutting out the vast wastage of information demanded by an Euler scheme. However there are computational difficulties of a different kind in full vortex cloud modelling which we have already discussed in Section 9.4.4 due to the proliferation of discrete vortices if no action is taken to economise. Thus for a body modelled by 50 surface elements taken through 200 time steps, 10^4 discrete vortex elements will have been created at the surface. If all of these were allowed to remain in the flow field 10^8 convections would need to be calculated at the next time step. In view of this several authors have considered the introduction of grids to capture or re-organise discrete vortices in order to gain economies. Two of these attributed to Spalart & Leonard (1981) and Stansby & Dixon (1983) will be described in Sections 12.2 and 12.3.

True Euler cell methods are outside the scope of this book but the boundary between the above mentioned grid management of vortex clouds and vortex in cell modelling is not always so marked as we would expect from the Lagrangian and Eulerian perspectives. Early contributions by Christiansen (1973) based on grid schemes in this no-mans-land will be referred to in Section 12.3. A study of laminar boundary layers by Lewis (1983b) making use of a vortex in cell model will be outlined in Sections 12.3 and 12.4 since this shows promise of future possibility for hybrid vortex cloud/cell schemes.

12.2   Cell-to-cell interaction method for speeding convection calculations

Spalart & Leonard (1981) have achieved considerable reductions in computation time for vortex cloud simulations of separated flows by grouping the discrete vortices into neighbourhoods and computing

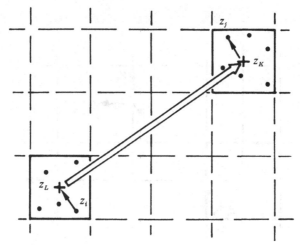

Fig. 12.1. Method for calculation of cell-to-cell interactions. Spalart & Leonard (1981).

the longer distance interactions from group to group rather than from vortex to vortex. A rectangular or square grid encompassing the active flow regime is the simplest way of accomplishing this, Fig. 12.1. The convective influence of the discrete vortex $\Delta\Gamma$ upon vortex $\Delta\Gamma_j$ may be approximated by a Taylor expansion of $1/(z_i - z_j)$ in the vicinity of $1/(z_K - z_L)$, where z_i and z_j are the complex coordinates of the vortices and z_K and z_L are the complex coordinates of their group cell centres. The coefficients of the series expansion depend only upon the fixed grid dimensions $(z_K - z_L)$ and the moments of the vorticity in each cell about its centre z_L as we will shortly demonstrate. In effect all the vortices in each group can be referred to the centre L of the cell and treated as a single vortex plus its first, second and higher order moments given by the Taylor expansion depending upon the accuracy demanded by the proximity of cells K and L. Thus many fewer interactions need be calculated and the coupling coefficients linking cell centres, being of fixed value, can be calculated once and for all at the outset. The penalty in storage is moderate and the 'link-list' technique (Hockney *et al.* (1974)) is used to manage this, no effort being spent on cells that are empty of vortices. If the cells L and K are adjacent the convection velocities are computed vortex to vortex. For distant cells one term only of the Taylor series is required. For a flow simulation involving 1000 vortices with 150 active cells, a 60%

471

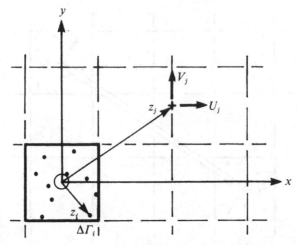

Fig. 12.2. Velocity induced at j due to a single cell centred on origin of (x, y) plane.

reduction in computational effort without loss of accuracy has been claimed for this technique.

Rather than attempt a full analysis here, we will illustrate the principles involved by considering first the velocity induced at z_j due to one discrete vortex $\Delta\Gamma_i$ at z_i belonging to a cell centred on the origin of the z plane, Fig. 12.2. The complex conjugate of the velocity at j may then be expressed

$$\bar{q}_{ij} = u_{ij} - iv_{ij} = \frac{i\,\Delta\Gamma_i}{2\pi(z_j - z_i)} \tag{12.1}$$

From Taylor's theorem or the binomial expansion we may then write

$$\bar{q}_{ij} = \frac{i\,\Delta\Gamma_i}{2\pi z_j(1 - z_i/z_j)} = \bar{q}_j\left[1 + \left(\frac{z_i}{z_j}\right) + \left(\frac{z_i}{z_j}\right)^2 + \ldots\right] \tag{12.2}$$

where $\bar{q}_j$ is the complex conjugate velocity at j induced by the discrete vortex $\Delta\Gamma_i$ shifted to the cell centre at the origin, namely

$$\bar{q}_j = u_j - iv_j = \left(\frac{\Delta\Gamma_i}{2\pi r_j}\right)ie^{-i\theta_j} \tag{12.3}$$

If we extend this to include an array of I discrete vortices contained

472

within the cell, the sum total induced velocity at j follows from

$$\bar{Q}_j = \sum_{i=1}^{I} \bar{q}_{ij}$$

$$= \frac{i \sum_{i=1}^{I} \Delta\Gamma_i}{2\pi z_j} + \frac{i}{2\pi z_j} \left[\frac{\sum_{i=1}^{I} \Delta\Gamma_i z_i}{z_j} + \frac{\sum_{i=1}^{I} \Delta\Gamma_i z_i^2}{z_j^2} + \ldots \right] \tag{12.4}$$

The first summation is equal to the total vortex strength Γ_0 contained within the cell

$$\Gamma_0 = \sum_{i=1}^{I} \Delta\Gamma_i \tag{12.5}$$

The other summations represent the first, second and higher order moments of the discrete vortices about the cell centre. For convenience we may define moment centres $Z_1, Z_2, \ldots$ for each term in the series as follows

$$Z_1 = \frac{\sum_{i=1}^{I} (\Delta\Gamma_i z_i)}{\Gamma_0}, \qquad Z_2 = \frac{\sum_{i=1}^{I} (\Delta\Gamma_i z_i^2)}{\Gamma_0} \text{ etc.} \tag{12.6}$$

Equation (12.4) then reduces for the ensemble to

$$\bar{Q}_j = U_j - iV_j$$

$$= \frac{i\Gamma_0}{2\pi z_j} \left[1 + \frac{Z_1}{z_j} + \frac{Z_2}{z_j^2} + \frac{Z_3}{z_j^3} + \ldots \right] \tag{12.7}$$

The computational saving comes from the fact that the moment terms Z_1, Z_2 etc. need only be calculated once for the cell and can be used unchanged to compute the velocity at any number of j locations. The term outside the square bracket represents the velocity Q_0 induced at j with all the discrete vortices shifted to the cell centre. The square bracket corrects this for the first, second and higher order moments due to the actual discrete vortex offsets z_i.

For simplicity let us consider the first term only in the expansion involving the centre of first moment $Z_1 = X_1 + iY_1$. This term reduces to

$$\frac{Z_1}{z_j} = a_1 + ib_1$$

where

$$a_1 = \frac{1}{r_j^2}(X_1 x_j + Y_1 y_j)$$

$$b_1 = \frac{1}{r_j^2}(Y_1 x_j - X_1 y_j)$$

$$(12.8)$$

Equation (12.7) then becomes for the ensemble of vortices

$$U_j - iV_j = (U_0 - iV_0)[1 + a_1 + ib_1]$$

so that finally we have the induced velocity components

$$U_j = (1 + a_1)U_0 + b_1 V_0$$
$$V_j = (1 + a_1)V_0 - b_1 U_0$$

$$(12.9)$$

where U_0, V_0 are the velocity components at j due to a concentrated vorticity Γ_0 at the cell centre, namely

$$U_0 = \frac{\Gamma_0 y_j}{2\pi r_j^2}, \qquad V_0 = -\frac{\Gamma_0 x_j}{2\pi r_j^2}$$

$$(12.10)$$

The analysis may be extended to higher orders with little difficulty if the nth term in (12.7) is expressed in the form

$$\frac{Z_n}{z_j^n} = \frac{X_n + iY_n}{(x_j + iy_j)^n} = a_n + ib_n$$

$$(12.11)$$

It is easier to write a computer algorithm to evaluate a_n and b_n sequentially for increasing values of n than to take the algebra further at this point. The recommended approach is first to evaluate all the required moments Z_n in the form $Z_n = X_n + iY_n$ from (12.6), which, for the nth term becomes

$$Z_n = \frac{1}{\Gamma_0} \sum_{i=1}^{I} \Delta \Gamma_i (x_i + iy_i)^n = X_n + iY_n$$

$$(12.12)$$

and then to evaluate equations (12.11). Pascal procedures to multiply or divide two complex numbers can then be used sequentially in a suitable algorithm to achieve maximum economy.

Predicted velocities due to an array of ten discrete vortices shown in Fig. 12.2 together with vortex data are given in Table 12.1 for location $x_j = 3.0$, $y_j = 1.0$ for up to ten orders of the series expansion. The point j chosen lies on the inner edge of a next but one cell, for which location five series terms are required to ensure

Table 12.1. *Discrete vortex data, Fig. 12.3 and predicted velocities at*
(3.0, 1.0)

i	x_i	y_i	$\Delta\Gamma_i$	Order	U_1	V_1
1	−0.25	−0.45	0.110	1	−0.006 119	−0.034 378
2	0.35	0.80	0.220	2	−0.005 021	−0.031 243
3	−0.47	0.65	0.520	3	−0.006 801	−0.031 272
4	0.95	−0.33	−0.590	4	−0.006 920	−0.032 018
5	0.70	0.10	0.510	5	−0.006 845	−0.032 017
6	−0.50	−0.90	−0.710	6	−0.006 814	−0.032 070
7	−0.68	0.22	−0.123	7	−0.006 781	−0.032 067
8	0.77	−0.85	0.243	8	−0.006 783	−0.032 058
9	0.15	0.22	0.400	9	−0.006 784	−0.032 058
10	−0.35	−0.10	−0.200	10	−0.006 784	−0.032 058
				exact	−0.006 784	−0.032 058

accuracy to four decimal places. The solution converges after the
ninth term to agree with the exact solution. Case studies such as the
above suggest the following rules:

(i) Adjacent cells – use vortex to vortex convection.
(ii) Next but one cell – use 5 terms of series vortex to vortex.
(iii) Next but two cells – use 4 terms.
(iv) Next but three cells – use 3 terms.
(v) Next but four cells – use 2 terms.
(vi) All other cells – use the first term.

These rules are independent of cell size and therefore have no
bearing upon grid selection. Too fine a grid structure (for example
200 cells for a field of 500 discrete vortices) will lead to little saving
in computation. Too coarse a structure will mean the more frequent
use of (i)–(v) above. A grid averaging say ten discrete vortices per
cell would seem a sensible compromise and the grid may be
redesigned to suit the flow field at each time step.

This simple analysis may be further modified as illustrated in Fig.
12.1 to refer to calculation points (x_j, y_j) also to the centre of cell K
by expanding the Taylor series in the vicinity of $1/(Z_K - Z_L)$ instead
of $1/(Z_j - Z_i)$ as we have just done. The velocities (U_0, V_0),
equations (12.10), are then dependent only upon the grid coordin-
ates Z_K and Z_L. $\bar{U}_0$, $\bar{V}_0$ coefficients for a unit strength vortex may
then be calculated once and for all at the outset resulting in yet
further savings at the expense of relatively little extra computer
storage.

12.3   Cloud-in-cell (CIC) method

Christiansen (1973) proposed an alternative convection scheme for
Lagrangian vortex cloud models involving temporary redistribution
of the scattered discrete vortices onto a fixed grid. The concepts,
developed in the 1960s for particle simulation in plasma physics, led
to the 'vortex-in-cell' and 'cloud-in-cell' methods of Birdsall & Fuss
(1969) and their later applications to vortex cloud modelling of
incompressible rotational flows by Baker (1979), Christiansen
(1973), Stansby & Dixon (1983) and others. An excellent review of
this and other important numerical techniques for improved and
economic vortex dynamics computations has been given by Leonard
(1980). More recently Smith & Stansby (1988) have shown the
computational attraction of using the vortex-in-cell method for
handling convection while using random walks to simulate viscous
diffusion, in this case for detailed resolution of the impulsive
starting flow around circular cylinders. In a later paper (1989) these
authors have shown how a turbulence model may then be incorpor-

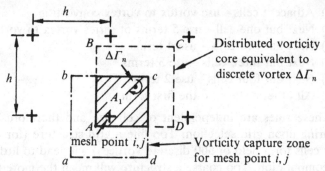

(a)   Vorticity capture area A_1 for mesh point i, j

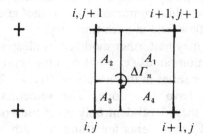

(b) Area weightings for discrete vortex at (x_n, y_n)

Fig. 12.3. Bi-linear interpolation (area weighting) method for redistribution of a discrete vortex at its local mesh points.

476

ated into vortex dynamics modelling through an 'effective viscosity' with application to high Reynolds number flows past circular cylinders.

We begin by superimposing a square mesh of side length h over the vortex cloud flow domain. Consider the discrete vortex $\Delta\Gamma_n$ located at (x_n, y_n) in cell (i, j), Fig. 12.3(a). Since $\Delta\Gamma_n$ is not in reality a discrete vortex but only a simplified representation of the local distributed vorticity, the CIC strategy is to split $\Delta\Gamma_n$ into four pieces, suitably weighted and relocated at the mesh corners. If each discrete vortex in the cell is treated likewise, the corner values may be accumulated. Thus a cloud of N discrete vortices in cell (i, j) will be replaced by only four vortices at the cell corners. Neighbouring cells will contribute to shared corners leading to yet greater reduction in the number of vortex convection calculations required. We will deal first with the distribution technique and then consider convection.

12.3.1 Vortex re-distribution to cell corners

A rational approach is to imagine each discrete vortex as having a square shaped core *ABCD*, also with grid scale dimensions h, containing uniformly distributed vorticity of strength

$$\omega_n = \Delta\Gamma_n/h^2 \tag{12.13}$$

As illustrated by Fig. 12.3(a), any of this core vorticity which lies within the region *abcd* defined by

$$
\left.
\begin{aligned}
x_i - \frac{h}{2} &\leqslant x \leqslant x_i + \frac{h}{2} \\[2mm]
y_i - \frac{h}{2} &\leqslant y \leqslant y_i + \frac{h}{2}
\end{aligned}
\right\}
\tag{12.14}
$$

namely the shaded area A_1, belongs to the mesh point $(i \cdot j)$. The discrete vortex contribution to this corner is thus given by

$$\Delta\Gamma_{n1} = \left(\frac{A_1}{h^2}\right)\Delta\Gamma_n$$

similar results being applicable to the remaining three mesh points of the cell.

A geometrically equivalent and numerically more convenient method for expressing the area weightings is shown in Fig. 12.3(*b*). The distributed vortex strength at the cell corners may then be expressed

$$\Delta\Gamma_{np} = \left(\frac{A_p}{A}\right)\Delta\Gamma_n \qquad p = 1\text{–}4 \tag{12.15}$$

where A_p is the area of the rectangle diagonally opposite to cell corner p and $A = h^2$ is the cell area.

12.3.2 Convection with grid distribution of vorticity

In many applications of this technique convection velocities have been derived from a solution of Poisson's equation for the stream function, namely

$$\nabla^2\psi = -\omega \tag{12.16}$$

where the velocity components are given by

$$u = \frac{\partial\psi}{\partial y}, \qquad v = -\frac{\partial\psi}{\partial x} \tag{12.17}$$

Following Stansby & Dixon (1983), values of ψ may be obtained at the mesh points by solving (12.16) expressed in finite difference form, namely

$$\frac{1}{h^2}(\psi_{i+1,j} - 2\psi_{ij} + \psi_{i-1,j})$$

$$+\frac{1}{h^2}(\psi_{i,j+1} - 2\psi_{ij} + \psi_{i,j-1}) = -\omega_{ij} \tag{12.18}$$

where the vorticity at mesh point (i, j) is given by

$$\omega_{ij} = \Gamma_{ij}/A^2 \tag{12.19}$$

Γ_{ij} is the accumulated vorticity following redistribution of all discrete vortices in cells sharing node (i, j). Equations (12.18) written for all (i, j) values form a banded matrix for which standard solution procedures are then available. Stansby & Dixon recommend an efficient method making use of Fast Fourier Transforms. Once the stream function has been determined velocities at the mesh points are calculated from the expressions

$$\left. \begin{aligned} u_{ij} &= \frac{1}{2h}(\psi_{i,j+1} - \psi_{i,j-1}) \\[2mm] v_{ij} &= -\frac{1}{2h}(\psi_{i+1,j} - \psi_{i-1,j}) \end{aligned} \right\} \tag{12.20}$$

and the vortex convection velocities are interpolated through the previous area weighting procedure.

$$\left. \begin{aligned} u_n &= \sum_{p=1}^{4} u(p)A_p/A \\[2mm] v_n &= \sum_{p=1}^{4} v(p)A_p/A \end{aligned} \right\} \tag{12.21}$$

Area weighting can be shown to be identical to bi-linear interpolation. It is of interest to note that the vortex redistribution method summarised by (12.15) is thus also equivalent in accuracy to bi-linear interpolation. Square cells were assumed above for simplicity but the analysis is easily extended to rectangular cells.

As an alternative to the above finite difference scheme, vortex to vortex Biot–Savart convective interactions can be calculated directly for the grid distribution of discrete vortices Γ_{ij} to provide the mesh point convection velocities (u_{ij}, v_{ij}). The bonus for a regular square shaped or rectangular grid is that a set of unit coupling coefficients may then be calculated once and for all at the outset linking mesh points and may be stored permanently in an array. Since these will be identical for any pair of mesh points distant (m, n) apart in the x and y directions additional economies of storage may be accomplished. Further insights into this strategy can be gained from the treatment of viscous diffusion by use of a diffusion coupling coefficient matrix given later in Section 12.4, where a more detailed description is provided. In effect this method which retains the use of normal vortex dynamics Biot–Savart law calculations, is an alternative to the other cell capturing technique of Spalart presented in Section 12.2 and could in some applications be more economic in computing time. It is however subject to the vortex smearing and interpolation errors implied by (12.15) and (12.21), whereas Spalart's method can be operated to any pre-

479

Table 12.2. *Convection paths for vortex* $\Delta\Gamma_A$ *of a vortex pair (See Fig. (8.4)*
by three methods

Method 1 Vortex-to-vortex (Biot–Savart) method		Method 2 Spalart's cell method		Method 3 Cloud-in-cell method (Christiansen)	
x	y	x	y	x	y
0.000 000	0.500 000	0.000 000	0.500 000	0.000 000	0.500 000
0.151 834	0.477 000	0.151 840	0.476 992	0.154 906	0.476 290
0.289 421	0.409 142	0.289 429	0.409 131	0.291 091	0.405 269
0.399 938	0.302 962	0.399 949	0.302 950	0.402 886	0.295 427
0.473 191	0.168 517	0.473 201	0.168 503	0.476 335	0.161 209
0.502 530	0.018 401	0.502 537	0.018 383	0.503 158	0.008 449
0.485 420	−0.133 444	0.485 425	−0.133 464	0.482 600	−0.145 547
0.423 618	−0.273 034	0.423 619	−0.270 052	0.415 072	−0.282 597
0.322 968	−0.387 619	0.322 969	−0.387 637	0.308 987	−0.396 258
0.192 812	−0.466 835	0.192 815	−0.466 852	0.177 586	−0.473 422
0.045 098	−0.503 614	0.045 102	−0.503 630	0.027 408	−0.505 279
−0.106 726	−0.494 791	−0.106 721	−0.494 809	−0.125 779	−0.490 971
−0.248 951	−0.441 334	−0.248 944	−0.441 354	−0.265 305	−0.430 494
−0.368 834	−0.348 215	−0.368 829	−0.348 241	−0.382 521	−0.330 795
−0.455 730	−0.223 918	−0.455 729	−0.223 951	−0.465 569	−0.203 403
−0.502 017	−0.079 639	−0.502 022	−0.079 577	−0.505 447	−0.056 777
−0.503 734	0.071 736	−0.503 750	0.071 697	−0.500 238	0.095 619
−0.460 894	0.216 792	−0.460 921	0.216 754	−0.449 680	0.238 484
−0.377 434	0.342 769	−0.337 473	0.342 735	−0.358 743	0.360 599
−0.260 828	0.438 677	−0.260 877	0.438 654	−0.236 953	0.451 920
−0.121 392	0.496 238	−0.121 448	0.496 228	−0.095 024	0.502 443

scribed level of accuracy by selecting sufficient terms in the fairly convergent Taylor expansion.

For a simple test of the accuracy of this and Spalart's method let us reconsider the self-convection of a vortex pair discussed in Section 8.2.1 (Fig. 8.4). Results are compared with vortex-to-vortex calculations (i.e. standard vortex dynamics) in Table 12.2 for second order Euler convection over ten time steps $\Delta t = 1.0$ for vortices of strength $\Delta\Gamma_A = \Delta\Gamma_B = 1.0$ distant $D = 1.0$ apart. (x, y) path coordinates are tabulated for vortex Γ_A only.

The grid region surrounding the vortex pair is that shown in Fig. 12.4 employing a 4×4 mesh structure. This imposes quite a severe

Cloud-in-cell (CIC) method

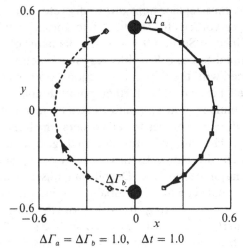

$$\Delta \Gamma_a = \Delta \Gamma_b = 1.0, \quad \Delta t = 1.0$$

Fig. 12.4. Self-convection of a vortex pair by the cloud-in-cell method with a 4 × 4 mesh.

test for both methods 2 and 3. Thus for the Spalart calculation, since the two vortices are always three cells apart, four moment terms are required in the Taylor series and an extremely accurate prediction of drift path is guaranteed. For the cloud-in-cell method the 4 × 4 grid may seem rather coarse bearing in mind that the vortex will cross one cell in only two time steps. Although the results tabulated above for method 3 are less accurate than method 2 they are nevertheless credible. A repeat of method 3 with an 8 × 8 mesh delivers much better results.

This test was based upon the convection of two vortices only. Studies undertaken by the author of the convection velocities for a cloud of 100 randomly scattered discrete vortices, on the other hand, have shown that accurate predictions by the cloud-in-cell method require that the number of cells should be no less than the number of discrete vortices. This is understandable in view of the vorticity smearing implied by the area weighting technique followed later by linear interpolation of the grid distribution of convection velocities. Finer grids will obviously lead to more accurate results but the outcome will then be increased rather than reduced computational effort. However we can argue with good reason that smeared vorticity is actually truer to reality than representation by discrete vortices. A fairer basis for testing the cellular method for vortex clouds is thus replacement of the discrete vortices by vortex blobs or Rankine vortices for the vortex-to-vortex calculation using the model outlined in Section 9.4.4. If the diameter of the Rankine core is made equal to the cell width, very much better comparisons are then obtained with the vortex convection velocities predicted by the cloud-in-cell method. The further point should be made that increasing the number of mesh points makes no increase in effort to disperse the N discrete vortices to cell corners which requires always $4N$ operations. The use of pre-calculated grid unit convection coefficients as an alternative to the solution of equations (12.18) also leads to major economies. Despite all this the use of direct vortex-to-vortex convections is probably to be preferred to the cloud-in-cell method, making use also of the alternative grid method of Spalart & Leonard described in Section 12.2 to speed calculations. Introduction of Rankine vortex cores into the latter is also perfectly acceptable.

There are two exceptions to this recommendation for which cell redistribution of the vorticity offers a distinct advantage, namely streamline plotting and the introduction of compressibility. The cloud-in-cell method gives the appearance or effect of smeared vorticity, since interpolated velocities (u_n, v_n) at any point within a cell vary smoothly across the cell and even across its boundaries unlike the velocities within a vortex cloud which become singular as (x_n, y_n) approaches any discrete vortex. Vortex cloud flow patterns are normally revealed by plotting short streak-lines traced out by the active discrete vortices. The cloud-in-cell technique will permit the plotting of streak- or stream-lines throughout the whole regime with finer detail.

Direct extension of pure Lagrangian vortex cloud modelling to high speed or compressible flows seems to offer little hope since fluid divergence due to density gradients cannot be pinned onto the discrete vortices, because it is spread throughout the whole flow field and is not by nature convective. The mesh method provides the means for introducing compressibility effects in a quite simple manner into Lagrangian vortex cloud modelling. This area of study has been barely touched upon at the time of writing but is one which offers future promise for dealing with high speed wake flows and practical gas or steam turbine cascade simulations with vortex wake interactions.

To conclude then, the grid method of Spalart & Leonard (1981) presented in Section 12.2 presents a real alternative to vortex-to-vortex convection calculations for a cloud of discrete vortices with significant savings in computational effort. Introduction of Rankine vortex cores is also perfectly acceptable. This strategy for smearing vorticity then leads to reasonable predictions of local velocity throughout the cloud. The mesh method of Christiansen is of less value for vortex cloud convection calculations unless data is required on a finer scale for more detailed flow plotting across the whole domain. On the other hand the mesh method holds the key to the introduction of other variables such as density variation present in compressible flows or the study of problems involving heat transfer with conduction and convection.

12.4   Cellular modelling of viscous boundary layers

A vortex dynamics model for flat plate boundary layers was considered in some detail in Section 9.4 resulting in reasonable predictions of the Blasius laminar boundary layer and the law-of-the-wall turbulent boundary layer profile. For adequate resolution however no less than 500 discrete vortices were required. In view of this Lewis (1982b) (1983b) investigated the use of a mesh system to reduce computational effort with application to laminar boundary layers. As illustrated by Fig. 12.5, a cellular control volume is placed over the active boundary layer region with M and N cells of equal size located along and normal to the wall respectively. Instead of random walks as in vortex cloud modelling, viscous diffusion between the cells may then be calculated analytically with some precision using the formulations derived in Section 9.2. The regular cell geometry then permits dramatic reduction in computational effort through the use of a unit diffusion matrix. This analysis will be developed over the next few sub-sections, beginning with the diffusion of a vortex sheet.

12.4.1   Numerical solution for a diffusing vortex sheet

As a first step towards simulation of a boundary layer, we shall consider a numerical simulation for the vortex sheet $\gamma(x)$ located initially along the x axis, the exact solution for which is given in Section 9.3 (9.19). For simplicity we will limit the range of vortex

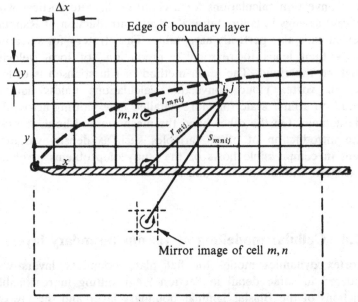

Fig. 12.5. Cellular mesh for analysis of a laminar boundary layer.

activity to $0 < x < X1$ and represent the sheet at $t = 0$ by M discrete vortices of strength $\gamma(x)\,\Delta x$ located on the wall at the mid points of the cell sides, Fig. 12.5. Thus for the time step Δt the vorticity diffusion into cell (i, j) will be given by the solution for a diffusing point vortex (9.7) with a summation of all of the M wall vortex elements.

$$\omega(i, j)_{\Delta t} = \frac{\gamma(x)\,\Delta x}{4\pi v\,\Delta t} \sum_{m=1}^{M} e^{\{-r_{mij}^2/4v\,\Delta t\}} \qquad (12.22)$$

where

$$r_{mij}^2 = \{x(m) - x(i)\}^2 + y(n)^2 \qquad (12.23)$$

As shown by Lewis (1982b), reasonable agreement with the exact solution is obtained for the central cells but vorticity values fall in magnitude towards either end of the x range due to end leakage (as also experienced in discrete vortex modelling, Fig. 9.5). However, knowing that the vorticity should be conserved at all x locations for the infinitely long diffusing vortex sheet, a vorticity conservation

484

Cellular modelling of viscous boundary layers

Table 12.3. Diffusion of a vorticity sheet

$\Delta t = 0.005, \; \nu = 1.0, \; \gamma = 2.0$				Exact solution
	Cell method			
y	$x = 0.05$	0.15	0.25	
0.05	7.041 307	7.041 307	7.041 309	7.041 309
0.15	2.590 351	2.590 352	2.590 351	2.590 352
0.25	0.350 566	0.350 566	0.350 566	0.350 566
0.35	0.017 453	0.017 453	0.017 453	0.017 454
0.45	0.000 320	0.000 320	0.000 320	0.000 320
0.55	0.000 002	0.000 002	0.000 002	0.000 002

condition may be imposed by the requirement that at $y = \infty$

$$U = \int_0^\infty \omega(y) \, dy$$

where $U = \frac{1}{2}\gamma(x)$ is the velocity induced at $y = \infty$. Thus to conserve vorticity numerically for each x_m grid location the above values may be scaled as follows.

$$\omega(i, j) := \frac{U}{\sum\limits_{n=1}^{N} \omega(i, n) \, \Delta y} \, \omega(i, n) \tag{12.24}$$

Application of this to a 6×6 square grid resulted in the prediction recorded in Table 12.3. In view of symmetry results are presented for only half of the grid. The above analysis with a single time step then delivers extremely accurate results.

As illustrated by Fig. 12.6 the diffusion of a vortex sheet over 20 successive time steps may also be predicted with accuracy by the cell method, in this case with a cell aspect ratio of $\Delta x / \Delta y = 10.0$ more suitable to later studies of boundary layers. In this case vorticity diffusion between all of the cells must also be accounted for at every time step except the first. In this model a reflection system is introduced to accomplish the wall boundary condition analogous to the discrete vortex simulation of Section 9.4.1. The vorticity diffused into cell (i, j) during time Δt, including contributions from all MN cells in the field including itself, will then be expressed

485

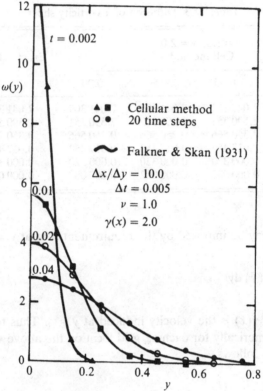

Fig. 12.6. Vorticity diffusion of a vortex sheet.

through

$$\omega(i,j)_{t+\Delta t} = \frac{\Delta x \, \Delta y}{4\pi\nu \, \Delta t} \sum_{m=1}^{M} \sum_{n=1}^{N} \omega(m,n)\{e^{\{-r_{mnij}^2/4\nu \, \Delta t\}}$$

$$+ e^{\{-s_{mnij}^2/4\nu \, \Delta t\}}\} \quad (12.25)$$

where, from Fig. 12.5

$$\left.\begin{array}{l} r_{mnij}^2 = [x(m) - x(i)]^2 + [y(n) - y(j)]^2 \\ s_{mnij}^2 = [x(m) - x(i)]^2 + [y(n) - y(j)]^2 \end{array}\right\} \quad (12.26)$$

Fig. 12.6 illustrates the extremely close agreement obtained if this procedure is applied over 20 time steps confirming the acceptability of this cellular method of viscous diffusion for transplantation into boundary layer analysis.

486

12.4.2 Diffusion coupling coefficient matrices

Evaluation of (12.25) involves $(MN)^2$ operations for each time step with a vast amount of repetition, since r_{mnij} and s_{mnij} are identical for cell pairs in similar juxtapositions. Lewis (1983b) therefore proposed the use of a diffusion matrix to evaluate these experimental terms once and for all at the commencement of a computation in the form of a diffusion coupling coefficient matrix defined for a cell of unit vortex strength, namely

$$S_{\text{mat}}(m, n) = \frac{\Delta x \, \Delta y}{4\pi v \, \Delta t} e^{\{-r_{mn}^2/4v \, \Delta t\}} \qquad (12.27)$$

where

$$r_{mn}^2 = \{x(m) - x(1)\}^2 + \{y(n) - y(1)\}^2 \qquad (12.28)$$

This technique is illustrated in Fig. 12.7 including also the use of a wall diffusing matrix for the case of boundary layer computations for which a new vortex sheet is created at the wall at the conclusion of each time step as we will see in the next section. The appropriate formulation is given by

$$W_{\text{mat}}(m, n) = \frac{\Delta x}{4\pi v \, \Delta t} e^{\{-r_{mn}^2/4v \, \Delta t\}} \qquad (12.29)$$

These matrices represent the vorticity found at location (m, n) after time Δt due to a unit vortex 0 located as illustrated in Fig. 12.7.

Now if the grid field is to be sufficiently large to contain all of the significant diffusing vorticity, it follows that vorticity diffused during the discrete time step Δt from a given cell will be of negligible value beyond some boundary such as *aefg*. In fact Δt may be selected with this in mind to achieve an acceptable diffusion zone which should contain no less than five cells in either direction for adequate representation of the diffusion function represented by (12.28). That illustrated in Fig. 12.7 is ideal. Furthermore the diffusion coupling coefficients are symmetrical about the central row and column so that the smaller matrix *abcd* is sufficient to contain all of the information needed to diffuse the vortex 0 into the region of significant diffusion *aefg*. In order to calculate the vorticity diffusion between all of the cells the 'template' *abcd* may simply be moved around the grid structure and centred on each cell in turn. The vorticity diffused from cell (m, n) into cell (i, j) is then given by

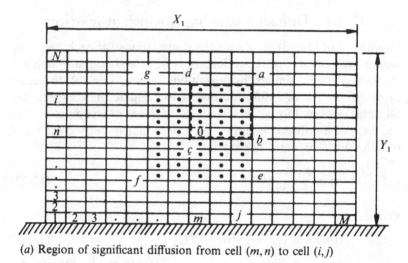

(a) Region of significant diffusion from cell (m, n) to cell (i, j)

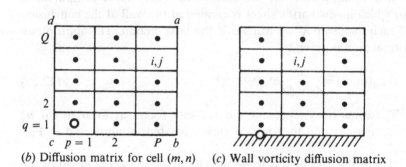

(b) Diffusion matrix for cell (m, n) (c) Wall vorticity diffusion matrix

O Diffusing vortex
• Diffused vorticity of significance

Fig. 12.7. Diffusion matrices for boundary layer calculations.

$\omega(m, n)S_{\text{mat}}(i, j)$, and the accumulated vorticity in cell (i, j) is thus

$$\omega(i, j)_{t+\Delta t} = \sum_{i=1}^{M} \sum_{j=1}^{N} \omega(m, n)S_{\text{mat}}(i, j) \qquad (12.30)$$

A method for selecting the appropriate time step for a prescribed size of diffusion matrix is as follows. Let us consider the matrix shown in Fig. 12.7(a) with P x-wise and Q y-wise elements. Focussing on the y direction let us prescribe that there should be five cells. At time t a unit vortex is located in cell $(1, 1)$ say. Then at time $t + \Delta t$ the scale of minimum diffused vorticity in cell $(1, Q)$ of the diffusion matrix will be

$$\varepsilon = \frac{\omega(1, Q)}{\omega(1, 1)} = e^{\{(Q-1)\,\Delta y\}^2/4v\,\Delta t}$$

This equation may be inverted to yield Δt for a prescribed value of ε (e.g. $\varepsilon = 10^{-6}$), namely

$$\Delta t = \{(Q - 1)\,\Delta y\}^2/\{4v \ln(1/\varepsilon)\} \tag{12.31}$$

Thus if we prescribe cell size Δy, allowable diffusion error ε and diffusion matrix dimension Q, the appropriate time step is determined. Should we alternatively wish to prescribe greater time steps, (12.31) may be rearranged to yield the required number of cells Q.

$$Q = 1 + \text{round}\left(\frac{2}{\Delta y}\right)\sqrt{[v\,\Delta t \ln(1/\varepsilon)]} \tag{12.32}$$

All this presupposes that $P\,\Delta x > Q\,\Delta y$ but a similar check in the x direction can be used to determine the appropriate value for P which should be no less than 3.

12.4.3   Boundary layer simulation by the cell method

Our next aim will be to simulate a steady boundary layer by allowing it to develop from the initial conditions of an irrotational mainstream $U(x)$ switched on at $t = 0$. The starting condition is thus equivalent to a vortex sheet $2U(x)$ on the wall surface which is allowed to diffuse over the initial time step Δt as just described. Thereafter a time stepping sequence is continued until the growing boundary layer settles down to its steady state condition. This procedure may be summarised as shown in the flow diagram below.

The procedures (iii) and (iv) in the flow diagram are as described in Section 12.4.2 in relation to simulation of a diffusing vortex sheet. For boundary layer flows the presence of an additionally imposed mainstream $U(x)$ introduces major convective motions which must be considered next in relation to procedure (v) of the flow diagram.

As illustrated by Fig. 12.8, the time convective process requires the calculation of (u, v) velocities for cell (i, j) under the influence

Use of grid systems in vortex dynamics

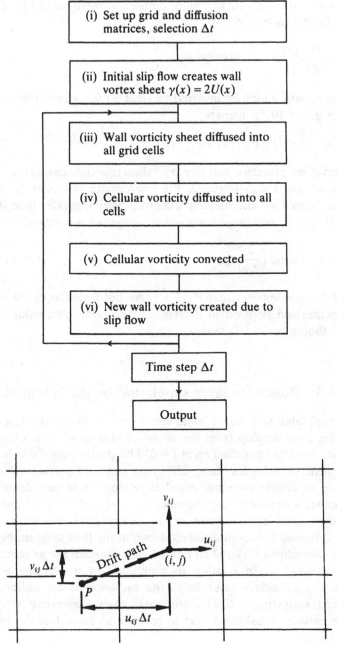

(i) Set up grid and diffusion matrices, selection Δt

(ii) Initial slip flow creates wall vortex sheet $\gamma(x) = 2U(x)$

(iii) Wall vorticity sheet diffused into all grid cells

(iv) Cellular vorticity diffused into all cells

(v) Cellular vorticity convected

(vi) New wall vorticity created due to slip flow

Time step Δt

Output

Fig. 12.8. Vorticity convection into cell $(i.j)$.

of $U(x)$ and all other cells, namely

$$\left.\begin{aligned}
u_{ij} &= U(i) + \sum_{m=1}^{M} \sum_{n=1}^{N} \frac{\omega(m, n)[y(i) - y(n)] \, \Delta x \, \Delta y}{\pi r_{mnij}{}^2} \\
v_{ij} &= -\sum_{m=1}^{M} \sum_{n=1}^{N} \frac{\omega(m, n)[x(i) - x(n)] \, \Delta x \, \Delta y}{2\pi r_{mnij}{}^2}
\end{aligned}\right\}
\tag{12.33}$$

We may then estimate the location P from which fluid is convected into location (x_i, y_i) during Δt, namely

$$\left.\begin{aligned}
x_P &= x(i) - u_{ij} \, \Delta t \\
y_P &= y(i) - v_{ij} \, \Delta t
\end{aligned}\right\}
\tag{12.34}$$

This will enable us to interpolate the vorticity ω_P and update cell (i, j). During the time step the average cell vorticity will thus be

$$\omega(i, j) = [\omega(i, j)_t + \omega_P]/2 \tag{12.35}$$

and this value should be used during the next diffusion process. At the end of the time step the updated cell vorticity is given by

$$\omega(i, j)_{t+\Delta t} = \omega_P \tag{12.36}$$

A good deal of calculation is involved in the evaluation of equations (12.33) for all cells which may be eliminated for laminar boundary layers by adopting the usual assumption that v_{ij} is zero. Equation (12.33a) may then be replaced by

$$u_{ij} = \int_0^{y_i} \omega \, dy \tag{12.37a}$$

A suitable numerical form yielding the velocity at the mid-cell points is then

$$\left.\begin{aligned}
u_{ij} &\approx \Delta y \left[\tfrac{1}{2}\omega(i, j) + \sum_{s=1}^{j-1} \omega(i, s) \right] && \text{for } j > 1 \\
&\approx \frac{\Delta y}{2} \omega(i, 1) && \text{for } j = 1 \\
v_{ij} &\approx 0
\end{aligned}\right\}
\tag{12.37b}$$

This restriction will of course prevent the onset of the Kelvin–Helmholtz instability (see Chapter 8) and thus limit the analysis to laminar shear flow. Use of the correct vorticity convection equations (12.33)–(12.36) on the other hand will permit the natural onset of turbulence to occur for high Reynolds numbers when convective motions predominate in the outer regions of the boundary layers.

Another defect of this approximate model is the invalidation of the vorticity convection equation (12.36) as we will now demonstrate, due to the implied assumption of constant static pressure across the boundary layer at any location x_i. Thus since stagnation pressure is conserved during the convective step,

$$p_{ij} + \tfrac{1}{2}\rho u_{ij}^2 = p_P + \tfrac{1}{2}\rho u_P^2$$

If this equation is differentiated with respect to y,

$$u_{ij}\omega(i, j) = u_P\omega_P \tag{12.38}$$

Thus, with this approximation, we see that the velocity weighted vorticity is conserved and the vorticity after convection must be derived by interpolating the quantity $u_P\omega_P$. Further detail has been given by Lewis (1982b) and useful formulations for deciding upon suitable grid dimensions will be given in the concluding section of this chapter.

12.4.4 The Blasius boundary layer

The above solution can be applied to boundary layers with arbitrary variation of the mainstream velocity $U(x)$. The Blasius solution for constant mainstream velocity, also modelled by vortex dynamics in Section 9.4, provides the simplest first case for consideration. Results are shown in Fig. 12.9 for the following data:

$$U = 50 \qquad v = 0.05 \qquad \Delta t = 0.05$$

$$X1 = 25.0 \qquad Y1 = 1.0$$

$$M = 10 \qquad N = 13$$

Both profile and vorticity were predicted with considerable accuracy. Various other predicted boundary layer parameters are compared with the Blasius solution in Table 12.4. Displacement thickness δ^* shows best agreement but momentum thickness θ and wall shear stress $\tau_0/\rho U^2$ were also predicted generally to within 2% of the exact solution. Since there were only 13 cells in the y

492

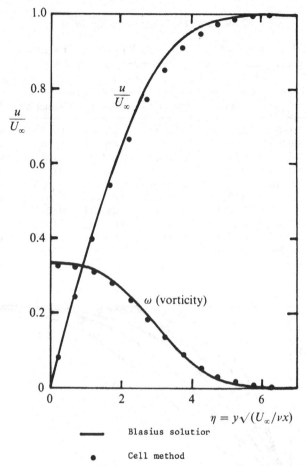

Fig. 12.9. The Blasius boundary layer.

direction interpolation of the data plotted in Fig. 12.9 was carried out to provide 100 values of $u(y)$ before calculating δ^* and θ.

12.4.5 Similarity boundary layers

Falkner & Skan (1931) and Hartree (1937) have shown that for mainstream velocities of the form

$$U(x) = U_1(x/X1)^m \tag{12.39}$$

similarity solutions can be derived from the boundary layer equa-

Table 12.4. *Flat plate boundary layer*

For $v = 0.05$, $U = 50.0$, $M = 10$, $N = 13$, $X1 = 25.0$, $Y1 = 1.0$

x	δ^*		θ		$\tau_0/\rho U^2$	
	Numerical	Blasius	Numerical	Blasius	Numerical	Blasius
1.25	0.059 021	0.061 16	0.025 959	0.023 474	0.012 239	0.009 390
3.75	0.103 037	0.105 94	0.426 603	0.040 661	0.006 346	0.005 422
6.25	0.134 965	0.136 77	0.054 600	0.052 494	0.004 546	0.004 200
8.75	0.161 050	0.161 83	0.064 584	0.062 113	0.003 721	0.003 549
11.25	0.183 520	0.183 49	0.073 208	0.070 426	0.003 231	0.003 130
13.75	0.203 433	0.202 86	0.080 832	0.077 861	0.002 897	0.002 831
16.25	0.221 355	0.220 53	0.087 650	0.084 643	0.002 652	0.002 604
18.75	0.237 590	0.236 89	0.093 770	0.090 992	0.002 463	0.002 425
21.25	0.252 285	0.252 19	0.099 252	0.096 794	0.002 315	0.002 278
23.75	0.265 480	0.266 61	0.104 137	0.102 329	0.002 198	0.002 154

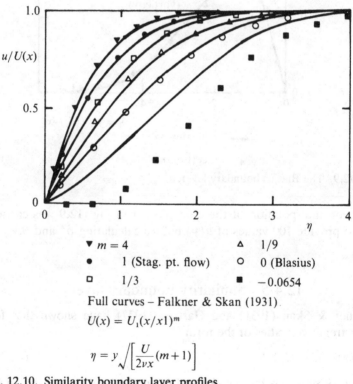

▼ $m = 4$ ▵ 1/9

● 1 (Stag. pt. flow)   ○ 0 (Blasius)

□ 1/3   ■ −0.0654

Full curves – Falkner & Skan (1931)

$$U(x) = U_1(x/x1)^m$$

$$\eta = y \sqrt{\left[\frac{U}{2vx}(m+1) \right]}$$

Fig. 12.10. Similarity boundary layer profiles.

Table 12.5. *Data for prediction of similarity boundary layers by the cell method*

m	M	N	$X1$	$Y1$	v	Δt	$U1$
−0.0654	10	13	25.0	1.0	0.05	0.05	50.0
0	10	13	25.0	1.0	0.05	0.05	50.0
1/9	10	13	25.0	1.0	0.05	0.05	50.0
1/3	10	13	25.0	1.0	0.05	0.05	50.0
1	10	13	25.0	0.6	0.05	0.0185	50.0
4	15	15	25.0	0.6	0.05	0.014	50.0

tions which may be expressed as a series of dimensionless velocity profiles, Fig. 12.10, of the form

$$\frac{u}{U} = f(\eta') \tag{12.40}$$

where the dimensionless coordinate η' is given by

$$\eta' = y \sqrt{\left[\frac{U}{2vx}(m+1)\right]} \tag{12.41}$$

Comparisons of the cellular method are shown in Fig. 12.10 for $m = -0.0654$, 0, 1/9, 1/3, 1 and 4, with input data as given in Table 12.5.

The main characteristic of this family of boundary layers with power law mainstream flow (12.39) is the production of similar profiles at all x locations. Lewis (1982b) found good agreement in all of the above cases for most of the cell region $0 < x < X1$, results shown in Fig. 12.10 being applicable to the mid position $x = X1/2$. This study represents an extreme test ranging from the fastest possible mainstream acceleration $m = 4$ application to potential flow around a plate end, to the mildly diffusing mainstream flow with $m = -0.0654$ for which the cell method predicted insipient separation. Extremely good results are obtained for acceleration flows and for the special case $m = 0$ which of course is the Blasius boundary layer. However the limiting similarity solution of Falkner & Skan predicts insipient separation for a value of $m = -0.0904$ in disagreement with the cell method. The probable explanation for this lies in both analyses. If a flow is near to separation the v velocity components and gradients may become of equal scale to the u values so that the convection assumptions of (12.37) and (12.38) are

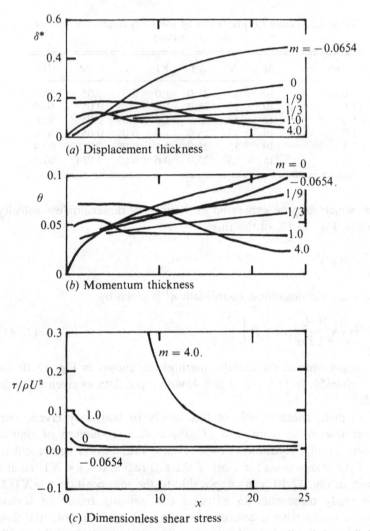

(a) Displacement thickness

(b) Momentum thickness

(c) Dimensionless shear stress

Fig. 12.11. Similarity boundary layer parameters predicted by cellular method.

invalidated. Even so the general shape of the separating boundary layer profile for $m = -0.0654$ is well predicted.

Other important boundary layer parameters predicted by the cell method are shown in Fig. 12.11 versus distance along the plate. For extreme acceleration $m = 4$ the boundary layer thicknesses δ^* and θ both in fact decrease proceeding downstream. An excellent test is presented by the case $m = 1$ which corresponds to flow away from a stagnation point for which the displacement and momentum thicknesses should both remain constant according to similarity solution theory. For $m < 1$ both δ^* and θ increase with x as expected.

12.4.6   Selection of grid data for cellular boundary layer computational schemes

Cell scales

The constraint required to maintain an accurate diffusion matrix determined in Section 12.4.2 resulted in fixture of the time step Δt for a prescribed cell dimension Δy. A suitable method for selecting Δx follows from convection considerations. The maximum drift δ occurs at the edge of the boundary layer, namely

$$\delta = U(x)\,\Delta t = k\,\Delta x = k(X1/M) \tag{12.42}$$

where k may be chosen at will. Thus if we wish to restrict maximum drift to one cell length, k will be chosen as unity. Where $U(x)$ is a function of x the maximum value should be chosen.

Field height

At the outset of a calculation we would like to select a field height $Y1$ which ensures that the boundary layer is adequately contained within the prescribed grid without vorticity leakage. If this condition is not fulfilled the vorticity conservation equation (12.24) will invalidate the computation.

Blasius (see Batchelor (1970)) defines the dimensionless thickness

$$\eta = y\sqrt{\left(\frac{U}{vx}\right)} \tag{12.43}$$

Flat plate laminar similarity profiles obey rules which result in universal profiles $u/U = f(\eta)$ as we have seen, Fig. 12.10. Thus if we define a minimum allowable dimensionless grid, say $\eta_1 = 6.0$, (12.42) yields a relationship for $X1$.

$$X1 = \frac{Y1^2 U}{v\eta_1^{\,2}} \tag{12.44}$$

Time step to ensure equal accuracy for convection and diffusion

Eliminating $X1$ from (12.42), we have

$$\Delta t = \frac{kY1^2}{Mv\eta_1^2} = \frac{kN^2 \Delta y^2}{Mv\eta_1^2} \qquad (12.45)$$

Δt is now the ideal time step to suit convection requirements. But we have already shown that Δt must obey (12.31) for satisfactory diffusion simulation. Eliminating Δt from these two equations we have finally a relationship between the appropriate numbers of cells in the x and y directions, namely

$$N = \text{round} \sqrt{\left[\frac{M\eta_1^2 (Q-1)^2}{4k \ln(1/\varepsilon)} \right]} \qquad (12.46)$$

This provides an alternative constraint on grid choice to (12.44) to ensure that both convection and diffusion are accurately modelled.

Alternatively for say high Reynolds number flows, we may pre-select $M, N, X1$ and $Y1$ and accept differing time steps Δt_d and Δt_c for convection and diffusion as discussed in Section 9.4.7. In this case $\Delta t_d / \Delta t_c$ must be an integer value say r and we would recycle the convective procedure (v) in the flow diagram of Section 12.4.3 r times for each diffusion procedure (iv).

APPENDIX

Computer programs

Program 1.1

Calculation of flow past a circular cylinder including surface
velocity, comparison with exact solution

```
program circle(input,output) ; {calculates flow past a circular cylinder.}

type
    index = 1..51 ;
    vector = array[index] of real ;
    matrix = array[index,index] of real ;
var
    k,m,n,i,j                                    : integer;
    coup                                         : matrix;
    xdata,ydata,ds,slope,sine,cosine,rhs,ans     : vector;
    x1,y1,x2,y2,radius,fi,dfi,pi,vel             : real;

            {*** Input data procedure ***}
procedure input_data;
begin
  writeln('number of pivotal points?'); read(m);
  writeln('mainstream velocity?'); read(vel);
  writeln('cylinder radius?'); read(radius);
  dfi := 2.0*pi/m;
  for n := 1 to m+1 do
    begin
    fi := (n-1)*dfi;
    xdata[n] := radius*(1.0-cos(fi)); ydata[n] := radius*sin(fi);
    end;
end;

            {*** Profile data preparation ***}
procedure data_preparation;
const
    ex = 0.000001;
var
    abscos,t : real; n : integer;
begin
  x1 := xdata[1]; y1 := ydata[1];
  writeln('pivotal points':24); writeln;
  writeln('x':11,'y':14); writeln;
  for n:=1 to m do
```

```
  begin
    x2 := xdata[n+1]; y2 := ydata[n+1];
    ds[n] := sqrt(sqr(x2-x1)+sqr(y2-y1));
    sine[n] := (y2-y1)/ds[n]; cosine[n] := (x2-x1)/ds[n];
    abscos := abs(cosine[n]);
    if abscos>ex then t := arctan(sine[n]/cosine[n]);
    if (abscos<ex) or (abscos=ex)
            then slope[n] := sine[n]/abs(sine[n])*pi/2.0;
    if cosine[n]>ex then slope[n] := t;
    if cosine[n]<(-ex) then slope[n] := t-pi;
    xdata[n] := (x1+x2)*0.5; ydata[n] := (y1+y2)*0.5;
    x1 := x2; y1 := y2;
    writeln(xdata[n]:14:6,ydata[n]:14:6);
  end; writeln;
end;


        {*** Coupling coefficients ***}
procedure coupling_coefficients;
var
  r,u,v,twopi : real; i,j : integer;
begin
  twopi := 2.0*pi;
  {*** Calculate the self inducing coupling coefficients ***}
  coup[1,1]    := -0.5-(slope[2]-slope[m]-2.0*pi)/(8.0*pi);
  coup[m,m]  := -0.5-(slope[1]-slope[m-1]-2.0*pi)/(8.0*pi);
  for i := 2 to m-1 do
  begin
    coup[i,i] := -0.5-(slope[i+1]-slope[i-1])/(8.0*pi);
  end;

  {*** Calculate coupling coefficients for j <> i ***}
  for i := 1 to m do for j := i to m do if j<>i then
    begin
      r := sqr(xdata[j]-xdata[i])+sqr(ydata[j]-ydata[i]);
      u := (ydata[j]-ydata[i])/(twopi*r);
      v := -(xdata[j]-xdata[i])/(twopi*r);
      coup[j,i] := (u*cosine[j]+v*sine[j])*ds[i];
      coup[i,j] := -(u*cosine[i]+v*sine[i])*ds[j];
    end;
end;


        {*** Matrix inversion ***}
procedure invert_matrix;
var
    a,b :real; pivot : vector; i,j,k : integer;
begin
for i := 1 to m do
  begin
```

```
a := coup[i,i]; coup[i,i] := 1.0;
for j := 1 to m do
  begin
    pivot[j] := coup[j,i]/a; coup[j,i] := pivot[j];
  end;
  for j := 1 to m do
  begin
  if i<>j then
    begin
      b := coup[i,j]; coup[i,j] := 0.0;
      for k := 1 to m do coup[k,j] := coup[k,j]-b*pivot[k];
    end;
  end;
end;
end;


        {*** Calculate right hand side values ***}
procedure right_hand_sides;
var i : integer;
begin
  for i := 1 to m do rhs[i] := -vel*cosine[i];
end;


  {*** Solve for surface vorticity element strengths ***}
procedure solution;
var exact : real; i,j : integer;
begin
  writeln('location','numerical':14,'exact':13); writeln;
  for i := 1 to m do
  begin
    {*** Multiply rhs column vector by inverted matrix ***}
    ans[i] := 0.0;
    for j := 1 to m do ans[i] := ans[i]+coup[i,j]*rhs[j];
    exact := 2.0*vel*sin((i-0.5)*dfi);      {Exact solution}
    writeln(i:5,ans[i]:18:6,exact:14:6);
  end;
end;


        {*** Main program ***}
begin
  pi := 4.0*arctan(1.0);
  input_data;
  data_preparation;
  coupling_coefficients;
  invert_matrix;
  right_hand_sides;
  solution;
end.
```

Appendix

Program 1.2

Calculation of the flow past a circular cylinder by the Douglas Neumann source panel method

```pascal
program source(input,output);
type
    index = 1..51 ;
    vector = array[index] of real ;
    matrix = array[index,index] of real ;

var
    k,m,n,i,j,ndivs                              : integer ;
    coup                                         : matrix ;
    xdata,ydata,ds,slope,sine,cosine,pivot,rhs,
    ans,source,delx,dely                         : vector ;
    u,v,r,x1,y1,x2,y2,rad,fi,dfi,pi,twopi,vel    : real ;

                    {*** procedures ***}
procedure inputdata;
var n : integer;
begin
  writeln('number of pivotal points?'); read(m);
  writeln('number of sub-elements?'); readln(ndivs);
  writeln('mainstream velocity?'); read(vel);
  writeln('cylinder radius?'); read(rad);
  dfi := 2.0*pi/m;
  for n := 1 to m+1 do
    begin
      fi := (n-1.0)*dfi;
      xdata[n] := rad*(1.0-cos(fi)); ydata[n] := rad*sin(fi);
    end;
end;

            {*** profile data preparation ***}
procedure data_preparation;
const
    ex = 0.000001;
var
    abscos,t  : real; n : integer;
begin
  x1 := xdata[1]; y1 := ydata[1];
  writeln('pivotal points':24); writeln;
  writeln('x':11,'y':14); writeln;
  for n:=1 to m do
    begin
      x2 := xdata[n+1]; y2 := ydata[n+1];
      delx[n] := (x2-x1)/ndivs; dely[n] := (y2-y1)/ndivs;
```

502

```
      ds[n] := sqrt(sqr(x2-x1)+sqr(y2-y1));
      sine[n] := (y2-y1)/ds[n]; cosine[n] := (x2-x1)/ds[n];
      abscos := abs(cosine[n]);
      if abscos>ex then t := arctan(sine[n]/cosine[n]);
      if (abscos<ex) or (abscos=ex)
                then slope[n] := sine[n]/abs(sine[n])*pi/2.0;
      if cosine[n]>ex then slope[n] := t;
      if cosine[n]<(-ex)  then slope[n] := t-pi;
      xdata[n] := (x1+x2)*0.5; ydata[n] := (y1+y2)*0.5;
      x1 := x2; y1 := y2;
      writeln(xdata[n]:14:6,ydata[n]:14:6);
    end; writeln;
end;


                {*** coupling coefficients ***}
procedure source_coupling_coefficients;
var i,j,k : integer;
begin
   for i := 1 to m do
   begin
     for j := 1 to m do if j<>i then
     begin
      u := 0.0; v := 0.0;
      for k := 1 to ndivs do
      begin
       x1 := xdata[i]+(k-0.5*(1+ndivs))*delx[i];
       y1 := ydata[i]+(k-0.5*(1+ndivs))*dely[i];
       r := sqr(xdata[j]-x1)+sqr(ydata[j]-y1);
       u := u+(xdata[j]-x1)/r;
       v := v+(ydata[j]-y1)/r;
      end;
      u := u/(twopi*ndivs); v := v/(twopi*ndivs);
      coup[j,i] := (-u*sine[j]+v*cosine[j])*ds[i];
     end;
   end;
   for i := 1 to m do
   begin
      coup[i,i] := 0.5;
   end;
end;


              {*** matrix inversion ***}
procedure invert_matrix;               (Same as for program 1.1)

              {*** calculate rhs values ***}
procedure right_hand_sides;
var i : integer;
begin
```

```
  for i := 1 to m do rhs[i] := vel*sine[i];
end;

    {*** solve for source element strengths ***}
procedure solve_for_source_strength;
var i,j : integer;
begin
  for i := 1 to m do
  begin
    source[i] := 0.0;
    for j := 1 to m do source[i] := source[i]+coup[i,j]*rhs[j];
  end;
end;

    {*** Perform second integral for surface velocity ***}
procedure calculate_surface_velocity;
var   exact : real; i,j,k : integer;
begin
 for i := 1 to m do
 begin
  {Contribution due to uniform stream}
  ans[i] := vel*cosine[i];
  {Contribution due to the surface source distribution}
  for j := 1 to m do if j<>i then
  begin
   u := 0.0; v := 0.0;
   for k := 1 to ndivs do
   begin
    x1 := xdata[j]+(k-0.5*(1+ndivs))*delx[j];
    y1 := ydata[j]+(k-0.5*(1+ndivs))*dely[j];
    r := sqr(x1-xdata[i])+sqr(y1-ydata[i]);
    u := u+(xdata[i]-x1)/r;
    v := v+(ydata[i]-y1)/r;
   end;
   u := u*source[j]*ds[j]/(twopi*ndivs);
   v := v*source[j]*ds[j]/(twopi*ndivs);
   ans[i] := ans[i]+u*cosine[i]+v*sine[i];
  end;
 end;
  {Now print out results}
  writeln('location','numerical':14,'exact':13); writeln;
 for i := 1 to m do
 begin
  exact := 2.0*vel*sin((i-0.5)*2.0*pi/m);
  writeln(i:5,ans[i]/vel:18:6,exact/vel:14:6);
 end;
end;
```

```
            {*** main program ***}
begin
  pi := 4.0*arctan(1.0) ; twopi := 2.0*pi;
  inputdata;
  data_preparation;
  source_coupling_coefficients;
  invert_matrix;
  right_hand_sides;
  solve_for_source_strength;
  calculate_surface_velocity;
end.
```

Program 1.3

Calculation of the flow past an ellipse, including surface velocity comparison
with exact solution and streamline pattern

```
program ellipse(input,output) ; {Calculates flow past an ellipse}
type
    vector  = array[1..51]  of real;
    vector1 = array[1..100] of real;
    matrix  = array[1..50,1..50] of real;
var
 k,m,number,diff,n,i,j                                    : integer;
 coup                                                     : matrix;
 xdata,ydata,ds,slope,sine,cosine,rhs,vorticity,ystart    : vector;
 x,y                                                      : vector1;
 major,ratio,theta,x1,y1,x2,y2,rad,fi,dfi,pi,Winf,
 Uinf,Vinf,alpha,xl,xr                                    : real;

            {*** Procedures ***}
procedure inputdata;
var i,n : integer;
begin
  writeln('number of pivotal points?'); read(m);
  writeln('mainstream velocity?'); read(Winf);
  writeln('angle of attack?'); readln(alpha);
  alpha := alpha*pi/180.0;
  Uinf := Winf*cos(alpha); Vinf := Winf*sin(alpha);
  writeln('major axis?'); read(major);
  writeln('minor axis/major axis?'); readln(ratio);
      {*** Data coordinates for ellipse ***}
  dfi := 2.0*pi/m;
  for n := 1 to m+1 do
    begin
      fi := (n-1)*dfi;
```

```
   xdata[n] := 0.5*major*(1.0-cos(fi));
   ydata[n] := ratio*0.5*major*sin(fi);
  end;


    {*** Set up specifications for streamlines ***}
  writeln('For forward difference enter 1');
  writeln('For central difference enter number of iterations');
  readln(diff); xl := 0.0; xr := 0.0;
  for i := 1 to m do if xdata[i]<xl then xl := xdata[i];
  for i := 1 to m do if xdata[i]>xr then xr := xdata[i];
  writeln('furthest data to left is at x =',xl:10:6);
  writeln('Enter x location on left to begin streamlines');
  readln(xl);
  writeln('furthest data to write is at x =',xr:10:6);
  writeln('Enter x location on right to end streamlines');
  readln(xr);
  writeln('Number of streamlines to be plotted'); readln(number);
  writeln('I.e. is roughly at y =',(ydata[1]+ydata[m])/2.0:10:6);
  writeln('Enter y positions for streamline starts');
  for n := 1 to number do
  begin
    write('Streamline',n:4,' of',number:4,' '); readln(ystart[n]);
  end;
  x[1] := xl;         {x starting position for streamlines}
  for i := 1 to m+1 do writeln(file1,xdata[i]:10:6,ydata[i]:10:6);
end;


    {*** Profile data preparation ***}          }
procedure data_preparation;                      }
    {*** Coupling coefficients ***}              } Same as for
procedure coupling_coefficients;                 } program1.1
    {*** Matrix inversion ***}                   }
procedure invert_matrix;                         }


    {*** Calculate right hand side values ***}
procedure right_hand_sides;
var i : integer;
begin
  for i := 1 to m do rhs[i] := -Uinf*cosine[i]-Vinf*sine[i];
end;


        {*** Multiply rhs column vector by inverted matrix ***}
procedure solution;
var
        exact,f,abyr : real; i,j : integer;
begin
  abyr := sqrt((1.0-ratio)/(1.0+ratio));
  writeln('location','numerical':14,'exact':13); writeln;
```

506

```
  for i := 1 to m do
  begin
    ans[i] := 0.0;
    for j := 1 to m do ans[i] := ans[i]+coup[i,j]*rhs[j];
      {*** Exact solution ***}
    theta := pi-(i-0.5)*2.0*pi/m;
    f := sqrt(1.0+sqr(srq(abyr))-2.0*sqr(abyr)*cos(2.0*theta));
    exact := 2.0*Winf*sin(theta-alpha)/f;
    writeln(i:5,ans[i]:18:6,exact:14:6);
  end;
end;


        {*** Induced u,v velocities at x,y due to surface vorticity ***}
procedure velocities(var u,v,x,y : real);
var  rsq,con,dx,dy : real; i : integer;
begin
  u := 0.0; v := 0.0;
  for i := 1 to m do
  begin
    dx := x-xdata[i]; dy := y-ydata[i];
    rsq := sqr(dx)+sqr(dy);
    con := ans[i]*ds[i]/2.0/pi/rsq;
    u := u+con*dy; v := v-con*dx;
  end;
  u := u+Uinf; v := v+Vinf;
end;


        {*** Calculation of streamlines ***}
procedure streamlines;
var
  con,u1,v1,u2,v2,dt,r : real; i,j,n : integer;
begin
  dt := (xr-xl)/50.0/Uinf;
  for n := 1 to number do
  begin
    j := 1;  y[1] := ystart[n];
    while ((j<200) and (x[j]<xr)) do
    begin
      velocities(u1,v1,x[j],y[j]);
      x[j+1] := x[j]+u1*dt;  y[j+1] := y[j]+v1*dt;
      if diff>1 then for k := 1 to (diff-1) do
      begin
        velocities(u2,v2,x[j+1],y[j+1]);
        u2 := 0.5*(u1+u2); v2 := 0.5*(v1+v2);
        x[j+1] := x[j]+u2*dt;  y[j+1] := y[j]+v2*dt;
      end;
      writeln(x[j+1]:10:6,y[j+1]:10:6); j := j+1;
    end;
```

```
   writeln; writeln(file1,j-1);
   for i := 1 to j-1 do writeln(file1,x[i]:10:6,y[i]:10:6);
   end;
end;


              {*** Main Program ***}
begin
pi := 4.0*arctan(1.0);
{*** Open and name one output file ***}
writeln('Name of output data file?'); readln(st1);
assign(file1,st1); rewrite(file1);
{*** Main computational sequence ***}
inputdata;
data_preparation;
coupling_coefficients;
invert_matrix;
right_hand_sides;
solution;
streamlines;
writeln('Solutions are in file ',st1); writeln;
end.
```

Program 2.1

Calculation of flow past a cylinder with bound circulation

```
program circle1(input,output);
type
    index = 1..51;
    vector = array[index] of real;
    matrix = array[index,index] of real;
var
    k,m,n,i,j                                   : integer;
    coup                                        : matrix;
    xdata,ydata,ds,slope,sine,cosine,rhs,ans    : vector;
    x1,y1,x2,y2,radius,fi,dfi,pi,vel,gamma       : real;

              {*** Input data procedure ***}
procedure input_data;
var n : integer;
begin
 writeln('number of pivotal points?'); read(m);
 writeln('mainstream velocity?'); read(vel);
 writeln('bound circulation strength?'); read(gamma);
 writeln('cylinder radius?'); read(radius);
 dfi := 2.0*pi/m;
```

```
  for n := 1 to m+1 do
    begin
     fi := (n-1)*dfi;
     xdata[n] := radius*(1.0-cos(fi)); ydata[n] := radius*sin(fi);
    end;
end;

procedure data_preparation;                (Same as for program 1.1)

procedure coupling_coefficients;           (Same as for program 1.1)

    {*** Procedure to add ds[i] to all matrix coefficients ***}
        {*** to allow for bound circulation ***}
procedure modify_matrix_for_bound_circ;
var i,j : integer;
begin
  for i := 1 to m do for j := 1 to m do coup[j,i] := coup[j,i]+ds[i];
end;

procedure invert_matrix;           (Same as for program 1.1)

        {*** Calculate right hand side values ***}
procedure right_hand_sides;
var i : integer;
begin
  for i := 1 to m do rhs[i] := -vel*cosine[i]+gamma;
end;

procedure solution;                        (Same as for program 1.1)

            {*** Main program ***}
begin
 pi := 4.0*arctan(1.0);
 input_data;
 data_preparation;
 coupling_coefficients;
 modify_matrix_for_bound_circ;
 invert_matrix;
 right_hand_sides;
 solution;
end.
```

Program_2.2

Flow past an ellipse with prescribed bound circulation.

```
program ellipse1(input,output) ;
type
    vector  = array[1..51] of real;
    matrix = array[1..50,1..50] of real;
var
 k,m,nsubs,n,i,j                              : integer;
 coup                                         : matrix;
 xdata,ydata,ds,slope,sine,cosine,rhs,ans     : vector;
 major,ratio,theta,x1,y1,x2,y2,r0,fi,pi,
 twopi,Winf,Uinf,Vinf,alpha,gamma,sum         : real;

                {*** Procedures ***}
procedure inputdata;
var n : integer;
begin
 writeln('number of pivotal points?'); read(m);
 writeln('number of sub-elements?'); read(nsubs);
 writeln('mainstream velocity?'); read(Winf);
 writeln('angle of attack?'); readln(alpha); alpha := alpha*pi/180.0;
 Uinf := Winf*cos(alpha); Vinf := Winf*sin(alpha);
 writeln('bound circulation?'); readln(gamma);
 writeln('major axis?'); read(major);
 writeln('minor axis/major axis?'); readln(ratio);
 for n := 1 to m+1 do        {*** Data coordinates for ellipse ***}
 begin
  fi := (n-1)*twopi/m;
  xdata[n] := 0.5*major*(1.0-cos(fi));
  ydata[n] := ratio*0.5*major*sin(fi);
 end;
end;


procedure data_preparation;        (Same as for program 1.1)

    {*** Coupling coefficients using sub-elements ***}
procedure coupling_coefficients;
var  r,u,v : real; i,j,k : integer;
begin
        {Self-induced coupling coefficients}
 coup[1,1] := -0.5-(slope[2]-slope[m]-2.0*pi)/(8.0*pi);
 coup[m,m] := -0.5-(slope[1]-slope[m-1]-2.0*pi)/(8.0*pi);
 for i := 2 to m-1 do
        coup[i,i] := -0.5-(slope[i+1]-slope[i-1])/(8.0*pi);
```

```
{remaining coupling coefficients using sub-elements}
for i := 1 to m do
begin
  for j := 1 to m do if i <> j then
  begin
   u := 0.0; v := 0.0;
   for k := 1 to nsubs do
   begin
    x1 := xdata[i]+(k-0.5*(1+nsubs))*ds[i]*cosine[i]/nsubs;
    y1 := ydata[i]+(k-0.5*(1+nsubs))*ds[i]*sine[i]/nsubs;
    r := sqr(xdata[j]-x1)+sqr(ydata[j]-y1);
    u := u+(ydata[j]-y1)/r; v := v-(xdata[j]-x1)/r;
   end;
   u := u/(twopi*nsubs); v := v/(twopi*nsubs);
   coup[j,i] :=  (u*cosine[j]+v*sine[j])*ds[i];
  end;
 end;
end;

{*** Back diagonal correction of coupling coefficient matrix ***}
procedure back_diagonal_correction;
var i,j : integer;
begin
  for i := 1 to m do
  begin
   sum := 0.0;
   for j := 1 to m do if j<>(m+1-i) then
      sum := sum+coup[j,i]*ds[j];
   coup[(m+1-i),i] := -sum/ds[m+1-i];
  end;
end;

   {*** Adjust coupling coefficient matrix for ***}
     {*** prescribed bound vortex strength ***}
procedure bound_vortex_correction;
var i,j : integer;
begin
  for i := 1 to m do for j := 1 to m do
     coup[i,j] := coup[i,j]+ds[j];
end;

procedure invert_matrix;          (Same as for program 1.1)

   {*** Calculate right hand side values ***}
procedure right_hand_sides;
var i : integer;
begin
```

511

```
    for i := 1 to m do rhs[i] := -Uinf*cosine[i]-Vinf*sine[i]+gamma;
end;
procedure solution;              (Same as for program 1.3)
            {*** Main program ***}
begin
 pi := 4.0*arctan(1.0) ; twopi := 2.0*pi;
 inputdata;
 data_preparation;
 coupling_coefficients;
 back_diagonal_correction;
 bound_vortex_correction;        {Add in extra equation for prescribed}
 invert_matrix;                  {bound vortex strength}
 right_hand_sides;
 solution;
end.
```

Program 2.3

Potential flow past an aerofoil

```
program aerofoil(input,output) ;
type
    vector = array[1..71] of real ;
    matrix = array[1..71,1..71] of real ;
var
    st1                                           : lstring(12);
    k,m,te,n,i,j,newcase                          : integer;
    coup                                          : matrix ;
    xdata,ydata,ds,slope,sine,cosine,pivot,
    rhs1,rhs2,ans1,ans2                           : vector;
    chord,a,b,x1,y1,x2,y2,pi,twopi,Winf,Uinf,Vinf,
    ans,Cp,gamma,gamma1,gamma2,alpha              : real;
    file1                                         : text;

            {*** procedures ***}
procedure inputdata;      {Real profile data points - xdata,ydata}
var n : integer;
begin                             {are the end coordinates of the elements}
  readln(file1,m); m := m-1; te := m div 2;   {te is trailing}
  for n := 1 to m+1 do                         {edge point}
    begin  readln(file1,xdata[n],ydata[n]); end;
  chord := xdata[te+1]-xdata[1];
end;

procedure input_flow_data;
begin
```

512

```
   writeln('mainstream velocity?'); read(Winf);
   writeln('angle of attack?'); readln(alpha);
   alpha := alpha*pi/180.0;
   Uinf := Winf*cos(alpha); Vinf := Winf*sin(alpha);
end;
procedure data_preparation;              (same as program 1.1)
procedure coupling_coefficients;         (-------- " --------)
procedure back_diagonal_correction;      (same as program 2.2)

procedure Kutta_condition;      {Trailing edge Kutta condition}
var i,j : integer;
begin
   {reduce matrix by one column and one row for Kutta condition}
      {First subtract column te+1 from te}
   for j := te to m do for i := 1 to m do
    if j>te then coup[i,j] := coup[i,j+1]
          else coup[i,j] := coup[i,j]-coup[i,j+1];
          {Now subtract row te+1 from row te}
   for i := te to m do for j := 1 to m do
    if i>te then coup[i,j] := coup[i+1,j]
          else coup[i,j] := coup[i,j]-coup[i+1,j];
          {Now adjust right hand sides to match}
   for i := te+1 to m do
   begin
    if i>te+1 then rhs1[i-1] := rhs1[i]
          else rhs1[i-1]     := rhs1[i-1]-rhs1[i];
    if i>te+1 then rhs2[i-1] := rhs2[i]
          else rhs2[i-1]     := rhs2[i-1]-rhs2[i];
   end;
   m := m-1;    {Matrix size is now reduced by one}
end;

procedure invert_matrix;       (same as program 1.1)

procedure right_hand_sides;  {*** calculate right hand side values ***}
var i : integer;
begin
   for i := 1 to m do rhs1[i] := -cosine[i]; rhs2[i] := -sine[i];
end;

   {*** Multiply rhs column vectors by inverted matrix ***}
procedure unit_solutions;
var
        exact : real; i,j : integer;
begin
   for i := 1 to m do
   begin
     ans1[i] := 0.0; ans2[i] := 0.0;
```

513

```
    for j := 1 to m do
    begin
      ans1[i] := ans1[i]+coup[i,j]*rhs1[j];
      ans2[i] := ans2[i]+coup[i,j]*rhs2[j];
    end;
  end;
  m := m+1;     {Revert to original matrix size}
  {Shift lower surface values along the array one place}
  for i := te to m-2 do
  begin
    ans1[m-i+te] := ans1[m-i+te-1]; ans2[m-i+te] := ans2[m-i+te-1];
  end;
  {replace lower trailing edge value by minus t.e. value}
  ans1[te+1] := -ans1[te]; ans2[te+1] := -ans2[te];
          {Sum up unit bound vortex strengths}
  gamma1 := 0.0; gamma2:= 0.0;
  for i := 1 to m do gamma1 := gamma1+ans1[i]*ds[i];
  for i := 1 to m do gamma2 := gamma2+ans2[i]*ds[i];
end;

procedure solution;
var i : integer;
begin
  writeln; writeln('Element ','Cp':9); writeln;
  for i := 1 to m do
  begin
    ans := Uinf*ans1[i]+Vinf*ans2[i]; Cp := 1.0-sqr(ans/Winf);
    writeln(i:4,Cp:16:6);
  end;
  gamma := Uinf*gamma1+Vinf*gamma2;
  writeln; writeln('Predicted circulation =',gamma:10:6);
  writeln; writeln('Cl =',(2.0*gamma)/(Winf*chord):10:6);
end;


                 {*** main program ***}
begin
  pi := 4.0*arctan(1.0) ; twopi := 2.0*pi;
  {*** open and name aerofoil profile input data file ***}
  writeln('Name of data input file?'); readln(st1);
  assign(file1,st1) ; reset(file1);
  inputdata;
  data_preparation;
  coupling_coefficients;
  right_hand_sides;
  back_diagonal_correction;
  Kutta_condition;
  invert_matrix;
  unit_solutions;
```

```
    newcase := 1;
    while newcase > 0 do
    begin
      input_flow_data; solution; writeln;
      writeln('For new flow data enter 1 otherwise 0'); readln(newcase);
    end;
  end.
```

Program 2.4

To calculate the potential flow through a turbomachine cascade

```
program bladerow(input,output) ;
type
    vector = array[1..71] of real ;
    matrix = array[1..71,1..71] of real ;
var
    st1                                              : lstring(12);
    k,m,te,n,i,j,newcase                             : integer;
    coup                                             : matrix;
    xdata,ydata,ds,slope,sine,cosine,pivot,
    rhs1,rhs2,ans1,ans2                              : vector;
    chord,pitch,stagger,a,b,x1,y1,x2,y2,pi,twopi,
    Winf,Uinf,Vinf,W1,ans,Cp,gamma,gamma1,gamma2,
    k1,k2,beta1,beta2,betainf                        : real;
    file1                                            : text;
              {*** procedures ***}
procedure inputdata;
var n : integer;
begin
    readln(file1,m); m := m-1; te := m div 2;    {Trailing edge}
    for n := 1 to m+1 do                  {location te}
      begin readln(file1,xdata[n],ydata[n]); end;
    chord := sqrt(sqr(xdata[te+1]-xdata[1])+sqr(ydata[te+1]-ydata[1]));
    writeln('Pitch/chord ratio?'); readln(pitch); pitch := pitch*chord;
    writeln('Stagger angle?'); readln(stagger);
    stagger := stagger*pi/180.0;
end;

procedure input_flow_data;
begin
    writeln('Inlet velocity?'); read(W1);
    writeln('Inlet angle?'); readln(beta1); beta1 := beta1*pi/180.0;
end;

procedure data_preparation;    (same as program 1.1)
```

515

```
procedure coupling_coefficients;
var
  r,u,v,a,b,k,sinh,cosh,e : real; i,j : integer;
begin
  for i := 1 to m do begin
    sine[i] := sin(stagger+slope[i]);
    cosine[i] := cos(stagger+slope[i]);
  end;
        {Self inducing coupling coefficients}
  coup[1,1] := -0.5-(slope[2]-slope[m]-2.0*pi)/(8.0*pi);
  coup[m,m] := -0.5-(slope[1]-slope[m-1]-2.0*pi)/(8.0*pi);
  for i := 2 to m-1 do
    coup[i,i] := -0.5-(slope[i+1]-slope[i-1])/(8.0*pi);
  for i := 1 to m do for j := i to m do if j<>i then
  if pitch/chord > 30.0 then        {Revert to single aerofoil for}
    begin                           {very wide blade spacing}
      r := sqr(xdata[j]-xdata[i])+sqr(ydata[j]-ydata[i]);
      u :=  (ydata[j]-ydata[i])/(r*twopi);
      v := -(xdata[j]-xdata[i])/(r*twopi);
      coup[j,i] :=  (u*cosine[j]+v*sine[j])*ds[i];
      coup[i,j] := -(u*cosine[i]+v*sine[i])*ds[j];
    end else          {Cascade coupling coefficients}
    begin
      a := ((xdata[i]-xdata[j])*cos(stagger)
         -(ydata[i]-ydata[j])*sin(stagger))*twopi/pitch;
      b := ((xdata[i]-xdata[j])*sin(stagger)
         +(ydata[i]-ydata[j])*cos(stagger))*twopi/pitch;
      e := exp(a); sinh := 0.5*(e-1.0/e); cosh := 0.5*(e+1.0/e);
      k := 0.5/pitch/(cosh-cos(b));
      coup[j,i] := (sinh*sine[j]-sin(b)*cosine[j])*k*ds[i];
      coup[i,j] := (-sinh*sine[i]+sin(b)*cosine[i])*k*ds[j];
    end;
end;


procedure back_diagonal_correction;     (Same as program 2.2)
procedure Kutta_condition                (Same as program 2.3)
procedure invert_matrix;                 (Same as program 1.1)
procedure right_hand_sides;              (Same as program 2.3)
procedure unit_solutions;                (Same as program 2.3)


procedure solution;
var i : integer;
begin
  k1 := (1.0-gamma2/2.0/pitch)/(1.0+gamma2/2.0/pitch);
  k2 := gamma1/pitch/(1.0+gamma2/2.0/pitch);
  beta2 := arctan(k1*sin(beta1)/cos(beta1)-k2);
  betainf := arctan(0.5*(sin(beta1)/cos(beta1)+sin(beta2)/cos(beta2)));
```

516

```
Winf := W1*cos(beta1)/cos(betainf);
Uinf := Winf*cos(betainf); Vinf := Winf*sin(betainf);
writeln; writeln('Element ','Cp':9);
writeln;
for i := 1 to m do
begin
    ans := Uinf*ans1[i]+Vinf*ans2[i];
    Cp := 1.0-sqr(ans/W1);
    writeln(i:4,Cp:16:6);
end;
gamma := Uinf*gamma1+Vinf*gamma2;
writeln; writeln('Predicted circulation =',gamma:10:6);
writeln; writeln('Cl =',(2.0*gamma)/(Winf*chord):10:6);
writeln; writeln('Beta2 =',beta2*180/pi:10:6);
end;


                {*** main program ***}
begin
  pi := 4.0*arctan(1.0) ; twopi := 2.0*pi;
  {*** open and name blade profile input data file ***}
  writeln('Name of data input file?'); readln(st1);
  assign(file1,st1) ; reset(file1);
  inputdata;
  data_preparation;
  coupling_coefficients;
  right_hand_sides;
  back_diagonal_correction;
  Kutta_condition;
  invert_matrix;
  unit_solutions;
  newcase := 1;
  while newcase > 0 do
  begin
    input_flow_data; solution; writeln;
    writeln('For new flow data enter 1 otherwise 0'); readln(newcase);
  end;
end.
```

Appendix

Program 4.1

Calculation of complete elliptic integrals of the first and
second kinds

```pascal
program elliptic(input,output);
{This program calculates elliptic integrals of the first and second kind and
files them. The user specifies the number of steps m and values are calculated
by the trapesium rule over the range fi = 0 to  (1-1/m)pi/2}

VAR
  pi,alpha,fi,dalpha,dfi,sumK,sumE,coefK,coefE,
  sqalpha,sqfi,dtor                          : real8;
  i,j,m,n                                    : integer;
  st1                                        : lstring(12);
  file1                                      : text;

begin
  pi := 4.0*arctan(1.0); dtor := pi/180.0;
  writeln('File name for K and E output?'); readln(st1);
  assign(file1,st1); rewrite(file1);
  writeln('Number of integration steps?'); readln(m);
  writeln('Number of fi steps from 0 to pi/2?'); readln(n);
  fi := 0.0;
  dfi := 90.0/n; dalpha := 90.0/m;
  for i := 1 to n do
  begin
    alpha := dalpha; sqfi := sqr(sin(fi*dtor));
    sumK := 0.0; sumE := 0.0;
    for j := 2 to m do
    begin
      sqalpha := sqr(sin(alpha*dtor));
      coefE := sqrt(1.0-sqfi*sqalpha);
      coefK := 1.0/coefE;
      sumK := sumK+coefK; sumE := sumE+coefE;
      alpha := alpha+dalpha;
    end;
    sumK := (sumK+0.5*(1.0+1.0/sqrt(1.0-sqfi)))*dalpha*dtor;
    sumE := (sumE+0.5*(1.0+sqrt(1.0-sqfi)))*dalpha*dtor;
    writeln(fi:14:6,sumK:14:6,sumE:14:6);
    writeln(file1,fi:14:6,sumK:14:6,sumE:14:6);
    fi := fi+dfi;
  end;
end.
```

Table: Complete Elliptic Integrals of the First and Second Kind

Ø	K	E	Ø	K	E
0.000000	1.570796	1.570796	23.000000	1.636517	1.509007
0.500000	1.570826	1.570766	23.500000	1.639523	1.506360
1.000000	1.570916	1.570677	24.000000	1.642604	1.503662
1.500000	1.571066	1.570527	24.500000	1.645761	1.500914
2.000000	1.571275	1.570318	25.000000	1.648995	1.498115
2.500000	1.571544	1.570049	25.500000	1.652307	1.495266
3.000000	1.571874	1.569720	26.000000	1.655697	1.492369
3.500000	1.572263	1.569332	26.500000	1.659166	1.489422
4.000000	1.572712	1.568884	27.000000	1.662716	1.486427
4.500000	1.573222	1.568376	27.500000	1.666347	1.483384
5.000000	1.573792	1.567809	28.000000	1.670059	1.480293
5.500000	1.574423	1.567183	28.500000	1.673855	1.477155
6.000000	1.575114	1.566497	29.000000	1.677735	1.473970
6.500000	1.575865	1.565752	29.500000	1.681700	1.470739
7.000000	1.576678	1.564948	30.000000	1.685750	1.467462
7.500000	1.577552	1.564084	30.500000	1.689888	1.464140
8.000000	1.578487	1.563162	31.000000	1.694114	1.460774
8.500000	1.579483	1.562181	31.500000	1.698430	1.457363
9.000000	1.580541	1.561142	32.000000	1.702836	1.453908
9.500000	1.581661	1.560044	32.500000	1.707334	1.450410
10.000000	1.582843	1.558887	33.000000	1.711925	1.446869
10.500000	1.584087	1.557673	33.500000	1.716610	1.443286
11.000000	1.585394	1.556400	34.000000	1.721391	1.439662
11.500000	1.586764	1.555069	34.500000	1.726269	1.435997
12.000000	1.588197	1.553681	35.000000	1.731245	1.432291
12.500000	1.589694	1.552235	35.500000	1.736321	1.428545
13.000000	1.591254	1.550732	36.000000	1.741499	1.424760
13.500000	1.592879	1.549172	36.500000	1.746780	1.420937
14.000000	1.594568	1.547555	37.000000	1.752165	1.417075
14.500000	1.596323	1.545881	37.500000	1.757657	1.413176
15.000000	1.598142	1.544150	38.000000	1.763256	1.409240
15.500000	1.600027	1.542364	38.500000	1.768965	1.405268
16.000000	1.601979	1.540522	39.000000	1.774786	1.401260
16.500000	1.603996	1.538623	39.500000	1.780720	1.397217
17.000000	1.606081	1.536670	40.000000	1.786769	1.393140
17.500000	1.608234	1.534661	40.500000	1.792936	1.389030
18.000000	1.610454	1.532597	41.000000	1.799222	1.384887
18.500000	1.612743	1.530479	41.500000	1.805629	1.380711
19.000000	1.615101	1.528306	42.000000	1.812160	1.376504
19.500000	1.617528	1.526080	42.500000	1.818817	1.372267
20.000000	1.620026	1.523799	43.000000	1.825602	1.367999
20.500000	1.622594	1.521465	43.500000	1.832518	1.363702
21.000000	1.625234	1.519079	44.000000	1.839567	1.359377
21.500000	1.627945	1.516639	44.500000	1.846751	1.355024
22.000000	1.630729	1.514147	45.000000	1.854075	1.350644
22.500000	1.633586	1.511603	45.500000	1.861539	1.346238

Table: Complete Elliptic Integrals of the First and Second Kind (cont.),

Ø	K	E	Ø	K	E
46.000000	1.869148	1.341806	68.000000	2.419842	1.136244
46.500000	1.876903	1.337350	68.500000	2.440135	1.131733
47.000000	1.884809	1.332870	69.000000	2.460999	1.127250
47.500000	1.892868	1.328367	69.500000	2.482462	1.122797
48.000000	1.901083	1.323842	70.000000	2.504550	1.118378
48.500000	1.909459	1.319296	70.500000	2.527296	1.113992
49.000000	1.917998	1.314730	71.000000	2.550731	1.109643
49.500000	1.926704	1.310144	71.500000	2.574893	1.105333
50.000000	1.935581	1.305539	72.000000	2.599820	1.101062
50.500000	1.944633	1.300917	72.500000	2.625553	1.096834
51.000000	1.953865	1.296278	73.000000	2.652138	1.092650
51.500000	1.963280	1.291623	73.500000	2.679625	1.088513
52.000000	1.972882	1.286954	74.000000	2.708068	1.084425
52.500000	1.982677	1.282270	74.500000	2.737525	1.080388
53.000000	1.992670	1.277574	75.000000	2.768063	1.076405
53.500000	2.002864	1.272866	75.500000	2.799753	1.072478
54.000000	2.013267	1.268147	76.000000	2.832673	1.068610
54.500000	2.023882	1.263417	76.500000	2.866911	1.064802
55.000000	2.034715	1.258680	77.000000	2.902565	1.061059
55.500000	2.045773	1.253934	77.500000	2.939744	1.057383
56.000000	2.057062	1.249182	78.000000	2.978569	1.053777
56.500000	2.068588	1.244424	78.500000	3.019179	1.050244
57.000000	2.080358	1.239661	79.000000	3.061729	1.046786
57.500000	2.092379	1.234895	79.500000	3.106396	1.043409
58.000000	2.104658	1.230127	80.000000	3.153385	1.040114
58.500000	2.117203	1.225358	80.500000	3.202930	1.036907
59.000000	2.130021	1.220589	81.000000	3.255303	1.033789
59.500000	2.143123	1.215821	81.500000	3.310823	1.030767
60.000000	2.156516	1.211056	82.000000	3.369848	1.027844
60.500000	2.170209	1.206295	82.500000	3.432887	1.025024
61.000000	2.184213	1.201538	83.000000	3.500422	1.022313
61.500000	2.198538	1.196788	83.500000	3.573138	1.019715
62.000000	2.213195	1.192046	84.000000	3.651856	1.017237
62.500000	2.228194	1.187312	84.500000	3.737613	1.014884
63.000000	2.243549	1.182589	85.000000	3.831742	1.012664
63.500000	2.259272	1.177878	85.500000	3.935998	1.010582
64.000000	2.275376	1.173179	86.000000	4.052758	1.008648
64.500000	2.291876	1.168496	86.500000	4.185351	1.006870
65.000000	2.308787	1.163828	87.000000	4.338654	1.005259
65.500000	2.326124	1.159178	87.500000	4.520223	1.003825
66.000000	2.343905	1.154547	88.000000	4.742717	1.002584
66.500000	2.362147	1.149936	88.500000	5.029861	1.001552
67.000000	2.380870	1.145348	89.000000	5.434910	1.000752
67.500000	2.400094	1.140783	89.500000	6.127779	1.000214

Computer programs

Program 4.2

Flow past a body of revolution

```
program axisym(input,output) ;
{Potential flow past a body of revolution in a uniform stream W = 1.0}
TYPE
    index1 = 1..81 ; index2 = 1..200;
    vector1 = array[index1] of real;
    vector2 = array[index2] of real;
    matrix  = array[index1,index1] of real;
VAR
    cons,con,pi,dtor,u,v,W                              : real;
    i,j,m,n,numberofKE,iloc                             : integer;
    E,K,fi                                              : vector2;
    xdata,rdata,sine,cosine,ds,slope,curve,rhs,ans      : vector1;
    coup                                                : matrix ;
    st1,st2                             : packed array[1..12] of char;
    file1,file2                                         : text;

function sign(a : real) : real;
begin
  sign := a/abs(a);
end;

procedure open_and_read_file;
var i : integer;
begin
  writeln('Name of elliptic integral data ',
                'file *** should normally be "kande"');
  readln(st1); assign(file1,st1); reset(file1); i := 0;
  while not eof(file1) do
    begin
    i := i+1;
    readln(file1,fi[i],K[i],E[i]);
    fi[i] := fi[i]*dtor;
    end;
  numberofKE := i; con := (numberofKE-1)/fi[numberofKE];
writeln; writeln('Elliptic integrals read from file ',st1); writeln;
writeln('Name of input body x,r data file?'); readln(st2);
assign(file2,st2); reset(file2);
readln(file2,m);
for i := 1 to m do readln(file2,xdata[i],rdata[i]);
m := m-1;
end;
```

521

```
procedure data_preparation;
var
  ex : real; i : integer;
begin
  ex := 0.00001;
  for i := 1 to m do
  begin
   ds[i] := sqrt(sqr(xdata[i+1]-xdata[i])+sqr(rdata[i+1]-rdata[i]));
   sine[i] :=  (rdata[i+1]-rdata[i])/ds[i];
   cosine[i] := (xdata[i+1]-xdata[i])/ds[i];
   if abs(cosine[i])<ex then slope[i] := sign(sine[i])*pi/2.0;
   if cosine[i]>ex  then slope[i] := arctan(sine[i]/cosine[i]);
   if cosine[i]<-ex then slope[i] := arctan(sine[i]/cosine[i])
                                  + pi*sign(sine[i]);
   xdata[i] := (xdata[i+1]+xdata[i])*0.5;
   rdata[i] := (rdata[i+1]+rdata[i])*0.5;
  end;
end;

procedure velocities(x1,r1,xm,rm : real);
var
  fi1,K1,E1,x,r,a          : real;
begin
 x := (x1-xm)/rm; r := r1/rm;
 a := x*x+sqr(r-1.0);
 fi1 := arctan(sqrt(4.0*r/a));
 cons := 0.5/(pi*rm*sqrt(x*x+sqr(r+1.0)));
 if fi1<=fi[numberofKE] then
 begin  {Use look-up tables}
  iloc := round(fi1*con)+1;
  K1 := K[iloc]; E1 := E[iloc];
 end else
 begin {Use assymptotic exressions}
  K1 := ln(4.0/cos(fi1));
  E1 := 1.0+0.5*(K1-1.0/1.2)*sqr(cos(fi1));
 end;
 u := -cons*(K1-(1.0+2.0*(r-1.0)/a)*E1);
 v := cons*x/r*(K1-(1.0+2.0*r/a)*E1);
end;

procedure coupling_coefficients;
var i,j : integer;
begin
 for i := 1 to m do
 for j := 1 to m do if j <> i then
 begin
  velocities(xdata[j],rdata[j],xdata[i],rdata[i]);
```

```
    coup[j,i] := (u*cosine[j]+v*sine[j])*ds[i];
   end;
   for i := 2 to m-1 do curve[i] := (slope[i+1]-slope[i-1])/8.0/pi;
   curve[1] := curve[2]; curve[m] := curve[m-1];
   for i := 1 to m do
   begin
    cons := 4.0*pi*rdata[i]/ds[i];
    coup[i,i] := -0.5-(ln(2.0*cons)-0.25)/cons*cosine[i]-curve[i];
   end;
  end;


               {*** Matrix inversion ***}
  procedure invert_matrix;
  VAR
     a,b   : real; i,j,n : integer; pivot : vector1;
  begin
  for i := 1 to m do
    begin
    a := coup[i,i]; coup[i,i] := 1.0;
    for j := 1 to m do
      begin
      pivot[j] := coup[j,i]/a; coup[j,i] := pivot[j];
      end;
      for j := 1 to m do
      begin
      if i<>j then
         begin
         b := coup[i,j]; coup[i,j] := 0.0;
         for n := 1 to m do coup[n,j] := coup[n,j]-b*pivot[n];
         end;
      end;
    end;
  end;


  procedure right_hand_sides;
  var i : integer;
  begin
   for i := 1 to m do rhs[i] := -cosine[i];
  end;


  procedure solve;
  var i,j : integer;
  begin
   writeln('Element':9,'Vel/W':9); writeln;
  for i := 1 to m do
    begin
    ans[i] := 0.0;
    for j := 1 to m do ans[i] := ans[i]+coup[i,j]*rhs[j];
```

```
   writeln(i:6,ans[i]:14:6);
  end;
 end;

              {*** Main program ***}
begin
 pi := 4.0*arctan(1.0); dtor := pi/180.0;
 open_and_read_file;
 data_preparation;
 coupling_coefficients;
 invert_matrix;
 right_hand_sides;
 solve;
end.
```

Program 4.3
Flow past an axisymmetric cowl or duct

```
program duct(input,output);
TYPE
    index1 = 1..81 ; index2 = 1..200;
    vector1 = array[index1] of real;  vector2 = array[index2] of real;
    matrix = array[index1,index1] of real;
VAR
 a,cons,con,pi,dtor,u,v,W,sum                   : real;
 i,j,m,n,numberofKE,iloc,te                     : integer;
 E,K,fi                                         : vector2;
 xdata,rdata,sine,cosine,ds,slope,curve,rhs,ans : vector1;
 coup                                           : matrix ;
 st1,st2                            : packed array[1..12] of char;
 file1,file2                                    : text;

function sign(a : real) : real;
begin  sign := a/abs(a); end;

procedure open_and_read_files;        (Same as program 4.2)

procedure data_preparation;
var    ex,t,abscos : real; i : integer;
begin
 ex := 0.00001;
 for i := 1 to m do
 begin
  ds[i] := sqrt(sqr(xdata[i+1]-xdata[i])+sqr(rdata[i+1]-rdata[i]));
  sine[i] :=   (rdata[i+1]-rdata[i])/ds[i];
  cosine[i] := (xdata[i+1]-xdata[i])/ds[i];
```

```
    abscos := abs(cosine[i]);
    if abscos>ex then t := arctan(sine[i]/cosine[i]);
    if abscos<ex then slope[i] := sign(sine[i])*pi/2.0;
    if cosine[i]>ex then slope[i] := t;
    if (cosine[i]<-ex) and (i>te) then slope[i] := t-pi;
    if (cosine[i]<-ex) and (i<te) then slope[i] := t+pi;
    xdata[i] := (xdata[i+1]+xdata[i])*0.5;
    rdata[i] := (rdata[i+1]+rdata[i])*0.5;
    end;
    for i := 2 to m-1 do curve[i] := (slope[i+1]-slope[i-1])/8.0/pi;
    curve[1] := (slope[2]-slope[m]-2.0*pi)/(8.0*pi);
    curve[m] := (slope[1]-slope[m-1]-2.0*pi)/(8.0*pi);
    curve[te] := 0.0; curve[te+1] := 0.0;
end;


procedure velocities(x1,r1,xm,rm : real);          )
                                                   )    Same as
procedure coupling_coefficients;                   )    Program 4.2
                                                   )
procedure right_hand_sides;                        )


procedure back_diagonal_correction;               (Same as Program 2.2)


    {Trailing edge Kutta condition}
procedure Kutta_condition;
var i,j : integer;
begin
    {reduce matrix by one column and one row for Kutta condition}
        {First subtract column te+1 from te}
    for j := te to m do
    begin
     for i := 1 to m do
     if j>te then coup[i,j] := coup[i,j+1]
     else coup[i,j] := coup[i,j]-coup[i,j+1];
    end;
            {Now subtract row te+1 from row te}
    for i := te to m do
    begin
     for j := 1 to m do
     if i>te then coup[i,j] := coup[i+1,j]
     else coup[i,j] := coup[i,j]-coup[i+1,j];
    end;
        {Now adjust right hand sides to match}
    for i := te+1 to m do
    if i>te+1 then rhs[i-1] := rhs[i] else rhs[i-1] := rhs[i-1]-rhs[i];
    m := m-1;
end;
```

525

```
procedure invert_matrix;          (Same as Program 1.1)

procedure solve;
var i,j : integer;
begin
  writeln('Element':9,'x':9,'r':14,'Vel/W':17,'Cp':14); writeln;
  for i := 1 to m do
   begin
    ans[i] := 0.0;
    for j := 1 to m do ans[i] := ans[i]+coup[i,j]*rhs[j];
   end;
  m := m+1;
  for i := te to m-2 do ans[m-i+te] := ans[m-i+te-1];
  ans[te+1] := -ans[te];
  for i := 1 to m do writeln(i:6,xdata[i]:16:6,rdata[i]:14:6,
              ans[i]:14:6,(1-sqr(ans[i])):14:6);
end;

        {*** Main program ***}
begin
  pi := 4.0*arctan(1.0); dtor := pi/180.0;
  open_and_read_files;
  data_preparation;
  te := m div 2;   {Trailing edge location}
  coupling_coefficients;
  right_hand_sides;
  back_diagonal_correction;
  Kutta_condition;
  invert_matrix;
  solve;
end.
```

Program 4.4

Flow through a contraction or diffuser

```
program contract(input,output) ;
TYPE
    index1 = 1..81 ; index2 = 1..200;
    vector1 = array[index1] of real;
    vector2 = array[index2] of real;
    matrix  = array[index1,index1] of real;
VAR
    cons,con,pi,dtor,u,v,W,xtube1,rtube1,xtube2,
    rtube2,fi1,K1,E1,pink,u1,v1,u2,v2,gamma1,gamma2          : real;
    i,j,m,n,numberofKE,iloc                                  : integer;
    E,K,fi                                                   : vector2;
```

```
xdata,rdata,sine,cosine,ds,slope,curve,rhs,ans          : vector1;
coup                                                    : matrix ;
st1,st2                                    : packed array[1..12] of char;
file1,file2                                             : text;

function sign(a : real) : real;
begin  sign := a/abs(a); end;

procedure open_and_read_file;
var i : integer;
begin
  writeln('Name of elliptic integral data ',
                'file *** should normally be "kande"');
  readln(st1); assign(file1,st1); reset(file1); i := 0;
  while not eof(file1) do
    begin
    i := i+1;
    readln(file1,fi[i],K[i],E[i]);
    fi[i] := fi[i]*dtor;
    end;
  numberofKE := i; con := (numberofKE-1)/fi[numberofKE];
  writeln; writeln('Elliptic integrals read from file ',st1); writeln;
  writeln('Name of input body x,r data file?'); readln(st2);
  assign(file2,st2); reset(file2);
  readln(file2,m);
  for i := 1 to m do readln(file2,xdata[i],rdata[i]);
  xtube1 := xdata[1]; rtube1 := rdata[1];
  xtube2 := xdata[m]; rtube2 := rdata[m];
  m := m-1;
  writeln('Inlet velocity?'); readln(W);
  gamma1 := -W; gamma2 := gamma1*sqr(rtube1/rtube2);
end;

procedure data_preparation;          (Same as program 4.2)

procedure look_up_and_interpolate;
  {Use of look-up tables to interpolate K(k) and E(k)}
var   ratio : real;
begin
  ratio := fi1*con; iloc := trunc(ratio)+1;
  if iloc >= numberofKE then iloc := numberofKE-1;
  K1 := K[iloc]+(K[iloc+1]-K[iloc])*(ratio+1-iloc);
  E1 := E[iloc]+(E[iloc+1]-E[iloc])*(ratio+1-iloc);
end;
```

527

```
procedure assymptotic_expressions;
begin {Assymptotic exressions for K(k) and E(k)}
   K1 := ln(4.0/cos(fi1));
   E1 := 1.0+0.5*(K1-1.0/1.2)*sqr(cos(fi1));
end;

procedure velocities(x1,r1,xm,rm : real);
var  x,r,a          : real;
begin
  x := (x1-xm)/rm; r := r1/rm; a := x*x+sqr(r-1.0);
  fi1 := arctan(sqrt(4.0*r/a));
  cons := 0.5/(pi*rm*sqrt(x*x+sqr(r+1.0)));
  {Elliptic integrals of the 1st and 2nd kinds}
  if fi1<=fi[numberofKE] then look_up_and_interpolate
  else assymptotic_expressions;
    u := -cons*(K1-(1.0+2.0*(r-1.0)/a)*E1);
    v := cons*x/r*(K1-(1.0+2.0*r/a)*E1);
end;

procedure third_kind(x,r,sqk : real);
var
   alpha,dalpha,sqalpha,sqn        : real;
   i                               : integer;
begin
  if abs(r-1.0)<0.0001 then pink := 0.0 else
  begin
   dalpha := pi/400; sqn := 4.0*r/sqr(r+1);
   pink := 0.0; alpha := dalpha;
   for i := 2 to 200 do
   begin
     sqalpha := sqr(sin(alpha));
     pink := pink+1.0/(1.0-sqn*sqalpha)/sqrt(1.0-sqk*sqalpha);
     alpha := alpha+dalpha;
   end;
   pink := (pink+0.5*(1.0+1.0/(1.0-sqn)/sqrt(1.0-sqk)))*dalpha;
  end;
end;

procedure uv_tube(xtube,rtube,xp,rp,gamma : real);
var x,r,a,b,sqk    : real;
begin
  x := (xp-xtube)/rtube; r := rp/rtube;
  {Find values of K(k) and E(k)}
  a := x*x+sqr(r-1.0); b := a+4.0*r; sqk := 4.0*r/b; b := sqrt(b);
  fi1 := arctan(sqrt(4.0*r/a));
  {Elliptic integrals of the 1st and 2nd kinds}
  if fi1<=fi[numberofKE] then look_up_and_interpolate
  else assymptotic_expressions;
```

```
{Now calculate pi(n,k), elliptic Int. of third kind}
third_kind(x,r,sqk);
if abs(r-1.0)<0.0001 then a := pi/2.0
        else if r<1.0 then a := pi else a := 0.0;
u := (a+x/b*(K1-(r-1.0)/(r+1.0)*pink))*gamma/(2.0*pi);
v := 2.0*gamma/pi/sqk/b*(E1-(1-0.5*sqk)*K1);
end;
```

procedure coupling_coefficients;   same as Program 4.2 except for
the expression for coup[i,i] which is changed to :-
coup[i,i] := +0.5-(ln(2.0*cons)-0.25)/cons*cosine[i]-curve[i];
                ⇑
                (note + sign for external boundary conditions)

```
               {*** Matrix inversion ***}
procedure invert_matrix;        (Same as program 1.1)

procedure right_hand_sides;
var i : integer;
begin
 for i := 1 to m do
 begin
  uv_tube(xdata[i],rtube1,xtube1,rdata[i],gamma1);
  u1:= u; v1 := -v;
  uv_tube(xtube2,rtube2,xdata[i],rdata[i],gamma2);
  u2 := u; v2 := v;
  rhs[i] := -(u1+u2)*cosine[i]-(v1+v2)*sine[i];
 end;
end;

procedure solve;
var i,j : integer;
begin
 writeln('Element':9,'x':9,'r':14,'Vel/W':17,'Cp':14); writeln;
 for i := 1 to m do
 begin
  ans[i] := 0.0;
  for j := 1 to m do ans[i] := ans[i]+coup[i,j]*rhs[j];
  writeln(i:6,xdata[i]:16:6,rdata[i]:14:6,
            ans[i]:14:6,(1-sqr(ans[i]/W)):14:6);
 end;
end;

              {*** Main program ***}
begin
pi := 4.0*arctan(1.0); dtor := pi/180.0;
open_and_read_file;
data_preparation;
```

529

```
coupling_coefficients;
invert_matrix;
right_hand_sides;
solve;
end.
```

Program 4.5

Potential flow past a body of revolution in a uniform stream W = 1.0, source panel method.

```
program dnaxisym(input,output);
TYPE
    index1 = 1..81 ; index2 = 1..200;
    vector1 = array[index1] of real;
    vector2 = array[index2] of real;
    matrix  = array[index1,index1] of real;
VAR
    cons,con,pi,dtor,twopi,u,v,u1,v1,W,ui,vi,q,xi,ri       : real;
    i,j,m,n,numberofKE,ndivs,iloc,self                     : integer;
    E,K,fi                                                 : vector2;
    xdata,rdata,delx,delr,sine,cosine,
    ds,slope,curve,rhs,ans,source                          : vector1;
    coup                                                   : matrix ;
    st1,st2                                    : packed array[1..12] of char;
    file1,file2                                            :   text;

function sign(a : real) : real;
begin  sign := a/abs(a); end;

procedure open_and_read_file;
var i : integer;
begin
  writeln('Number of sub-elements?'); readln(ndivs);
  if ndivs = 1 then ndivs := 2; ndivs := (ndivs div 2)*2;
  writeln('Name of elliptic integral data file'); readln(st1);
  assign(file1,st1); reset(file1); i := 0;
  while not eof(file1) do
    begin
    i := i+1; readln(file1,fi[i],K[i],E[i]);
    fi[i] := fi[i]*dtor;
    end;
  numberofKE := i; con := (numberofKE-1)/fi[numberofKE];
  writeln; writeln('Elliptic integrals read from file ',st1); writeln;
  writeln('Name of input body x,r data file?'); readln(st2);
  assign(file2,st2); reset(file2);
```

530

```
readln(file2,m); for i := 1 to m do readln(file2,xdata[i],rdata[i]);
m := m-1;
end;

procedure data_preparation;
var  ex : real; i : integer;
begin
 ex := 0.00001;
 for i := 1 to m do
 begin
  ds[i] := sqrt(sqr(xdata[i+1]-xdata[i])+sqr(rdata[i+1]-rdata[i]));
  delx[i] := (xdata[i+1]-xdata[i])/ndivs;
  delr[i] := (rdata[i+1]-rdata[i])/ndivs;
  sine[i] :=  (rdata[i+1]-rdata[i])/ds[i];
  cosine[i] := (xdata[i+1]-xdata[i])/ds[i];
  if abs(cosine[i])<ex then slope[i] := sign(sine[i])*pi/2.0;
  if cosine[i]>ex  then slope[i] := arctan(sine[i]/cosine[i]);
  if cosine[i]<-ex then slope[i] := arctan(sine[i]/cosine[i])
                              + pi*sign(sine[i]);
  xdata[i] := (xdata[i+1]+xdata[i])*0.5;
  rdata[i] := (rdata[i+1]+rdata[i])*0.5;
 end;
 for i := 2 to m-1 do curve[i] := (slope[i-1]-slope[i+1])/8.0/pi;
 curve[1] := curve[2]; curve[m] := curve[m-1];
end;
procedure velocities(x1,r1,xm,rm : real);
{u,v vels at (x1,y1) due to a unit strength ring source at (xm,rm)}
var
  fi1,K1,E1,x,r,a           : real;
begin
 x := (x1-xm)/rm; r := r1/rm; a := x*x+sqr(r-1.0);
 fi1 := arctan(sqrt(4.0*r/a));
 cons := 0.5/(pi*rm*sqrt(x*x+sqr(r+1.0)));
 if fi1<=fi[numberofKE] then
 begin  {Use look-up tables}
  iloc := round(fi1*con)+1;
  K1 := K[iloc]; E1 := E[iloc];
 end else
 begin {Use assymptotic exressions}
  K1 := ln(4.0/cos(fi1));
  E1 := 1.0+0.5*(K1-1.0/1.2)*sqr(cos(fi1));
 end;
  u1 := cons*2.0*x*E1/a;
  v1 := cons/r*(K1-(1.0-2.0*r*(r-1)/a)*E1);
end;
```

```
procedure coupling_coefficients;{for ring source with ndiv sub-elements}
var i,j,n integer;
begin
 for i := 1 to m do
 for j := 1 to m do
 begin
  u := 0.0; v := 0.0;
  for n := 1 to ndivs do
  begin
   xi := xdata[i]+(n-0.5*(1+ndivs))*delx[i];
   ri := rdata[i]+(n-0.5*(1+ndivs))*delr[i];
   velocities(xdata[j],rdata[j],xi,ri);
   u := u+u1; v := v+v1;
  end;
  u := u/ndivs; v := v/ndivs;
  coup[j,i] := (-u*sine[j]+v*cosine[j])*ds[i];
 end;
 for i := 1 to m do coup[i,i] := coup[i,i]+0.5;
end;


procedure invert_matrix;                    (Same as program 1.1)


procedure right_hand_sides;
var i : integer;
begin
 for i := 1 to m do rhs[i] := sine[i];
end;


procedure solve_for_source_strength;
var i,j : integer;
begin
for i := 1 to m do
 begin
   source[i] := 0.0;
   for j := 1 to m do source[i] := source[i]+coup[i,j]*rhs[j];
 end;
end;


procedure surface_velocity;
var i,j,n : integer;
begin
 writeln('Element':9,'Vel/W':9); writeln;
 for j := 1 to m do
 begin
  ans[j] := cosine[j];
  for i := 1 to m do
  begin
   u := 0.0; v := 0.0;
```

532

```
    for n := 1 to ndivs do
    begin
     xi := xdata[i]+(n-0.5*(1+ndivs))*delx[i];
     ri := rdata[i]+(n-0.5*(1+ndivs))*delr[i];
     velocities(xdata[j],rdata[j],xi,ri);
     u := u+u1; v := v+v1;
    end;
    u := u/ndivs; v := v/ndivs;
    q := (u*cosine[j]+v*sine[j])*ds[i]*source[i];
    ans[j] := ans[j]+q;
   end;
   writeln(j:6,ans[j]:14:6);
  end;
end;


            {*** Main program ***}
begin
 pi := 4.0*arctan(1.0); dtor := pi/180.0; twopi := 2.0*pi;
 open_and_read_file;
 data_preparation;
 coupling_coefficients;
 invert_matrix;
 right_hand_sides;
 solve_for_source_strength;
 surface_velocity;
end.
```

Program 5.1

Potential flow through an engine intake sucked from downstream
  by a cylindrical duct and located in a uniform stream W

```
program suckduct(input,output);
TYPE
    index1 = 1..81 ; index2 = 1..200;
    vector1 = array[index1] of real;
    vector2 = array[index2] of real;
    matrix  = array[index1,index1] of real;
VAR
    cons,con,pi,dtor,u,v,W,Vj,xtube,rtube,gamma,
    K1,E1,pink,fi1,sum,u1,v1                        : real;
    i,j,m,n,numberofKE,iloc,te,newcase              : integer;
    E,K,fi                                          : vector2;
    xdata,rdata,sine,cosine,ds,slope,curve,rhs,rhsW,
    rhsgamma,ans,ansW,ansgamma                      : vector1;
```

533

```
coup                                           : matrix ;
st1,st2                                        : packed array[1..12] of char;
file1,file2                                    : text;

function sign(a : real) : real;
begin  sign := a/abs(a); end;

procedure open_and_read_files;
var i : integer;
begin
  writeln('Name of elliptic integral data ',
                'file *** should normally be "kande"');
  readln(st1); assign(file1,st1); reset(file1); i := 0;
  while not eof(file1) do
    begin
    i := i+1; readln(file1,fi[i],K[i],E[i]);
    fi[i] := fi[i]*dtor;
    end;
  numberofKE := i; con := (numberofKE-1)/fi[numberofKE];
  writeln; writeln('Elliptic integrals read from file ',st1); writeln;
  writeln('Name of input body x,r data file?'); readln(st2);
  writeln; assign(file2,st2); reset(file2);
  readln(file2,m); for i := 1 to m do readln(file2,xdata[i],rdata[i]);
  te := (m-1) div 2; xtube := xdata[te+1]; rtube := rdata[te+1];
  m := m-1;
  writeln('Mainstream velocity?'); readln(W);
  writeln('Suction duct velocity?'); readln(Vj); gamma := W-Vj;
end;

procedure data_preparation:    (Same as Program 4..3 but
                                      including the following last line)
  if abs((xtube-xdata[te])/rtube)<0.01 then xtube := xtube+0.2*rtube;
procedure look_up_and_interpolate;            )
                                              )
procedure assymptotic_expressions;            )
                                              )          (Same as
procedure velocities(x1,r1,xm,rm: real);      )          program 4.4)
                                              )
procedure third_kind(x,r,sqk : real);         )
                                              )
procedure uv_tube(xtube,rtube,xp,rp,gamma : real);  )

procedure coupling_coefficients;          (Same as program 4.2)

procedure right_hand_sides;
var i : integer;
begin
  for i := 1 to m do
```

534

```
  begin
   rhsW[i] := -cosine[i];
   uv_tube(xtube,rtube,xdata[i],rdata[i],1.0);
   rhsgamma[i] := -u*cosine[i]-v*sine[i]-coup[i,te+1];
  end;
end;
```

```
procedure back_diagonal_correction;        (Same as program 4.2)
```

```
procedure Kutta_condition;        {Trailing edge Kutta condition}
var i,j : integer;
begin
    {reduce matrix by one column and one row for Kutta condition}
        {First subtract column te+1 from te}
   for j := te to m do
   begin
    for i := 1 to m do
    if j>te then coup[i,j] := coup[i,j+1]
    else coup[i,j] := coup[i,j]-coup[i,j+1];
   end;
           {Now subtract row te+1 from row te}
   for i := te to m do
   begin
    for j := 1 to m do
    if i>te then coup[i,j] := coup[i+1,j]
    else coup[i,j] := coup[i,j]-coup[i+1,j];
   end;
           {Now adjust right hand sides to match}
   for i := te+1 to m do
   if i>te+1 then
   begin
    rhsW[i-1] := rhsW[i]; rhsgamma[i-1] := rhsgamma[i];
   end else
   begin
    rhsW[i-1] := rhsW[i-1]-rhsW[i];
    rhsgamma[i-1] := rhsgamma[i-1]-rhsgamma[i];
   end; m := m-1;
end;
```

```
procedure invert_matrix;             (Same as program 1.1)
```

```
procedure unit_solutions;
var i : integer;
begin
for i := 1 to m do
  begin
   ansW[i] := 0.0; ansgamma[i] := 0.0;
    for j := 1 to m do
```

```
   begin
    ansW[i] := ansW[i]+coup[i,j]*rhsW[j];
    ansgamma[i] := ansgamma[i]+coup[i,j]*rhsgamma[j];
   end;
  end; m := m+1;
 for i := te to m-2 do
  begin
   ansW[m-i+te] := ansW[m-i+te-1];
   ansgamma[m-i+te] := ansgamma[m-i+te-1];
  end;
  ansW[te+1] := -ansW[te]; ansgamma[te+1] := -ansgamma[te]+1.0;
end;

procedure solution;
var i : integer;
begin
 writeln(m);
 writeln('Element':9,'x':9,'r':14,'Vel/W':17,'Cp':14); writeln;
 for i := 1 to m do
 begin
  ans[i] := W*ansW[i]+gamma*ansgamma[i];
  writeln(i:6,xdata[i]:16:6,rdata[i]:14:6,
                ans[i]:14:6,(1-sqr(ans[i])):14:6);
 end;
end;


                {*** Main program ***}
begin
 pi := 4.0*arctan(1.0); dtor := pi/180.0;
 open_and_read_files;
 data_preparation;
 coupling_coefficients;
 back_diagonal_correction;
 right_hand_sides;
 Kutta_condition;
 invert_matrix;
 unit_solutions;
 solution;
 newcase := 1;
 repeat
  writeln('For new case enter 1 otherwise 0'); readln(newcase);
  if newcase=1 then
  begin
   writeln('Mainstream velocity?'); readln(W);
   writeln('Suction duct velocity?'); readln(Vj);
   gamma := W-Vj; for i := 1 to 5 do writeln;
   solution;
```

```
    end;
  until newcase = 0;
end.
```

Program 5.2

Potential flow through a ducted propeller in a uniform stream W

```
program ductprop(input,output) ;
TYPE
    index1 = 1..81 ; index2 = 1..200;
    vector1 = array[index1] of real;
    vector2 = array[index2] of real;
    matrix  = array[index1,index1] of real;
VAR
  cons,con,pi,dtor,u,v,W,Vj,xtube,rtube,gamma,K1,E1,pink,
  Ctduct,Ctprop,ratio,hubtip,tc,diameter,fi1,sum,u1,v1        : real;
  i,j,m,mhub,mduct,n,numberofKE,iloc,te,newcase               : integer;
  E,K,fi                                                      : vector2;
  xdata,rdata,sine,cosine,ds,slope,curve,rhs,rhsW,
  rhsgamma,ans,ansW,ansgamma,Cp                               : vector1;
  coup                                                        : matrix ;
  st1,st2                                   : packed array[1..12] of char;
  file1,file2                                                 : text;

function sign(a : real) : real;
begin sign := a/abs(a); end;

procedure open_and_read_files;
var i : integer;
begin
  writeln('Name of elliptic integral data file?'); readln(st1);
  assign(file1,st1); reset(file1); i := 0;
  while not eof(file1) do
    begin
    i := i+1;
    readln(file1,fi[i],K[i],E[i]);
    fi[i] := fi[i]*dtor;
    end;
  numberofKE := i; con := (numberofKE-1)/fi[numberofKE];
  writeln; writeln('Elliptic integrals read from file ',st1); writeln;
  writeln('Name of input body x,r data file?'); readln(st2);
  writeln; assign(file2,st2); reset(file2);
  readln(file2); readln(file2,mhub);
  for i := 1 to mhub do readln(file2,xdata[i],rdata[i]);
  readln(file2,mduct); m := mhub+mduct;  for i := mhub+1 to m do
  readln(file2,xdata[i],rdata[i]);
```

```
te := ((mduct-1) div 2)+mhub; rtube := rdata[mhub+1];
for i := mhub+1 to m do if rdata[i]<rtube then
    begin rtube := rdata[i]; j := i; end;
writeln('Minimum radius is ',rtube:10:6,' at x =',xdata[j]:10:6);
diameter := 2.0*rtube;
writeln('Tip clearance?'); readln(tc); rtube := rtube-tc;
writeln('Axial location of propeller?'); readln(xtube);
writeln('Mainstream velocity?'); readln(W);
writeln('Propeller thrust coefficient?'); readln(Ctprop);
writeln('Hub/tip ratio?'); readln(hubtip);
Vj := W*sqrt(1.0+Ctprop/(1.0-sqr(hubtip)));
gamma := W-Vj;
end;

procedure data_preparation;
var
    ex,dels,t,abscos : real; i,imod : integer;
begin
    ex := 0.00001; i := 0;
    for imod := 1 to m-1 do if imod<>mhub then
    begin
        if imod<mhub then i := imod else i := imod-1;
        dels :=
        sqrt(sqr(xdata[imod+1]-xdata[imod])+sqr(rdata[imod+1]-rdata[imod]));
        ds[i] := dels;
        sine[i] :=  (rdata[imod+1]-rdata[imod])/dels;
        cosine[i] := (xdata[imod+1]-xdata[imod])/dels;
        xdata[i] := (xdata[imod+1]+xdata[imod])*0.5;
        rdata[i] := (rdata[imod+1]+rdata[imod])*0.5;
        abscos := abs(cosine[i]);
        if abscos>ex then t := arctan(sine[i]/cosine[i]);
        if abscos<ex then slope[i] := sign(sine[i])*pi/2.0;
        if cosine[i]>ex then slope[i] := t;
        if (cosine[i]<-ex) and (i>te) then slope[i] := t-pi;
        if (cosine[i]<-ex) and (i<te) then slope[i] := t+pi;
    end;
    m := m-2; mhub := mhub-1; mduct := mduct-1; te := te-1;
    if abs((xtube-xdata[te])/rtube)<0.01 then xtube := xtube+0.2*rtube;
    {Hub surface curvatures}  for i := 2 to mhub-1 do curve[i] := (slope[i+1]-
slope[i-1])/8.0/pi;
    curve[1] := curve[2]; curve[mhub] := curve[mhub-1];
        {Duct surface curvatures}
    for i := mhub+2 to m-1 do curve[i] := (slope[i+1]-slope[i-1])/8.0/pi;
    curve[mhub+1] := (slope[mhub+2]-slope[m]-2.0*pi)/(8.0*pi);
    curve[m] := (slope[mhub+1]-slope[m-1]-2.0*pi)/(8.0*pi);
    curve[te] := 0.0; curve[te+1] := 0.0;
end;
```

```
procedure look_up_and_interpolate;                    )
                                                      )
procedure assymptotic_expressions;                    )
                                                      )    (Same as
procedure velocities(x1,r1,xm,rm : real);             )    program 4.4)
                                                      )
procedure third_kind(x,r,sqk : real);                 )
                                                      )
procedure uv_tube(xtube,rtube,xp,rp,gamma : real);    )

procedure coupling_coefficients;              (Same as program 4.2)

procedure right_hand_sides;                   (Same as program 5.1)

   {Back diagonal correction}
procedure back_diagonal_correction;
var i,j : integer;
begin
 for i := mhub+1 to m do
 begin
  sum := 0.0;
  for j := mhub+1 to m do
          if j<>(m-i+1) then sum := sum-coup[j,i]*ds[j];
  coup[(m-i+1),i] := sum/ds[m-i+1];
  end;
end;

procedure Kutta_condition;                    (Same as program 5.1)

procedure invert_matrix;                      (Same as program 1.1)

procedure unit_solutions;                     (Same as program 5.1)

procedure solution;                           (Same as program 5.1)

procedure duct_thrust;
var i : integer;
begin
 Ctduct := 0.0;
 for i := mhub+1 to m do Ctduct := Ctduct+Cp[i]*sine[i]*rdata[i]*ds[i];
 Ctduct := -Ctduct*8.0/sqr(diameter);
 ratio := Ctprop/(Ctprop+Ctduct);
 writeln;
 writeln('Ct duct = ',Ctduct:10:6,' Thrust ratio =',ratio:10:6);
 writeln;
end;
```

539

```
        {*** Main program ***}
begin
 pi := 4.0*arctan(1.0); dtor := pi/180.0;
 open_and_read_files;
 data_preparation;
 coupling_coefficients;
 back_diagonal_correction;
 right_hand_sides;
 Kutta_condition;
 invert_matrix;
 unit_solutions;
 solution;
 duct_thrust;
 newcase := 1;
 repeat
   writeln('For new case enter 1 otherwise 0'); readln(newcase);
   if newcase=1 then
   begin
     writeln('Mainstream velocity?'); readln(W);
     writeln('Propeller thrust coefficient?'); readln(Ctprop);
     Vj := W*sqrt(1.0+Ctprop/(1.0-sqr(hubtip)));
     gamma := W-Vj; for i := 1 to 5 do writeln;
     solution;
     duct_thrust;
   end;
 until newcase = 0;
 writeln('Edit first integer on file from 2 to 2*(number of cases)');
end.
```

Program 8.1

Program for experimentation with convection of vortex clouds.

```
program convect(input,output);
type
        vector1   = array[1..20] of real;
var
 pi,dt,r,r1                                             :   real;
 m,n,step,nsteps,Z,it,its                               :  integer;
 gamma,xa,ya,xb,yb,ua,va,ub,vb                          :  vector1;
 xd,yd                                        : array[1..200,1..20] of real;
 st1,st2,st3,st4                              : packed array[1..12] of char;
 file1,file2,file3,file4                                :   text;
```

```
procedure open_files;
var n : integer;
begin
  writeln('Name of input data file?'); readln(st1);
  writeln('Name of output data file?'); readln(st2);
  assign(file1,st1); reset(file1); assign(file2,st2); rewrite(file2);
  st3 := 'finalsol  '; assign(file3,st3); rewrite(file3);
  st4 := 'reverse   '; assign(file4,st4); rewrite(file4);
  readln(file1,Z); writeln('Vortex data on file ',st1);
  writeln('Gamma':12,'x':13,'y':14);
  for n := 1 to Z do
  begin
    readln(file1,gamma[n],xa[n],ya[n]);
    writeln(gamma[n]:14:6,xa[n]:14:6,ya[n]:14:6);
  end;
end;

procedure uv_vels(var x,y,u,v  : vector1);
var  rsq : real; m,n : integer;
begin
  for m := 1 to Z do
  begin
    u[m] := 0.0; v[m] := 0.0;
    for n := 1 to Z do if n<>m then
    begin
      rsq := sqr(x[m]-x[n])+sqr(y[m]-y[n]);
      u[m] := u[m]+(y[m]-y[n])/rsq; v[m] := v[m]+(x[n]-x[m])/rsq;
    end;
    u[m] := u[m]/(2.0*pi); v[m] := v[m]/(2.0*pi);
  end;
end;

procedure file_final_coordinates;
var n : integer;
begin
  writeln(file4,Z);
  for n := 1 to Z do
  writeln(file4,gamma[n]:14:6,xd[nsteps+1,n]:14:6,yd[nsteps+1,n]:14:6);
  writeln(file3,1); writeln(file3,Z);
  for n := 1 to Z do
  writeln(file3,xd[nsteps+1,n]:14:6,yd[nsteps+1,n]:14:6);
end;

                {Main program}
begin
  pi := 4.0*arctan(1.0);
  open_files;
  r1 := sqrt(sqr(xa[1])+sqr(ya[1]));
```

```
writeln('Size of time steps?'); readln(dt);
writeln('Number of time steps?'); readln(nsteps);
writeln('Number of iters? 1 = forward diff, 2 = central diff. etc.');
readln(its);
writeln(file2,Z);
for n := 1 to Z do
begin
  xd[1,n] := xa[n]; yd[1,n] := ya[n];
end;
for step := 1 to nsteps do
begin
  writeln('step ',step);
  {do the first forward convection to from xa,ya to xb,yb}
  uv_vels(xa,ya,ua,va);
  for n := 1 to Z do
  begin
    xb[n] := xa[n]+ua[n]*dt; yb[n] := ya[n]+va[n]*dt;
  end;
{now do (its-1) iteration towards true central difference convection}
  for it := 2 to its do
  begin
    uv_vels(xb,yb,ub,vb);
    for n := 1 to Z do
    begin
      xb[n] := xa[n]+0.5*(ua[n]+ub[n])*dt;
      yb[n] := ya[n]+0.5*(va[n]+vb[n])*dt;
    end;
  end;
  {store results in double dim. arrays xd,yd for the moment}
  for n := 1 to Z do
  begin
    xa[n] := xb[n]; ya[n] := yb[n];
    xd[step+1,n] := xb[n]; yd[step+1,n] := yb[n];
  end;
end;
{now file the drift paths xd,yd}
for n := 1 to Z do
begin
  writeln(file2,nsteps+1); writeln(nsteps+1);
  for step := 1 to nsteps+1 do
    writeln(xd[step,n]:14:6,yd[step,n]:14:6);
  for step := 1 to nsteps+1 do
    writeln(file2,xd[step,n]:14:6,yd[step,n]:14:6);
end;
writeln('final solution in file ',st3);
writeln('data for reversibility check is in file ',st4);
file_final_coordinates;
end.
```

Program 9.1

Program to generate a set of random numbers and sort them into bins.

```
program ranbox(input,output);
type
  vector1 = array[1..101] of real;
  vector2 = array[1..100] of integer;
var
  df              : real;
  p               : real8;
  m,nbins,i,j     : integer;
  f               : vector1;
  bin             : vector2;
  st1             : lstring(12);
  file1           : text;
begin
  st1 := 'ranout'; assign(file1,st1); rewrite(file1);
  writeln('Input a real number 0.0 to 10.0'); readln(p);
  writeln('How many random numbers shall we generate and sort?');
  readln(m);
  writeln('How many bins shall we sort them into - (up to 100)?');
  readln(nbins); f[1] := 0.0; df := 1.0/nbins;
  for j := 1 to nbins do begin f[j+1] := j*df; bin[j] := 0; end;
  for i := 1 to m do
  begin
    p := exp(5.0*ln(p+1.10316)); p := p-trunc(p);
    writeln(p);
    for j := 1 to nbins do
      if (f[j]<p) and (p <= f[j+1]) then bin[j] := bin[j]+1;
  end;
  writeln('For':8,m:5,'random numbers spread into':27, nbins:4,'bins':5);
  writeln; writeln('bin':9,'range':10,'number':11); writeln;
  writeln(file1,'For':8,m:5,'random numbers spread into':27, nbins:4,'bins':5);
  writeln(file1);
  writeln(file1,'bin':9,'range':10,'number':11); writeln(file1);
  for j := 1 to nbins do writeln(j:8,f[j]:8:4,f[j+1]:7:4,bin[j]:5);
  for j := 1 to nbins do
          writeln(file1,j:8,f[j]:8:4,f[j+1]:7:4,bin[j]:5);
  writeln('Results are in file ',st1);
end.
```

Program 9.2

Diffusion of a point vortex

```pascal
program pointv(input,output);
type
   vector1 = array[1..1000] of real8;
   vector2 = array[1..100] of integer;
   vector3 = array[1..101] of real;
var
   r,dr,dx,dy,dt,viscosity,radius,coef,pi,twopi        :  real;
   m,nbins,i,j,step,nsteps,nrads                       :  integer;
   x,y,P,Q                                             :  vector1;
   bin                                                 :  vector2;
   rad,vorticity,rms,exact                             :  vector3;
   st1                                                 :  lstring(12);
   file1                                               :  text;

procedure input_data;
begin
   st1 := 'ranout'; assign(file1,st1); rewrite(file1);
   writeln('Number of vortex elements?');  readln(m);
   writeln('How many radial strips - (up to 100)?'); readln(nrads);
   writeln('Number of time steps?'); readln(nsteps);
   writeln('Time step?'); readln(dt);
   writeln('Viscosity?'); readln(viscosity);
   writeln('Target area radius?'); readln(radius);
   dr := radius/nrads; coef := sqrt(4.0*viscosity*dt);
   writeln('Input a real number 0.0 to 10.0'); readln(P[1]);
end;

procedure set_radial_bins;
var i : integer;
begin
  rad[1] := 0.0;
  for i := 1 to nrads do
  begin
    rad[i+1] := rad[i]+dr;
  end; writeln;
end;

procedure random_numbers;
var i : integer;
begin
   if step>1 then P[1] := Q[m];
   Q[1] := exp(5.0*ln(P[1]+1.10316));
   for i := 2 to m do
   begin
```

```
      P[i] := exp(5.0*ln(Q[i-1]+1.10316)); P[i] := P[i]-trunc(P[i]);
      Q[i] := exp(5.0*ln(P[i]+1.10316)); Q[i] := Q[i]-trunc(Q[i]);
    end;
  end;

procedure offsets;
var i : integer;
begin
  for i := 1 to m do
  begin
    dr := coef*sqrt(ln(1.0/P[i]));
    dx := dr*cos(twopi*Q[i]); dy := dr*sin(twopi*Q[i]);
    x[i] := x[i]+dx; y[i] := y[i]+dy;
  end;
end;

procedure bin_sort;
var i,j : integer;
begin
  for j := 1 to nrads do bin[j] := 0;
  for i := 1 to m do
  begin
    r := sqrt(sqr(x[i])+sqr(y[i]));
    for j := 1 to nrads do
      if (rad[j]<r) and (rad[j+1]>r) then bin[j] := bin[j]+1;
  end;
end;

procedure vorticity_distribution;
var  argu : real; i : integer;
begin
  writeln('Step',step:5,'  t =',step*dt:9:4); writeln;
  writeln('Vorticity':58); writeln;
  writeln('Bin':6,'Elements':12,'Radius range':19,
                        ' Numerical':16,'Exact':9);
  writeln;
  for i := 1 to nrads do
  begin
    vorticity[i] := bin[i]/(pi*m*(sqr(rad[i+1])-sqr(rad[i])));
    rms[i] := sqrt(0.5*(sqr(rad[i])+sqr(rad[i+1])));
    argu := rms[i]*rms[i]/(4.0*viscosity*step*dt);
    if argu>40.0 then exact[i] := 0.0 else
          exact[i] := 1.0/(pi*4.0*viscosity*step*dt*exp(argu));
    writeln(i:5,bin[i]:10,rad[i]:12:4,
          'to':4,rad[i+1]:9:4,vorticity[i]:12:6,exact[i]:12:6);
  end;
end;
```

```
procedure output_data;
var i : integer;
begin
  write(file1,'Number of vortex elements =',m:5);
  writeln(file1,'    Viscosity =',viscosity:10:6);
  writeln(file1); writeln(file1,'Step',nsteps:5,'  t =',step*dt:9:4);
  writeln(file1); writeln(file1,'Vorticity':67); writeln(file1);
  writeln(file1,'Bin':6,'Elements':12,
        'Radius range':19,'rms rad':13,' Numerical':13,'Exact':9);
  writeln(file1);
  for i := 1 to nrads do
    writeln(file1,i:5,bin[i]:10,rad[i]:12:4,'to':4,
        rad[i+1]:9:4,rms[i]:10:6,vorticity[i]:12:6,exact[i]:12:6);
end;

begin
  pi := 4.0*arctan(1.0); twopi := 2.0*pi;
  input_data;
  set_radial_bins;
  for step := 1 to nsteps do
  begin
    random_numbers;
    offsets;
    bin_sort;
    vorticity_distribution;
  end;
  output_data;
  writeln('Output in file ',st1);
end.
```

Bibliography

Abbott, I. H., Von Doenhoff, A. E. & Stivers, L. S. (1945). *Summary of airfoil data.* NACA Rept. No. 824.

Abbott, I. H. & Von Doenhoff, A. E. (1959). *Theory of wing sections.* Dover Publications, Inc., New York.

Abernathy, F. H. & Kronauer, R. E. (1962). The formation of vortex streets. *J. Fluid Mech.,* **13**, 1–20.

Ackeret, J. (1942). Zum enwurf Dichtstehendes Schaufelgitter Schweiz. *Bauzeitung,* **120**, No. 9. Also available as: *The design of closely spaced blade grids.* Min. of Aviation translation RTP 2007.

Bagley, J. A., Kirby, N. B. & Harcer, P. J. (1961). *A method of calculating the velocity distribution on annular aerofoils in incompressible flow.* R. & M. No. 3146.

Bagley, J. A. & Purvis, H. G. F. (1972). *Experimental pressure distributions on an annular aerofoil in cross-flow.* RAE TR 72098.

Baker, G. R. (1979). The 'cloud-in-cell' technique applied to the roll up of vortex sheets. *J. Comp. Phys.* **31**, 76–95.

Balabaskaran, V. (1982). Aerodynamic investigations of a ducted propeller. Ph.D. Thesis, Indian Institute of Technology, Madras.

Basu, B. C. & Handcock, G. J. (1978). The unsteady motion of a two-dimensional aerofoil in incompressible inviscid flow. *J. Fluid Mech.,* **87**, 159–78.

Batchelor, G. K. (1970). *An introduction to fluid dynamics.* Cambridge University Press.

Bearman, P. W. & Graham, J. M. R. (1979). *Vortex shedding from bluff bodies in oscillatory flow.* A report on Euromech 119.

Bettess, P. & Downie, M. J. (1988). Wave loading on tubular structure members due to appurtenances. *Proceedings of International Conference on Computer Modelling in Ocean Engineering,* Venice.

Birdsall, C. K. & Fuss, D. (1969). Clouds-in-clouds, clouds-in-cell physics for many-body plasma simulation. *J. Comp. Phys.* **3**, No. 4, 494–511.

Birkhoff, G. & Fisher, J. (1959). Do vortex sheets roll up? *Rendi. Circ. Mat.* Palermo, Ser. **2**, **8**, 77–80.

Blasius, H. (1908). Grenzschichten in Flussigkeiten mit Kleiner Reibung. *Z. Math. u. Phys.,* **56**, 1. NACA Tech. Memo. No. 1256.

Bragg, S. L. & Hawthorne, W. R. (1950). Some exact solutions of the flow through annular cascade actuator discs. *J. Aero. Sci.,* **17**, No. 4.

Bristow, D. R. (1974). *A solution to the inverse problem for incompressible, axisymmetric potential flow.* AIAA Paper 74–520.

Bromwich, T. J. I'a. (1908). *An introduction to the theory of infinite series.* Macmillan, New York.

Bibliography

Burrill, L. C. (1955). The optimum diameter of marine propellers: A new design approach. *Trans. N.E.C.I.E.S.*, **72**.

Busemann, A. (1928). Lift ratio of radial-flow centrifugal pumps with logarithmic spiral blades. *Math. Mech.* **8**(5), 372–84.

Carr, L. W., McAlister, K. W. & McCrosbay, W. J. (1977). *Analysis of the development of dynamic stall based on oscillating aerofoil experiments.* NASA TN D-8382.

Carter, A. D. S. (1949). *The low speed performance of related aerofoils in cascade.* A.R.C., C.P. No. 29.

Cebeci, T. & Bradshaw, P. (1977). *Momentum transfer in boundary layers.* McGraw-Hill, New York.

Cheng, K. Y. (1981). Horizontal axis wind turbine and profile aerodynamics. Ph.D. Thesis, University of Newcastle upon Tyne.

Chorin, A. J. (1973). Numerical study of slightly viscous flow. *J. Fluid Mech.*, **57**, 785–96.

Chorin, A. J. (1978). Vortex sheet approximation of boundary layers. *J. Comp. Phys.* **27**, 428–442.

Chorin, A. J. & Bernard, P. S. (1973). Discretisation of a vortex sheet with an example of roll up. *J. Comp. Phys.* **13**, 423.

Christiansen, J. P. (1973). Numerical simulation of hydrodynamics by the method of point vortices. *J. Comp. Phys.*, **13**, 363–79.

Chuen-Yen Chow (1979). *An introduction to computational fluid mechanics.* John Wiley & Sons. London.

Clauser, F. H. (1956). The turbulent boundary layer. *Adv. App. Mech.* **4**, Academic Press, London.

Clements, R. R. (1973). An inviscid model of two-dimensional vortex shedding. *J. Fluid Mech.*, **57**, 321.

Clements, R. R. (1977). *Flow representation, including separated regions, using discrete vortices.* AGARD Lecture Series No. 86.

Clements, R. R. & Maull, D. J. (1975). The representation of sheets of vorticity by discrete vortices. *Prog. Aero. Sci.*, **16**(2), 129–46.

Critzos, C. C., Heyson, H. H. & Boswinkle, R. W. (1954). *Aerodynamic characteristics of NACA 0012 airfoil section.* NACA Tech. 3361.

Czibere, T. (1962). *An iterative method of computing blade profiles of straight and radial vaned grates.* Ganz-Mavag Bulletin No. 32, Budapest, Paper 16.

Czibere, T. (1963). Uber die Berechnung der Schaufelprofile von Stromungsmachinen mit halbaxialer Durschtromung. *Acta Technica Hung.* **44**, 149–93.

Denton, J. D. (1974). *A time marching method for two- and three-dimensional blade to blade flows.* ARC R&M No. 3775.

Denton, J. D. (1976). *Extension of the finite area time marching method to three dimensions.* A.R.C. R&M No. 36906. Ae. Th. **33**.

Denton, J. D. (1982). *An improved time marching method for turbomachinery flow calculations.* ASME Paper 82-GT-239.

Dixon, S. L. (1975). *Fluid mechanics, thermodynamics of turbomachinery.* Pergamon Press, Oxford.

Downie, M. J. (1981). An inviscid model for the fluid forces induced by vortex shedding from a circular cylinder. Ph.D. thesis, Royal Military College of Science, Schrivenham.

Dwight, H. B. (1963). *Tables of integrals and other mathematical data.*

Bibliography

Macmillan, New York.

Einstein, A. (1956). *Investigation on the theory of Brownian motion.* Dover, New York.

Fairbairn, G. W. (1976). Three-dimensional flows through the rotor passages of mixed-flow fans. Ph.D. Thesis, University of Newcastle upon Tyne.

Fairbairn, G. W. & Lewis, R. I. (1982). Three-dimensional effects due to the through-flow eddy in a mixed-flow fan rotor. *Int. Conf. on Fan Design and Applications, B.H.R.A. Fluid Engineering,* Paper J2.

Falkner, V. M. & Skan, S. W. (1931). Some approximate solutions of the boundary layer equations. *Phil. Mag.,* **12**, 865. (See also ARC R.&M. No. 1314 (1930)).

Fink, P. T. & Soh, W. K. (1976). A new approach to roll-up calculations of vortex sheets. *Proc. Roy. Soc.,* A, **362**, 195.

Fisher, E. H. (1975). Performance of mixed-flow pumps and fans. Ph.D. thesis, University of Newcastle upon Tyne.

Fisher, E. H. (1980). A study of diffuser/rotor interaction in a centrifugal compressor. Dept. of Mech. Eng. Report No. Tb. 52, University of Newcastle upon Tyne.

Fisher, E. H. (1986). A singularity solution to the problem of blade to blade flow in a mixed-flow rotor. *International Seminar on Engineering Applications of the Surface and Cloud Vorticity Methods.* Technical University of Wroclaw, Poland.

Fisher, E. H. & Inoue, M. (1981). A study of diffuser/rotor interaction in a centrifugal compressor. *J. Mech. Eng. Sci.,* **23**, No. 3.

Fisher, E. H., & Lewis, R. I. (1971a). *Exact solutions for conical mixed-flow rotors. Part I Symmetrical profiled blades.* N.E.L. Report No. 498, D.T.I.

Fisher, E. H. & Lewis, R. I. (1971b). *Exact solutions for conical mixed-flow rotors. Part II Cambered blades of finite thickness.* N.E.L. Report No. 524, D.T.I.

Fuzy, O. (1970). Existence and use of the singularity carrier auxiliary curve in airfoil cascades. *Periodica Polytechnica, Budapesti Muszaki Egyetem, Mech. Eng.,* **14**, No. 3.

Garrick, I. E. (1944). *On the plane potential flow past a lattice of arbitrary aerofoils.* NACA Rep. 788.

George, M. F. (1976). The aerodynamic performance of a propeller in a non-axisymmetrical duct. Ph.D. Thesis, Dept. of Aeronautics, Glasgow University.

George, M. F. (1978). Linearised theory applied to annular aerofoils of non-axisymmetrical shape in a non-uniform flow. *R.I.N.A.* **120**, 201–10.

Gerard, J. H. (1967). Numerical computation of the magnitude and frequency of the lift on a circular cylinder. *Phil. Trans. Roy. Soc.* A.**261**, 137–62.

Gibson, I. S. (1972). Application of vortex singularities to ducted propellers. Ph.D. Thesis, University of Newcastle upon Tyne.

Gibson, I. S. & Lewis, R. I. (1973). Ducted propeller analysis by surface vorticity and actuator disc theory. *Symposium on Ducted Propellers.* R.N.I.A., Paper No. 1.

Giesing, J. P. (1964). *Extension of the Douglas Neumann program to*

Bibliography

problems of lifting, infinite cascades. Report No. L.B31653, D.T.M.B.

Glauert, H. (1948). *The elements of aerofoil and airscrew theory.* Cambridge University Press.

Glover, E. J. (1970). Slipstream deformation and its influence on marine propeller design. Ph.D. Thesis, University of Newcastle upon Tyne.

Gostelow, J. P. (1964). *Potential flow through cascades. A comparison between exact and approximate solutions.* A.R.C. CP.807.

Gostelow, J. P. (1984). *Cascade aerodynamics.* Pergamon Press, Oxford.

Graham, D. G. (1972). Some effects of sweep in annular and straight cascades. Ph.D. Thesis, University of Newcastle upon Tyne.

Graham, D. G. & Lewis, R. I. (1970). Analysis and computer program for calculation of frictionless flow through axial cascades. Report No. Tb. 14, Dept. of Mechanical Engineering, University of Newcastle upon Tyne.

Graham, D. & Lewis, R. I. (1974). A surface vorticity analysis of three-dimensional flow through strongly swept turbine cascades. *J. Mech. Eng. Sci.,* **16,** No. 6.

Graham, J. M. R. (1985a). Application of discrete vortex methods to the computation of separated flows. *Int. Symposium on Computational Fluid Dynamics,* Reading.

Graham, J. M. R. (1985b). Numerical simulation of steady and unsteady flow about sharp-edged bodies. *IAHR Symposium on Separated Flow around Marine Structures,* Trondheim, Norway.

Graham, J. M. R. (1988). Numerical simulation of unsteady separated flows with application to a rapidly moving spoiler. Imperial College Report.

Hama, F. R. & Burke, E. R. (1960). On the rolling up of a vortex sheet. University of Maryland, Tech. Note No. BN. 220.

Hansen, R. J. & Hoyt, J. G. (1984). Laminar-to-turbulent transition on a body of revolution with an extended favourable pressure gradient forebody. *J. Fluid Eng.* **106,** 202–10.

Hartree, D. R. (1937). On an equation occurring in Falkner and Skan's approximate treatment of the equations of the boundary layer. *Proc. Camb. Phil. Soc.,* **33,** Part II, 223.

Herrig, L. J., Emery, J. C. & Erwin, J. R. (1951). *Systematic two-dimensional cascade tests of NACA 65-series compressor blades at low speeds.* NACA R.M. L51G31.

Hess, J. L. (1962). Calculation of potential flow about bodies of revolution having axes perpendicular to the free-stream direction. *J. Aerospace Sci.,* 726–42.

Hess, J. L. (1971). Numerical solution of the integral equation for Neumann problem with application to aircraft and ships. Douglas Aircraft Company Engineering Paper No. 5987.

Hess, J. L. (1974). The problem of three-dimensional lifting potential flow and its solution by means of surface singularity distribution. *Comp. Meth. App. Mech. & Eng.* **4,** 283–319.

Hess, J. L. (1976). On the problem of shaping an axisymmetric body to obtain low drag at large Reynolds numbers. *J. Ship Res.* **20,** 51.

Hess, J. L. & Smith, A. M. O. (1962). Calculation of non-lifting potential flow about arbitrary three-dimensional bodies. Douglas Aircraft Company Rep. No. ES40622.

Bibliography

Hess, J. L. & Smith, A. M. O. (1966). Calculation of potential flow about arbitrary bodies. *Prog. Aero. Sci.* **8**, 1–138, Pergamon Press, Oxford.

Hill, J. M. (1971). Sweep and dihedral effects in turbine cascades. Ph.D. Thesis, University of Newcastle upon Tyne.

Hill, J. M. (1975). A simple computer solution for general turbine cascade flows based upon actuator disc theory. *Int. J. Mech. Sci.*, **17**, 551–5.

Hill, J. M. & Lewis, R. I. (1974). Experimental investigations of strongly swept turbine cascades with low speed flow. *J. Mech. Eng. Sci.*, **16**, No. 1.

Hill, R. S., Shim, K. C. & Lewis, R. I. (1986). Sources of excitation in tube banks due to vortex shedding. *Proc. Inst. Mech. Eng.*, **200**, No. C4.

Hill, V. P. (1975). Surface vorticity techniques applied to the annular aerofoil in non-axisymmetric flow. Ph.D. Thesis, University of Newcastle upon Tyne.

Hill, V. P. (1978). A surface vorticity theory for propeller ducts and turbofan engine cowls in non-axisymmetric incompressible flow. *I. Mech. E., J. Mech. Eng. Sci.*, **20**, No. 4.

Hockney, R. W., Goel, S. P. & Eastwood, J. W. (1974). Quite High Resolution Computer Model of a Plasma. J. Phys. **14**, 148.

Horlock, J. H. (1952). *Some actuator disc theories for the flow of air through axial turbomachines.* A.R.C., R&M. 3030.

Horlock, J. H. (1958). *Axial Flow Compressors. Fluid Mechanics and Thermodynamics.* Butterworth Publications Ltd, London.

Horlock, J. H. (1978). *Actuator disc theory.* McGraw-Hill International, New York.

Hourigan, K., Stokes, A. N., Thompson, M. C. & Welsh, M. C. (1986). Flow induced acoustic resonances for a bluff body in a duct: a numerical study. *9th Australiasian Fluid Mechanics Conference,* Auckland. pp. 504–7.

Hourigan, K., Thompson, M. C., Stokes, A. N. & Welsh, M. C. (1987). Prediction of flow/acoustic interactions using vortex models: duct with baffles. *Conference on Computational Techniques and Applications: CTAC 87,* Sydney.

Howe, M. S. (1975). Contributions to the theory of aerodynamic sound, with application to excess jet noise and the theory of the flute. *J. Fluid Mech.*, **71**, 625–73.

Howell, A. R. (1948). The theory of arbitrary aerofoils in cascades. *Phil. Mag.* 39299.

Hunt, B. (1978). The panel method for subsonic aerodynamic flows: A survey of mathematical formulations and numerical models and an outline of the new British Aerospace scheme. *Von Karmen Inst. for Fluid Dynamics, Computational Fluid Dynamics, Lecture Series, No. 4.*

Hunt, B. & Hewitt, B. L. (1986). *Developments in Boundary Element Methods—4,* Chapter 8, (Ed. Banerjee & Watson), Elsevier.

Inoue, M. (1980). Centrifugal compressor diffuser studies. Ph.D. Thesis, Cambridge University.

Jacob, K. & Riegels, F. W. (1963). The calculation of the pressure distributions over aerofoil sections of finite thickness with and without

551

Bibliography

flaps and slats. *Z. Flugwiss*, **11**, 9, 357–67. Available as R.A.E. Library translation No. 1101, 1965).

Kacker, S. C., Pennington, B. & Hill, R. S. (1974). Fluctuating lift coefficient for a circular cylinder in a cross flow. *J. Mech. Eng. Sci.* **16**, No. 4.

Karman, Th. von (1911). Uber den Mechanismus des Widerstandes, den ein bewegter Korper in einer Flussigkeit erfahrt. *Gottingen Nachrichten Maths.-Phys. Kl.*, 509–17.

Karman, Th. von (1930). *Calculation of the flow field around airships.* NACA TM 574.

Kellogg, O. D. (1929). *Foundations of potential theory.* Frederick Ungar Publishing Company, New York, 1929. Also Dover Publications, New York.

Kerwin, J. E. & Lee, Chang-Sup. (1978). Prediction of steady and unsteady marine propeller performance by numerical lifting-surface theory. *S.N.A.M.E., Annual Meeting*, Paper No. 8.

Kuchemann, D. & Weber, J. (1953). *Aerodynamics of propulsion.* McGraw-Hill, New York.

Kuwahara, K. (1973). Numerical study of flow past an inclined flat plate by an inviscid model. *Kn. Phys. Soc. Japan*, **35**, No. 5, 1545–51.

Lamb, H. (1945). *Hydrodynamics.* Cambridge University Press.

Leonard, A. (1980). Vortex methods for flow simulation. *J. Comp. Phys.* **37**, No. 3, 289–335.

Levine, P. (1958). Incompressible flow about axially symmetric bodies in ducts. *J. Aero. Sci.*, **25**, No. 1.

Lewis, R. I. (1964a). Internal Aerodynamics of turbo-machines. *Proc. I. Mech. Eng.* **179**, Pt. I.

Lewis, R. I. (1964b). A theoretical investigation of the rotational flow of incompressible fluids through axial turbomachines with tapered annulus walls. *Int. J. Mech. Sci.*, **6**, 51–75.

Lewis, R. I. (1978). Turbomachinery fluid dynamic problems and techniques. *8th National Conference on Fluid Mechanics and Fluid Power*, PSG College of Technology, Coimbatore, India.

Lewis, R. I. (1980). Simplification to surface vorticity aerofoil and cascade theory to facilitate computation. Report No, Tb. 55, Department of Mechanical Engineering, University. of Newcastle upon Tyne.

Lewis, R. I. (1981). Surface vorticity modelling of separated flows from two-dimensional bluff bodies of arbitrary shape. *J. Mech. Eng. Sci.* **23**, No. 1.

Lewis, R. I. (1982a). *A method for inverse aerofoil and cascade design by surface vorticity.* ASME Paper No. 82-GT-154.

Lewis, R. I. (1982b). A Cellular method for unsteady boundary layer calculations. Internal report No. Tb.57, Department of Mechanical Engineering, University of Newcastle upon Tyne.

Lewis, R. I. (1983). Simultaneous analysis of boundary layer and potential flows. *Proceedings of 7th Conference on Fluid Machinery*, Akademiai Kiado, Budapest.

Lewis, R. I. (1984a). Velocities induced by a ring vortex, using look-up tables. University of Newcastle upon Tyne, Department of Mechanical Engineering, Internal Report No. Tb. 72.

Lewis, R. I. (1984b). An introduction to Martensen's method for plane two-dimensional potential flows. Department of Mechanical Engineering, University of Newcastle upon Tyne, Internal Report No. Tb.63.

Lewis, R. I. (1984c). Calculation of Joukowski aerofoils as a check on the Martensen method. Department of Mechanical Engineering, University of Newcastle upon Tyne, Internal Report No. Tb.64.

Lewis, R. I. (1986). The vorticity method – a natural approach to flow modelling. Keynote Address to the *International Seminar on Engineering Applications of the Surface and Cloud Vorticity Methods*, Wroclaw, Poland, Vol. 46, No. 9, pp. 1–36.

Lewis, R. I. (1987a). User manual for MIXEQU. A program for the analysis or design of mixed-flow cascades and blades rows by Martensen's method. (Author's software).

Lewis, R. I. (1987b). User instructions for PROMOD. A data preparation program for Martensen analysis of aerofoils and blade rows. (Author's software).

Lewis, R. I. (1987c). Recent developments and engineering applications of the vortex cloud method. *Comp. Meth. Appl. Mech. & Eng.*, **64**, 153–76.

Lewis, R. I. (1988). Application of the vortex cloud method to cascades. *Int. J. Turbo & Jet Engines*.

Lewis, R. I. & Balabaskaran, V. (1983). Aerodynamic investigations of a Kort nozzle ducted propeller. *Proceedings 7th Conference on Fluid Machinery*, Akademiai Kiado, Budapest.

Lewis, R. I., Fairbairn, G. W. (1980). Analysis of the through-flow relative eddy of mixed-flow turbomachines. *Int. J. Mech. Sci.*, **22**, 535–49.

Lewis, R. I., Fisher, E. H. & Saviolakis, A. (1972). *Analysis of mixed-flow rotor cascades*. Aero. Res. Council, R.&M No. 3703.

Lewis, R. I. & Hill, J. M. (1971). The influence of sweep and dihedral in turbomachinery blade rows. *J. Mech. Eng. Sci.*, **13**, No. 4.

Lewis, R. I. & Horlock, J. H. (1961). Non-uniform three-dimensional and swirling flows through diverging ducts and turbomachines. *Int. J. Mech. Sci.*, **3**, 170–96.

Lewis, R. I. & Horlock, J. H. (1969). Flow disturbances due to blade thickness in turbomachines. *J. Mech. Eng. Sci.*, **11**, No. 1.

Lewis, R. I. & Lo, Y. C. (1986). Investigation of the performance of airbrakes for a wind turbine. *International Seminar on Engineering Applications of the Surface and Cloud Vorticity Methods*. Technical University of Wroclaw, Poland.

Lewis, R. I. & Mughal, H. U. (1986). Application of mixed-flow surface vorticity and actuator disc theories to analysis of the Newcastle fan rig. *International Seminar on Engineering Applications of the Surface and Cloud Vorticity Methods*, Technical University of Wroclaw, Poland.

Lewis, R. I., Mughal, H. U. (1989). *A vortex singularity boundary integral solution of meridional flows in turbomachinery*. Hydroturbo 89, Brno.

Lewis, R. I., Porthouse, DTC (1983a). Numerical simulation of stalling flows by an integral equation method. *AGARD Meeting. Viscous effects in turbomachines*, AGARD-CPP-351, Copenhagen.

Lewis, R. I. & Porthouse, D. T. C. (1983b). Recent advances in the theoretical simulation of real fluid flows. *Trans. N.E.C.I.*, **99**, No. 3.

Bibliography

Lewis, R. I. & Ryan, P. G. (1972). Surface vorticity theory for axisymmetric potential flow past annular aerofoils and bodies of revolution with application to ducted propellers and cowls. *J. Mech. Eng. Sci.*, **14**, No. 4.

Lewis, R. I. & Shim, K. C. (1986). Vortex cloud analysis of the average and fluctuating flow past a cylinder in a uniform stream. *International Seminar on Enginering Applications of the Surface and Cloud Vorticity Methods*, Technical University of Wroclaw, Poland, Vol. 46, No. 9, pp. 145–61.

Lewis, R. I. & Sorvatziotis, H. (1987). Some contributions to the surface and cellular vorticity analysis of axisymmetric flow past a body of revolution. *Proceedings of the Eighth Conference on Fluid Machinery, Vol. 1,* Akademiai Kiado, Budapest.

Ling, Guo-Can (1986). Numerical study of secondary separation and circulation reduction in the wake flow around a circular cylinder. *Internal Seminar on Engineering Applications of the Surface and Cloud Vorticity Methods,* Wroclaw, Poland, Vol. 51, No. 14, pp. 103–117.

Ling Guo-Can, Bearman, P. W., Graham, J. M. R. (1986). A further simulation of starting flow around a flat plate by a discrete vortex model. *Internal Seminar on Engineering Applications of the Surface and Cloud Vorticity Methods.* Wroclaw, Poland, Vol. 51, No. 14, pp. 118–138.

Ling, Guo-Can, Yin, Xie-Yuan (1984). Vortex motion in the early stages of unsteady flow around a circular cylinder. *Proc. Ind. Acad. Sci.,* **7**, Pt. 2, 119–35.

Mair, W. A. & Maull, D. J. (1971). Bluff Bodies and vortex shedding – A report on Euromech 17. *J. Fluid Mech.* **45**, 209.

Mani, R. & Acosta, A. J. (1968). Quasi two-dimensional flows through a cascade. *J. Eng. Pwr. Trans. A.S.M.E.,* **90**, 119.

Marble, F. E. (1948). The flow of a perfect fluid through an axial turbomachine with prescribed blade loading. *J. Aero. Sci.,* **15**, No. 8.

Marsh, H. (1966). *A digital computer program for the through-flow fluid mechanics of an arbitrary turbomachine using a matrix method.* ARC R&M No. 3508.

Martensen, E. (1959). Berechnung der Druckverteilung an Gitterprofilen in ebener Potentialstromung mit einer Fredholmschen Integralgleichung. *Arch. Rat. Mech., Anal.* **3**, 235–70.

Maskell, E. L. (1972). *On the Kutta–Joukowski conditions in two-dimensional unsteady flow.* R.A.E. Tech. Memo. Aero. 1451.

Maull, D. J. (1986). *Flow models using vortex dynamics – work in the United Kingdom.* AGARD Advisory Report No. 239.

Mellor, G. (1956). The 65-series cascade data. Gas Turbine Lab. M.I.T. (unpublished).

Merchant, W. & Collar, A. R. (1941). *Flow of an ideal fluid past a cascade of blades. (Part II).* A.R.C., R&M. 1893.

Miley, S. J. (1982). A catalogue of low Reynolds number airfoil data for wind turbine applications. Department of Aerospace Engineering, Texas A&M. University, College Station, Texas, RFP-3387, UC-60.

Milne-Thomson, L. M., (1955). *Theoretical Hydrodynamics.* Macmillan & Co. London.

Montgomery, S. R. (1958). Three-dimensional flow in com-

Bibliography

pressor cascades. Rep. No. 48, Gas Turbine Laboratory, M.I.T.

Montgomery, S. R. (1959). *Spanwise variations of lift in compressor cascades. J. Mech. Eng. Sci.,* **1,** No. 3.

Morgan, W. B. (1961). A theory of the ducted propeller with a finite number of blades. Ph.D. Thesis, University of California.

Morgan, W. B. & Caster, E. B. (1968). Comparison of theory and experiment on ducted propellers. *7th Symposium on Naval Hydrodynamics.* Office of Naval Research, Dept. of U.S. Navy.

Mughal, H. U. (1989). Extension of Vorticity Techniques to Meridional Flows in Turbomachines. Ph.D. Thesis, University of Newcastle upon Tyne.

Navier, M. (1827). Memoire sur les lois du mouvement des fluides. *Mem. Acad. Sci.,* **6,** 387.

Nyiri, A. (1964). Calculation of the Meridional Flow Pattern of Hydraulic Machines. *Kulonnyomat ez acta Technica* 45/1-2. Szamabol, Budapest.

Nyiri, A. (1970). Determination of the theoretical characteristics of hydraulic machines, based on potential theory. *Acta Technica Academiae Scientiarum Hungaricae;* Tomus 69 (3–4), pp. 243–273.

Nyiri, A. (1972). Potential Flow around the blades of hydraulic machines. *Proc. of Second International J.S.M.E. Symposium Fluid Machinery and Fluidics, Vol. 1., Fluid Machinery*—1. Tokyo.

Nyiri, A. & Baranyi, L. (1983). Numerical method for calculating the flow around a cascade of aerofoils. *Proc. of VII International J.S.M.E. Symposium Fluid Machinery and Fluidics,* Tokyo, Vol. 2.

Oberkampf, W. L. & Watson, L. E. (1974). Incompressible potential flow solutions for arbitrary bodies of revolution. *AIAA Journal,* **12,** 409–11.

Parker, R. (1966). Resonant effects in the wake shedding from parallel plates: some experimental observations. *J. Sound & Vibration,* **4,** 67–72.

Parsons, J. S. & Goodson, R. E. (1972). The optimum shaping of axisymmetric bodies for minimum drag in incompressible flow. Purdue University Report ACC-72-6.

Peraire, J., Vahdati, M., Morgan, K. & Zienkiewiez, O. C. Z. (1986). Adaptive remeshing for compressible flow computations. *Inst. Num. Meth. Eng.,* University of Swansea, Report No. CR/R/544/86.

Pien, P. C. (1961). The calculation of marine propellers based on lifting-surface theory. *J. Ship Res.*

Poisson, S. D. (1831). Memoire sur les equations general des l'equilibre et du mouvement des corps solides elastiques et des fluides. *J. Ecole Polytech.,* Vol. B, p. 1.

Pollard, D. & Horlock, J. D. (1963). *A theoretical investigation of the effect of change in axial velocity on the potential flow through a cascade of aerofoils.* A.R.C. CP. No. 619.

Pollard, D. (1964). Low speed performance of two-dimensional cascades of aerofoils. Ph.D. thesis, University of Liverpool.

Pollard, D. (1965). The extension of Schlichting's analysis to mixed-flow cascades. *Proc. I. Mech. E.,* **180,** Pt. 3J. 86–95.

Pollard, D. & Gostelow, J. P. (1967). Some experiments at low speed on compressor cascades. *J. Eng. Pwr., Trans. A.S.M.E.,* **89,** 427.

Porthouse, D. T. C. (1983). Numerical simulation of aerofoil and bluff body flows by vortex dynamics. Ph.D. Thesis, University of Newcastle upon Tyne.

Bibliography

Porthouse, D. T. C. & Lewis, R. I. (1981). Simulation of viscous diffusion for extension of the surface vorticity method to boundary and separated flows. *J. Mech. Eng. Sci., I. Mech. E.*, **23,** No. 3, 157–67.

Potts, I. (1987). The importance of S-1 stream surface twist in the analysis of inviscid flow through swept linear turbine cascades. Paper C258/87, *International conference on turbomachinery-efficiency prediction and improvement,* Cambridge University, I. Mech. E.

Prandtl, L. & Tietjens, O. G. (1931). *Hydro- und Aeromechanik,* 2. Berlin. New York. Also available as Dover Publication, New York (1934).

Railly, J. W. (1951). The flow of an incompressible fluid through an axial turbo-machine with any number of rows. *Aero. Quart.,* **3.**

Railly, J. W. (1965). Ackeret method for the design of tandem cascades. *Engineer,* London, pp. 1086–89.

Railly, J. W. (1967). The design of mixed flow cascades by Ackeret's method. *Engineer,* London, 224 (5827), pp. 405–16.

Railly, J. W., Houlton, J. M. & Murugesan, K. (1969). A solution of the direct problem of flow in an arbitrary mixed-flow turbomachine. N.E.L. Report No. 413, Ministry of Technology.

Rhoden, H. G. (1956). *Effects of Reynolds number on the flow of air through a cascade of compressor blades.* ARC, R&M. No, 2919.

Ribaut, M. (1968). Three-dimensional calculation of flow in turbomachines with the aid of singularities. *J. Eng. Power,* pp. 258–64.

Riegels, F. (1948). Das Umstromungsproblem bei inkompressiblen Potentialstromungen. *Ing. Arch. Bd.* **16,** 373–6 and Bd. 17, 94–106.

Riegels, F. (1949). Formeln und Tabellen fur ein in der raumlichen Potentialtheorie auftretendes elliptisches Integral. *Arch. Math.* **1,** 117.

Riegels, F. (1952). Die stromung um schlanke, fast drehsymmetrische Korper. *Mitteilungen an der Max-Plank-Institut für Stromungsforschung. No. 5.* Also R.A.E. Library Translation 779, (1958).

Roberts, K. V. & Christiansen, J. P. (1972). Topics in computational fluid mechanics. *Comp. Phys. Comm.* 3, Suppl., 14–32.

Rohatynski, R. (1986). The integral representation of three-dimensional potential velocity fields. *International Seminar on Engineering Applications of the Surface and Cloud Vorticity Methods.* Technical University of Wroclaw, Poland, No. 46, Series 9, pp. 37–95.

Rosenhead, L. (1931). The formation of vortices from a surface of discontinuity. *Proc. Roy. Soc. A.,* **134,** 170–92.

Ryall, D. L. & Collins, I. F. (1967). *Design and test of series of annular aerofoils.* Min. Tech., A.R.C., R&M. No. 3492.

Ryan, P. G. (1970). Surface vorticity distribution techniques applied to ducted propeller flows. Ph.D. Thesis, University of Newcastle upon Tyne.

Ryan, P. G. & Glover, E. J. (1972). A ducted propeller design method: A new approach using surface vorticity distribution techniques and lifting line theory. *Trans. R.I.N.A.,* **114.**

Sarpkaya, T. (1975). An inviscid model of two-dimensional vortex shedding for transient and asymptotically steady separated flow over an inclined plate. *J. Fluid Mech.,* **68,** Part I, 109–28.

Sarpkaya, T. (1979). Vortex induced oscillations. *A.S.M.E. J. Appl. Mech.,* **46,** 241–58.

Bibliography

Sarpkaya, T. & Shoaff, R. L. (1979a). A discrete-vortex analysis of flow about stationary and transversely oscillating circular cylinders. Naval Postgraduate School, Monterey, Calif, NPS-69SL79011.

Sarpkaya, T. & Shoaff, R. L. (1979b). An inviscid model of two-dimensional vortex shedding for transient and asymptotically steady separated flow over a cylinder. *AIAA 17th Aerospace Sciences Meeting*, New Orleans, Paper No. 79-0281.

Schlichting, H. (1955a). *Boundary layer theory.* Pergamon, Oxford.

Schlichting, H. (1955b). Berechnung der reibungslosen inkompressiblen stromung fur ein vorgegebenes ebenes schaufelgitter, *V.D.I.—Forschungshaft 447, Edition B,* **21**. (Also available as NACA/TIL Misc. 128 translation, Langley Research Centre, 1958).

Scholz, N. (1951). On the calculation of the potential flow around aerofoils in cascade. *J. Aero. Sci. Bd.* **18,** 68–9.

Semple, W. G. (1977). *A note on the relationship of the influence of sources and vortices in incompressible and linearised compressible flow.* British Aircraft Corp. Report AE/A/541. (1958).

Shaalan, M. R. A. & Horlock, J. H. (1966). *The effect of change in axial velocity on the potential flow in cascades.* A.R.C. Report No. 28, 611, P.A. 1194.

Shim, K. C. (1985). Fluctuating phenomena in tube banks in cross-flow. Ph.D. Thesis, University of Newcastle upon Tyne.

Shim, K. C., Hill, R. S. & Lewis, R. I. (1988). Fluctuating lift forces and pressure distributions due to vortex shedding in tube banks. *Int. J. Heat & Fluid Flow,* **9,** No. 2.

Smith, A. M. O. (1962). Incompressible flow about bodies of arbitrary shape. *I.A.S. National Summer Meeting,* Paper No. 62–143.

Smith, A. M. O. & Pierce, J. (1958). *Exact solutions of the Neumann problem. Calculation of non-circulatory plane and axially symmetric flows about or within arbitrary boundaries.* Douglas Report No. ES.26988. Also available in condensed form in *Proc. Third U.S. National Congress of Appl. Mech.,* 1958, A.S.M.E., pp. 807–815.

Smith, A. M. O. & Hess, J. L. (1966). Calculation of potential flow about arbitrary bodies. *Prog. Aero. Sci. 8.* Pergamon Press.

Smith, D. J. L., & Johnson, I. H. (1966). Investigations on an experimental single stage turbine of conservative design – Part I. *A rational aerodynamics design procedure.* N.G.T.E. Report No. R.283.

Smith, L. H., & Yeh, H. (1963). Sweep and dihedral effects in axial flow turbomachinery. *A.S.M.E. J. Basic Eng.* **85,** 401–16. (Also ASME Paper No. 62-WA-102, 1962).

Smith, P. A. & Stansby, P. K. (1988). Impulsively started flow around a circular cylinder by the vortex method. *J. Fluid Mech.,* **198,** 45–77.

Smith, P. A. & Stansby, P. K. (1989a). An efficient surface algorithm for random-particle simulation of vorticity and heat transport. *J. Comp. Phys.,* **81,** No. 2.

Smith, P. A. & Stansby, P. K. (1989b). Post critical flow around a circular cylinder by the vortex method. *J. Fluid Struct.* **3.**

Sorvatziotis, H. (1985). Extension of surface vorticity method to thick shear layers in axisymmetric flows. M.Sc. Thesis, University of Newcastle upon Tyne.

Soundranayagam, S. (1971). The effects of axial velocity variation in

Bibliography

aerofoil cascades. *J. Mech. Eng. Sci.*, **13**, No. 2, 92–9.

Spalart, P. R. (1984). Two recent extensions of the vortex method. *AIAA 22nd Aerospace Sciences Meeting*, Reno, Nevada, AIAA-84-0343.

Spalart, P. R. & Leonard, A. (1981). Computation of separated flows by a vortex tracing algorithm. *AIAA 14th Fluid and Plasma Dynamics Conference*, Palo Alto, California, AIAA-81-1246.

Stansby, P. K. & Dixon, A. G. (1983). Simulation of flow around cylinders by a Lagrangian vortex scheme. *Appl. Ocean Res.* **5**, No. 3, 167–78.

Stenning, A. H. & Kriebel, A. R. (1957). *Stall propagation in a cascade of aerofoils.* A.S.M.E. Paper No. 57-5A-29.

Stratford, B. S. (1959). The prediction of separation of the turbulent boundary layer. *J. Fluid Mech.* **5**.

Strouhal, V. (1878). Uber eine besondere Art der Tonerregung. *Ann. Phys. und. Chemie. Nav. Series* **5**, 216–51.

Thompkins, W. T. & Oliver, D. A. (1976). Three-dimensional flow calculation of a transonic compressor rotor. *AGARD Conference Proc. No. 195*, DFULR Research Centre, Koln.

Thwaites, B. (1949). Approximate calculation of the laminar boundary layer. *Aero. Quart.*, **1**.

Traupel, W. (1945). Calculation of potential flow through blade grids. *Sulzer Review*, No. 1.

Turbal, V. K. (1973). Theoretical solution of the problem on the action of a non-axisymmetrical ducted propeller system in a non-uniform flow. *Symposium on ducted propellers, R.I.N.A.*, Paper No. 2.

Vallentine, H. R. (1967). *Applied hydrodynamics.* Butterworth, London.

Van Manen, J. D. & Oosterveld, M. W. C. (1966). Analysis of ducted propeller design. *Trans. Soc. N.A.M.E.*, **74**, 552–62.

Vezza, M. & Galbraith, R. A. McD. (1985). A method for predicting unsteady potential flow about an aerofoil. *Int. J. Num. Meth. Fluid*, **5**, 347–56.

Weber, J. (1955). *The calculation of the pressure distribution on the surface of thick cambered wings and the design of wings with given pressure distribution.* A.R.C., R&M. No. 3026.

Weissinger, J. & Maass, D. (1968). Theory of the ducted propeller. A review. *7th Symposium on Naval Hydrodynamics*, Office of Naval Research, Dept. of U.S. Navy.

Welsh, M. C., Stokes, A. N. & Parker, R. (1984). Flow-resonant sound interaction in a duct containing a plate. Part I: Semi-circular leading edge. *J. Sound Vib.*, **95**, 305–23.

White, F. M. (1974). *Viscous fluid flow.* McGraw Hill, New York.

Whitney, W. T., Szanca, E. M., Moffat, T. P. & Monroe, D. E. (1967). *Cold air investigation of a turbine for high temperature engine application – turbine design and overall stator performance.* NASA TN D-3751.

Wilkinson, D. H. (1967a). *A numerical solution of the analysis and design problems for the flow past one or more aerofoils or cascades.* A.R.C., R&M. No. 3545.

Wilkinson, D. H. (1967b). *The calculation of rotational, non-solenoidal flows past two-dimensional aerofoils and cascades.* English Electric, M.E.L., Report No. W/M(6C), p. 1291.

Wilkinson, D. H. (1969). *The analysis and design of blade shape for* radial, mixed and axial turbomachines with incompressible flow. M.E.L. Report No. W/M(3F), English Electric Co., Whetstone, Leicester.

Wu, Chung-Hua (1952). *A general theory of three-dimensional flow in subsonic and supersonic turbomachines of axial, radial and mixed-flow types.* NACA TN.2604.

Young, C. (1969). *An investigation of annular aerofoils for turbofan engine cowls.* RAE Technical Report 69285.

Young, C. (1971). *An analysis of the drag of two annular aerofoils.* R.A.E. Tech. Rep. 71126.

Young, L. (1958). Runners of experimental turbo-machines. *Engineering,* London, **185,** 376.

Zedan, M. F. & Dalton, C. (1978). Potential flow around axisymmetric bodies: direct and inverse problems. *AIAA J.,* **16,** No. 3.

Zedan, M. F. & Dalton, C. (1981). Design of low drag axisymmetric shapes by the inverse method. *J. Hydro.* **15,** 48–54.

Index

Index

Index

Index

Index

Stoke's equation, in axisymmetric flow, 268

Stoke's stream function, 215, 268

streamlines
  plotting, in CIC method, 482
  plotting, in surface vorticity schemes, 32–5

streamwise vorticity, in meridional flows, 217, 218, 229

stream function
  use in CIC method, 478, 479
  use in vortex dynamics, 435

Strouhal number, for circular cylinder wake flow, 423

sub-elements
  source panel method, 28, 29
  surface vorticity method, 30, 55, 60
  use in inverse design method, 295–7
  vortex cloud modelling, 346, 356ff

sucked duct rig, for engine intake testing, 192ff

surface vorticity method, outlined, 8ff

sweep, of blade rows and aerofoils, 248ff

S1, S2 Wu surfaces, 99, 100, 191, 233, 249, 251

tandem aerofoil and cascade, 96–8, 309

Taylor expansion, in cell-to-cell method, 472–5

through-flow analysis, 99, 191

three-dimensional flows, 39, 234ff

thrust coefficient, or ducted propeller, 201

thrust ratio, of ducted propeller, 201

time marching analysis, 191

trailing edge, 59ff (see also Kutta–Joukowski condition)

trailing vorticity, 235, 237ff, 254

transition, of boundary layer, 438, 440

turbine nozzle cascade, 455–8

vector mean velocity, of a cascade, 79, 453

viscous diffusion
  in a boundary layer, 381
  of point vortex, 366, 371
  of vortex sheet, 374–6

vortex array, 79

vortex cloud method
  applied to an aerofoil, 428ff
  applied to a cascade, 451ff
  basic theory, 316ff
  convection schemes, 410, 411
  flow diagram of computational scheme, 405
  full vortex cloud analysis, 404ff

potential flow due to a vortex cloud, 407

shedding of surface vorticity, 409

vorticity conservation in vortex cloud modelling, 408

with fixed separation points, 355ff

vortex-body interaction
  circulation correction for discrete vortex close to body, 396–8
  convection of a vortex close to a body, 339ff
  in vortex cloud modelling, 412

vortex cylinder, semi-infinite
  applied to flow through an annulus, 174–6
  applied to flow through a contraction, 170–2
  free vortex ducted propeller model wake flow, 198ff
  non-free vortex ducted propeller, 204ff, 218ff
  sucked duct or pipe flow rig, 192–5
  velocities induced by, 167ff

vortex dynamics, 316ff, 354ff (see vortex cloud method)

vortex-in-cell method, 476ff

vortex panels, 8

vortex pair
  Euler (reversible) convection of in vortex dynamics, 322ff
  convection by the CIC method, 480

vortex cell corner redistribution, in CIC method, 477, 478

vortex street
  interaction with acoustic resonance for plate in duct, 463
  vortex cloud method, 361

vorticity
  bound and shed, 40, 41
  separation from a sharp edge, 360–2
  separation from a smooth surface, 359, 409
  smoke-ring and streamwise, 217, 218, 229

vorticity convection, 14–16

vorticity, line, 9

vorticity production
  at a body surface, 14, 15, 378, 409, 410
  at a sharp edged separation point, 355, 356
  due to wall slip flow in vortex cloud modelling, 378ff
  errors in production due to close proximity of discrete vortex to wall, 382–4

565

Index